ED YONG

Ed Yong is a science journalist who reports for *The Atlantic*, and is based in Washington DC. His work has also featured in *National Geographic*, the *New Yorker*, *Wired*, *Nature*, *New Scientist*, *Scientific American*, and many more.

Praise for *I Contain Multitudes*:

'Beyond fascinating. An amazing book. It'll change the way you think about the world. It'll change who you think you are' Helen Macdonald, author of *H is for Hawk*

'Masterful . . . smart, canny, and often quite beautiful . . . In the realm of the life sciences there is a noble tradition of popular books that have reshaped our view of the world . . . In my own reading I think of Schrödinger's *What is Life*, Dawkins' *The Selfish Gene* and, more recently, Nick Lane's *The Vital Question*. Among such company, *I Contain Multitudes* can hold its head high' *Guardian*

'*I Contain Multitudes* has a terrific story to tell . . . Ed Yong is infectiously enthusiastic about microbes and he describes them with verve . . . Even the book's endnotes are rich with interesting asides, swarming with interesting sidelights, a teeming microbial world' *New York Times*

'I very much enjoyed and admired *I Contain Multitudes*' Bill Bryson, *Observer*, Book of the Year

'An account that is far, far more entertaining than an exploration of microbes should be' Tom Whipple, *The Times*, Book of the Year

'[An] utterly absorbing and hugely important book . . . [Yong] is an extraordinarily adept guide. Writing with lightness and panache, he has a knack of explaining complex science in terms that are both easy to understand and totally enthralling . . . *I Contain Multitudes* is popular science writing at its best. Reading this book will make you view the world differently' *Literary Review*

'Ed Yong has written a riveting account of the microbes that make the world work. *I Contain Multitudes* will change the way you look at yourself – and just about everything else' Elizabeth Kolbert, author of the Pulitzer Prize-winning *The Sixth Extinction*

'A marvellous book! Ed Yong's brilliant gift for storytelling and precise writing about science converge in *I Contain Multitudes* to make the invisible and tiny both visible and mighty. A unique, entertaining, and smart read' Jeff Vandermeer

'With a simply wonderful book, Ed Yong opens the doorway to a hidden world around and inside us. He's smart, he's witty, and he's at the cutting edge. You could not get a better guide' Tim Harford, author of *The Undercover Economist Strikes Back* and *Messy*

'Ed Yong has done something beautiful, and unlikely: he's rendered the unseen world of bacteria thrilling, captivating and highly entertaining. This is a much-needed guide to the hidden kingdom that dominates life on Earth. It cuts through all the buzzwordy hooey and flakey hype of microbiomes with a scientifically steady hand, but told with an infectious sense of awe' Adam Rutherford, broadcaster and author of *Creation*

'Ed Yong is one of our finest young explainers of science – wicked smart, broadly informed, sly, savvy, so illuminating. And this is an encyclopedia of fascinations – a teeming intellectual ecosystem, a keen book on the intricacies of the microbiome and more' David Quammen, author of *The Song of the Dodo* and *Spillover*

ED YONG

I Contain Multitudes

The Microbes within us and a Grander View of Life

9 10

Vintage
20 Vauxhall Bridge Road,
London SW1V 2SA

Vintage is part of the Penguin Random House group of companies
whose addresses can be found at global.penguinrandomhouse.com

First published in Vintage in 2017
First published in hardback by The Bodley Head in 2016

penguin.co.uk/vintage

A CIP catalogue record for this book is
available from the British Library

ISBN 9781784700171

Printed and bound by Clays Ltd, Elcograf S.p.A.

For Mum

CONTENTS

PROLOGUE: A TRIP TO THE ZOO

Baba does not flinch. He is unfazed by the throng of excited kids who have gathered around him. He is unperturbed by the Californian summer heat. He does not mind the cotton swabs that brush his face, body and paws. His nonchalance makes sense, for his life is safe and cushy. He lives in San Diego Zoo, wears an impregnable suit of armour, and is currently curled around the waist of a zookeeper. Baba is a white-bellied pangolin – an utterly endearing animal that looks like a cross between an anteater and a pine cone. He's about the size of a small cat. His black eyes have a doleful air, and the hair that frames his cheeks looks like unruly mutton chops. His pink face ends in a tapering toothless snout that's well adapted for slurping up ants and termites. His stocky front legs are tipped with long, curved claws for clinging to tree trunks and tearing into insect nests, and he has a long tail for hanging off tree branches (or friendly zookeepers).

But his most distinctive features, by far, are his scales. His head, body, limbs and tail are covered in them – pale orange, overlapping plates that create an extremely tough defensive coat. They are made of the same material as your nails – keratin. Indeed, they look and feel a lot like fingernails, albeit large, varnished, and badly chewed ones. Each one is flexibly but firmly attached to his body, so they sink down and spring back as I run my hand down his back. If I stroked him in the opposite direction, I'd probably cut myself – many of the scales are sharp-edged. Only Baba's face, belly and paws are unprotected, and if he chose to, he could easily defend them by rolling up into a ball. It's this ability that gives his kind their name: pangolin comes from the Malay word *pengguling*, meaning 'something that rolls up'.

Baba is one of the zoo's ambassador animals – exceptionally docile and well-trained individuals who take part in public activities. Keepers frequently take him to nursing homes and children's hospitals to brighten up the days of sick people, and to teach them about unusual animals. But today, he gets the day off. He just sits around the keeper's midriff, like the world's strangest cummerbund, while Rob Knight gently dabs a cotton swab against the side of his face. 'This is one of the species that I've been captivated by since I was a kid – just that something like that exists,' he says.

Knight, a tall, lanky New Zealander with buzzcut hair, is a scholar of microscopic life, a connoisseur of the invisible. He studies bacteria and other microscopic organisms – microbes – and he is specifically enthralled by those that live in or on the bodies of animals. To study them, he must first collect them. Butterfly collectors use nets and jars; Knight's tool of choice is the cotton swab. He reaches over with a small bud and rolls it over Baba's nose for a couple of seconds, long enough to infuse the end with pangolin bacteria. Thousands, if not millions, of microscopic cells are now entangled in the white fuzz. Knight moves delicately so as not to perturb the pangolin. Baba couldn't look less perturbed if he tried. I get the feeling that if a bomb went off next to him his only reaction would be to fidget slightly.

Baba is not just a pangolin. He is also a teeming mass of microbes. Some of them live inside him, mostly in his gut. Others live on the surface, on his face, belly, paws, claws, and scales. Knight swabs each of these places in turn. He has swabbed his own body parts on more than one occasion, for he too hosts his own community of microbes. So do I. So does every beast in the zoo. So does every creature on the planet, except for a few lab animals that scientists have deliberately bred to be sterile.

All of us have an abundant microscopic menagerie, collectively known as the *microbiota* or *microbiome*.[1] They live on our surface, inside our bodies, and sometimes inside our very cells. The vast majority of them are bacteria, but there are also other tiny organisms including fungi (such as yeasts) and archaea, a mysterious group that we will meet again later. There are viruses too, in unfathomable numbers – a *virome* that infects all the other microbes and occasionally the host's cells. We can't see any of these minuscule specks. But if our own cells were to mysteriously disappear, they would perhaps be detectable as a ghostly microbial shimmer, outlining a now-vanished animal core.[2]

In some cases, the missing cells would barely be noticeable. Sponges are among the simplest of animals, with static bodies never more than a few cells thick, and they are also home to a thriving microbiome.[3] Sometimes, if you look at a sponge under a microscope, you will barely be able to see the animal for the microbes that cover it. The even simpler placozoans are little more than oozing mats of cells; they look like amoebae but they are animals like us, and they also have microbial partners. Ants live in colonies that can number in their millions, but every single ant is a colony unto itself. A polar bear, trundling solo through the Arctic, with nothing but ice in all directions, is completely surrounded. Bar-headed geese carry microbes over the Himalayas, while elephant seals take them into the deepest oceans. When Neil Armstrong and Buzz Aldrin set foot on the Moon, they were also taking giant steps for microbe-kind.

When Orson Welles said 'We're born alone, we live alone, we die alone', he was mistaken. Even when we are alone, we are never alone. We exist in symbiosis – a wonderful term that refers to different organisms living together. Some animals are colonised by microbes while they are still unfertilised eggs; others pick up their first partners at the moment of birth. We then proceed through our lives in their presence. When we eat, so do they. When we travel, they come along. When we die, they consume us. Every one of us is a zoo in our own right – a colony enclosed within a single body. A multi-species collective. An entire world.

These concepts can be hard to grasp, not least because we humans are a global species. Our reach is boundless. We have expanded into every corner of our blue marble, and some of us have even left it. It can be weird to consider existences that play out in an intestine or in a single cell, or to think about our body parts as rolling landscapes. And yet, they assuredly are. The Earth contains a variety of different ecosystems: rainforests, grasslands, coral reefs, deserts, salt marshes, each with its own particular community of species. But a single animal is full of ecosystems too. Skin, mouth, guts, genitals, any organ that connects with the outside world: each has its own characteristic community of microbes.[4] All of the concepts that ecologists use to describe the continental-scale ecosystems that we see through satellites also apply to ecosystems in our bodies that we peer at with microscopes. We can talk about the diversity of microbial species. We can draw food webs, where different organisms eat and feed each other. We can single out keystone microbes that exert a disproportionate influence on their environment – the equivalents of sea otters or wolves. We can treat disease-causing microbes – pathogens – as invasive creatures, like cane toads or fire ants. We can compare the gut of a person with inflammatory bowel disease to a dying coral reef or a fallow field: a battered ecosystem where the balance of organisms has gone awry.

These similarities mean that when we look at a termite or a sponge or a mouse, we are also looking at ourselves. Their microbes might be different to ours, but the same principles govern our alliances. A squid with luminous bacteria that glow only at night can tell us about the daily ebbs and flows of bacteria in our guts. A coral reef whose microbes are running amok because of pollution or overfishing hints at the turmoil that occurs in our guts when we swallow unhealthy food or antibiotics. A mouse whose behaviour changes under the sway of its gut microbes can show us something about the tendrils of influence that our own companions insinuate into our minds. Through microbes, we find unity with our fellow creatures, despite our incredibly different lives. None of those lives is lived in isolation; they always exist in a microbial context, and involve constant negotiations between species big and small. Microbes move between animals, too, and between our bodies and the soils, water, air, buildings, and other environments around us. They connect us to each other, and to the world.

All zoology is really ecology. We cannot fully understand the lives of animals without understanding our microbes and our symbioses with them. And we cannot fully appreciate our own microbiome without appreciating how those of our fellow species enrich and influence their lives. We need to zoom out to the entire animal kingdom, while zooming in to see the hidden ecosystems that exist in every creature. When we look at beetles and elephants, sea urchins and earthworms, parents and friends, we see individuals, working their way through life as a bunch of cells in a single body, driven by a single brain, and operating with a single genome. This is a pleasant fiction. In fact, we are legion, each and every one of us. Always a 'we' and never a 'me'. Forget Orson Welles, and heed Walt Whitman: 'I am large, I contain multitudes.'[5]

1. LIVING ISLANDS

The Earth is 4.54 billion years old. A span of time that big is too mind-boggling to comprehend, so let's collapse the planet's entire history into a single calendar year.[1] Right now, as you're reading this page, it is 31 December, just before the stroke of midnight. (Thankfully, fireworks were invented nine seconds ago.) Humans have only existed for 30 minutes or fewer. The dinosaurs ruled the world until the evening of 26 December, when an asteroid hit the planet and wiped them out (except for the birds). Flowers and mammals evolved earlier in December. In November, plants invaded the land and most of the major animal groups appeared in the seas. Plants and animals are all made up of many cells, and similar multicellular organisms had certainly evolved by the start of October. They may have appeared before that – the fossils are ambiguous and open to interpretation – but they would have been rare. Before October, almost every living thing on the planet consisted of single cells. They would have been invisible to the naked eye, had eyes existed. They had been that way ever since life first emerged, some time in March.

Let me stress: all the visible organisms that we're familiar with, everything that springs to mind when we think of 'nature', are latecomers to life's story. They are part of the coda. For most of the tale, microbes were the only living things on Earth. From March to October in our imaginary calendar, they had the sole run of the planet.

During that time, they changed it irrevocably. Bacteria enrich soils and break down pollutants. They drive planetary cycles of carbon, nitrogen, sulphur and phosphorus, by converting these elements into compounds that can be used by animals and plants and then returning them to the world by decomposing organic bodies. They were the first organisms to make their own food, by harnessing the sun's energy in a process called photosynthesis. They released oxygen as a waste product, pumping out so much of the gas that they permanently changed the atmosphere of our planet. It is thanks to them that we live in an oxygenated world. Even now, the photosynthetic bacteria in the oceans produce the oxygen in half the breaths you take, and they lock away an equal amount of carbon dioxide.[2] It is said that we are now in the Anthropocene: a new geological period characterised by the enormous impact that humans have had on the planet. You could equally argue that we are still living in the Microbiocene: a period that started at the dawn of life itself and will continue to its very end.

Indeed, microbes are everywhere. They live in the water of the deepest oceanic trenches and in the rocks below. They persist in belching hydrothermal vents, boiling springs, and Antarctic ice. They can even be found in clouds, where they act as seeds for rain and snow. They exist in astronomical numbers. Actually, they far exceed astronomical numbers: there are more bacteria in your gut than there are stars in our galaxy.[3]

This is the world in which animals originated, one smothered in and transformed by microbes. As palaeontologist Andrew Knoll once said, 'Animals might be evolution's icing, but bacteria are really the cake.'[4] They have always been part of our ecology. We evolved among them. Also, we evolved *from* them. Animals belong to a group of organisms called *eukaryotes*, which also includes every plant, fungus and alga. Despite our obvious variety, all eukaryotes are built from cells that share the same basic architecture, which distinguishes them from other forms of life. They pack almost all their DNA into a central nucleus, a structure that gives the group its name – 'eukaryote' comes from the Greek for 'true nut'. They have an internal 'skeleton' that provides structural support and shuttles molecules from place to place. And they have mitochondria – bean-shaped power stations that supply cells with energy.

All eukaryotes share these traits because we all evolved from a single ancestor, around two billion years ago. Before that point, life on Earth could be divided into two camps or *domains*: the bacteria, which we already know about, and the archaea, which are less familiar and have a fondness for colonising inhospitable and extreme environments. These two groups both consisted of single cells that lack the sophistication of eukaryotes. They had no internal skeleton. They lacked a nucleus. They had no energy-providing mitochondria, for reasons that will soon become abundantly clear. They also looked superficially similar, which is why scientists originally believed that archaea *were* bacteria. But appearances are deceptive; archaea are as different from bacteria in biochemistry as PCs are from Macs in operating systems.

For roughly the first 2.5 billion years of life on Earth, bacteria and archaea charted largely separate evolutionary courses. Then, on one fateful occasion, a bacterium somehow merged with an archaeon, losing its free-living existence and becoming entrapped forever within its new host. That is how many scientists believe eukaryotes came to be. It's our creation story: two great domains of life merging to create a third, in the greatest symbiosis of all time. The archaeon provided the chassis of the eukaryotic cell while the bacterium eventually transformed into the mitochondria.[5]

All eukaryotes descend from that fateful union. It's why our genomes contain many genes that still have an archaeal character and others that more resemble those of bacteria. It's also why all of us contain mitochondria in our cells. These domesticated bacteria changed everything. By providing an extra source of energy, they allowed eukaryotic cells to get bigger, to accumulate more genes, and to become more complex. This explains what biochemist Nick Lane calls the 'black hole at the heart of biology'. There's a huge void between the simpler cells of bacteria and archaea and the more complex ones of eukaryotes, and life has managed to cross that void exactly once in four billion years. Since then, the countless bacteria and archaea in the world, all evolving at breakneck speed, have never again managed to produce a eukaryote. How could that possibly be? Other complex structures, from eyes to armour to many-celled bodies, have evolved on many independent occasions but the eukaryotic cell is a one-off innovation. That's because, as Lane and others argue, the merger that created it – the one between an archaeon and a bacterium – was so breathtakingly improbable that it has never been duplicated, or at least never with success. By forging a union, those two microbes defied the odds and enabled the existence of all plants, animals, and anything visible to the naked eye – or anything with eyes, for that matter. They're the reason I exist to write this book and you exist to read it. In our imaginary calendar, their merger happened some time in the middle of July. This book is about what happened afterwards.

After eukaryotic cells evolved, some of them started cooperating and clustering together, giving rise to multicellular creatures, like animals and plants. For the first time, living things became big – so big that they could host huge communities of bacteria and other microbes in their bodies.[6] Counting such microbes is difficult. It's commonly said that the average person contains ten microbial cells for every human one, making us rounding errors in our own bodies. But this 10-to-1 ratio, which shows up in books, magazines, TED talks, and virtually every scientific review on this topic, is a wild guess, based on a back-of-the-envelope calculation that became unfortunately enshrined as fact.[7] The latest estimates suggest that we have around 30 trillion human cells and 39 trillion microbial ones – a roughly even split. Even these numbers are inexact, but that does not really matter: by any reckoning, we contain multitudes.

If we zoomed in on our skin, we would see them: spherical beads, sausage-like rods, and comma-shaped beans, each just a few millionths of a metre across. They are so small that, despite their numbers, they collectively weigh just a few pounds in total. A dozen or more would line up cosily in the width of a human hair. A million could dance on the head of a pin.

Without access to a microscope, most of us will never directly glimpse these miniature organisms. We only notice their consequences, and especially the negative ones. We can feel the painful cramp of an inflamed gut, and hear the sound of an uncontrollable sneeze. We can't see the bacterium *Mycobacterium tuberculosis* with our naked eyes, but we can see the bloody spittle of a tuberculosis patient. *Yersinia pestis*, another bacterium, is similarly invisible to us, but the plague epidemics that it causes are all too obvious. These disease-causing microbes – pathogens – have traumatised humans throughout history, and have left a lingering cultural scar. Most of us still see microbes as germs: unwanted bringers of pestilence that we must avoid at all costs. Newspapers regularly churn out scare stories in which everyday items, from keyboards to mobile phones to doorknobs, turn out to be – gasp! – covered in bacteria. Even more bacteria than on a toilet seat! The implication is that these microbes are contaminants, and their presence a sign of filth, squalor, and imminent disease. This stereotype is grossly unfair. Most microbes are not pathogens. They do not make us sick. There are fewer than 100 species of bacteria that cause infectious diseases in humans;[8] by contrast, the thousands of species in our guts are mostly harmless. At worst, they are passengers or hitchhikers. At best, they are invaluable parts of our bodies: not takers of life but its guardians. They behave like a hidden organ, as important as a stomach or an eye but made of trillions of swarming individual cells rather than a single unified mass.

The microbiome is infinitely more versatile than any of our familiar body parts. Your cells carry between 20,000 and 25,000 genes, but it is estimated that the microbes inside you wield around 500 times more.[9] This genetic wealth, combined with their rapid evolution, makes them virtuosos of biochemistry, able to adapt to any possible challenge. They help to digest our food, releasing otherwise inaccessible nutrients. They produce vitamins and minerals that are missing from our diet. They break down toxins and hazardous chemicals. They protect us from disease by crowding out more dangerous microbes or killing them directly with antimicrobial chemicals. They produce substances that affect the way we smell. They are such an inevitable presence that we have outsourced surprising aspects of our lives to them. They guide the construction of our bodies, releasing molecules and signals that steer the growth of our organs. They educate our immune system, teaching it to tell friend from foe. They affect the development of the nervous system, and perhaps even influence our behaviour. They contribute to our lives in profound and wide-ranging ways; no corner of our biology is untouched. If we ignore them, we are looking at our lives through a keyhole.

This book will open the door fully. We are going to explore the incredible universe that exists within our bodies. We'll learn about the origins of our alliances with microbes, the counter-intuitive ways in which they sculpt our bodies and shape our everyday lives, and the tricks we use for keeping them in line and ensuring a cordial partnership. We'll look at how we inadvertently disrupt these partnerships and, in doing so, jeopardise our health. We'll see how we might reverse these problems by manipulating the microbiome for our benefit. And we'll hear the stories of the gleeful, imaginative, driven scientists who have dedicated their lives to understanding the microbial world, often in the face of scorn, dismissal, and failure.

We won't focus only on humans, either.[10] We'll see how microbes have bestowed on animals extraordinary powers, evolutionary opportunities, and even their own genes. The hoopoe, a bird with a pickaxe profile and a tiger's colours, paints its eggs with a bacteria-rich fluid that it secretes from a gland beneath its tail; the bacteria release antibiotics that stop more dangerous microbes from infiltrating the eggs and harming the chicks. Leafcutter ants also carry antibiotic-producing microbes on their bodies, and use these to disinfect the fungi that they cultivate in underground gardens. The spiky, expandable pufferfish uses bacteria to make tetrodotoxin – an exceptionally lethal substance which poisons any predator that tries to eat it. The Colorado potato beetle, a major pest, uses bacteria in its saliva to suppress the defences of the plants that it eats. The zebra-striped cardinalfish houses luminous bacteria, which it uses to attract its prey. The ant lion, a predatory insect with fearsome jaws, paralyses its victims with toxins produced by the bacteria in its saliva. Some nematode worms kill insects by vomiting toxic glowing bacteria into their bodies;[11] others burrow into plant cells, and cause vast agricultural losses, using genes stolen from microbes.

Our alliances with microbes have repeatedly changed the course of animal evolution and transformed the world around us. It is easiest to appreciate how important these partnerships are by considering what would happen if they broke. Imagine if all microbes on the planet suddenly disappeared. On the upside, infectious diseases would be a thing of the past, and many pest insects would be unable to eke out a living. But that's where the good news ends. Grazing mammals, like cows, sheep, antelope, and deer would starve since they are utterly dependent on their gut microbes to break down the tough fibres in the plants they eat. The great herds of Africa's grasslands would vanish. Termites are similarly dependent on the digestive services of microbes, so they would also disappear, as would the larger animals that depend on them for food, or on their mounds for shelter. Aphids, cicadas, and other sap-sucking bugs would perish without bacteria to supplement the nutrients that are missing from their diets. In the deep oceans, many worms, shellfish, and other animals rely on bacteria for all of their energy. Without microbes, they too would die, and the entire food webs of these dark, abyssal worlds would collapse. Shallower oceans would fare little better. Corals, which depend on microscopic algae and a surprisingly diverse collection of bacteria, would become weak and vulnerable. Their mighty reefs would bleach and erode, and all the life they support would suffer.

Humans, oddly, would be fine. Unlike other animals, for whom sterility would mean a quick death, we would get by for weeks, months, even years. Our health might eventually suffer, but we'd have more pressing concerns. Waste would rapidly build up, for microbes are lords of decay. Along with other grazing mammals, our livestock would perish. So would our crop plants; without microbes to provide plants with nitrogen, the Earth would experience a catastrophic de-greening. (Since this book focuses entirely on animals, I offer my sincerest apologies to enthusiasts of botany.) 'We predict complete societal collapse only within a year or so, linked to catastrophic failure of the food supply chain,' wrote microbiologists Jack Gilbert and Josh Neufeld, after running through this thought experiment.[12] 'Most species on Earth would become extinct, and population sizes would be reduced greatly for the species that endured.'

Microbes matter. We have ignored them. We have feared and hated them. Now, it is time to appreciate them, for our grasp of our own biology is greatly impoverished if we don't. In this book, I want to show you what the animal kingdom really looks like, and how much more wondrous it becomes when you see it as the world of partnerships that it actually is. This is a version of natural history that deepens the more familiar one, the one laid down by the greatest naturalists of the past.

In March 1854, a 31-year-old British man named Alfred Russel Wallace began an epic eight-year trek through the islands of Malaysia and Indonesia.[13] He saw fiery-furred orang-utans, kangaroos that hopped in trees, resplendent birds of paradise, giant birdwing butterflies, the babirusa pig whose tusks grow up through its snout, and a frog that glides from tree to tree on parachute-like feet. Wallace netted, grabbed, and shot the wonders he saw, eventually amassing an astonishing collection of over 125,000 specimens: shells; plants; thousands of insects, pinned in trays; birds and mammals, skinned, stuffed, or preserved in spirits. But unlike many of his contemporaries, Wallace also labelled everything meticulously, noting *where* each specimen was collected.

That was crucial. From these details, Wallace extracted patterns. He noticed a lot of variation in the animals that live in a certain place, even among those of the same species. He saw that some islands were home to unique species. He realised that as he sailed east from Bali to Lombok – a distance of just 22 miles – the animals of Asia suddenly gave way to the very different fauna of Australasia, as if these two islands were separated by an invisible barrier (which would later be called the Wallace Line). For good reason, Wallace is today heralded as the father of biogeography – the science of where species are, and where they are not. But as David Quammen writes in *The Song of the Dodo*: 'As practiced by thoughtful scientists, biogeography does more than ask *Which species?* and *Where*? It also asks *Why*? And, what is sometimes even more crucial, *Why not*?'[14]

The study of microbiomes begins in exactly this way: cataloguing the ones that are found on different animals, or on different body parts of the same animal. Which species live where? Why? And why not? We need to know their biogeography before we can gain deeper insights into their contributions. Wallace's observations and specimens led him towards *the* defining insight of biology: that species change. *'Every species has come into existence coincident both in space and time with a pre-existing closely allied species,'* he wrote, repeatedly and sometimes in italics.[15] As animals compete, the fittest individuals survive and reproduce, passing their advantageous traits to their offspring. That is, they evolve, by means of natural selection. This was as important an epiphany as science has ever produced, and it all began with a restless curiosity about the world, a desire to explore it, and an aptitude for noticing what lives where.

Wallace was just one of many naturalist explorers who traipsed around the world and catalogued its riches. Charles Darwin endured a five-year, round-the-world voyage aboard the HMS *Beagle*, in which he would discover the fossilised bones of giant ground sloths and armadillos in Argentina, and encounter the giant tortoises, marine iguanas, and diverse mockingbirds of the Galapagos Islands. His experiences and collections planted the intellectual seeds of the same idea that had independently germinated in Wallace's mind – the theory of evolution, which would become inextricably linked with his name. Thomas Henry Huxley, who became known as 'Darwin's bulldog' for his ferocious advocacy of natural selection, sailed to Australia and New Guinea and studied their marine invertebrates. The botanist Joseph Hooker meandered his way to Antarctica, collecting plants along the way. More recently, E. O. Wilson, after studying the ants of Melanesia, wrote the textbook on biogeography.

It is often assumed that these legendary scientists focused entirely on the visible worlds of animals and plants, ignoring the hidden worlds of microbes. That is not entirely true. Darwin certainly collected microbes – he called them 'infusoria' – that blew onto the deck of the *Beagle*, and he corresponded with the leading microbiologists of the day.[16] But there was only so much he could do with the tools available to him.

By contrast, today's scientists can collect samples of microbes, break them apart, extract their DNA, and identify them by sequencing their genes. In this way, they can do exactly what Darwin and Wallace did. They can collect specimens from different locations, identify them, and ask the fundamental question: what lives where? They can do biogeography – just on a different scale. The gentle caress of a cotton bud replaces the swing of a butterfly net. A read-out of genes is like a flick through a field guide. And an afternoon at the zoo, walking from cage to cage, can be like the voyage of the *Beagle*, sailing from island to island.

Darwin, Wallace and their peers were particularly fascinated by islands, and for good reason. Islands are where you go if you want to find life at its most outlandish, gaudy, and superlative. Their isolation, restricted boundaries, and constrained size allow evolution to go to town. The patterns of biology resolve into sharper focus more readily than they would do on the extensive, contiguous mainland. But an island doesn't have to be a land mass surrounded by water. To microbes, every host is effectively an island – a world surrounded by void. My hand, reaching out and stroking Baba at San Diego Zoo, is like a raft, conveying species from a human-shaped island to a pangolin-shaped one. An adult being ravaged by cholera is like Guam being invaded by foreign snakes. No man is an island? Not so: we're all islands from a bacterium's point of view.[17]

Each of us has our own distinctive microbiome, sculpted by the genes we inherited, the places we've lived in, the drugs we've taken, the food we've eaten, the years we've lived, the hands we've shaken. Microbially, we are similar but different. When microbiologists first started cataloguing the human microbiome in its entirety they hoped to discover a 'core' microbiome: a group of species that everyone shares. It's now debatable if that core exists.[18] Some species are common, but none is everywhere. If there is a core, it exists at the level of *functions,* not organisms. There are certain jobs, like digesting a certain nutrient or carrying out a specific metabolic trick, that are always filled by *some* microbe – just not always the same one. You see the same trend on a bigger scale. In New Zealand, kiwis root through leaf litter in search of worms, doing what a badger might do in England. Tigers and clouded leopards stalk the forests of Sumatra but in cat-free Madagascar that same niche is filled by a giant killer mongoose called the fossa; meanwhile, in Komodo, a huge lizard claims the top predator role. Different islands, different species, same jobs. The islands in question could be huge land masses, or individual people.

In fact, every individual is more like an archipelago – a *chain* of islands. Each of our body parts has its own microbial fauna, just as the various Galapagos islands have their own special tortoises and finches. The human skin microbiome is the domain of *Propionibacterium, Corynebacterium,* and *Staphylococcus,* while *Bacteroides* lords over the gut, *Lactobacillus* dominates the vagina, and *Streptococcus* rules the mouth. Every organ is also variable in itself. The microbes that live at the start of the small intestine are very different from those in the rectum. Those in dental plaque vary above and below the gumline. On the skin, microbes in the oily lakes of the face and chest differ from those in the hot and humid jungles of the groin and armpit, or those colonising the dry deserts of the forearms and palms. Speaking of palms, your right hand shares just a sixth of its microbial species with your left hand.[19] The variations that exist between body parts dwarf those that exist between people. Put simply, the bacteria on your forearm are more similar to those on my forearm than to those in your mouth.

The microbiome varies in time as well as space. When each baby is born, it leaves the sterile world of its mother's womb and is immediately colonised by her vaginal microbes; almost three-quarters of a newborn's strains can be traced directly back to its mother. Then follows an age of expansion. As the baby picks up new species from its parents and environment, its gut microbiome becomes gradually more diverse.[20] The dominant species rise and fall: as the baby's diet changes, milk-digesting specialists like *Bifidobacterium* give way to carbohydrate-eaters like *Bacteroides*. And as the microbes change, so do their antics. They start making different vitamins and they unlock the ability to digest a more adult diet.

This period is turbulent but follows predictable stages. Imagine watching a forest recently scoured by fire, or a fresh island newly risen from the sea. Both would quickly be colonised by simple plants like lichens and mosses. Grasses and small shrubs would follow. Taller trees would arrive later. Ecologists call this *succession*, and it applies to microbes too. It takes anywhere from one to three years for a baby's microbiome to reach an adult state. Then, a lasting stability. The microbiome may vary from day to day, from sunrise to sunset, or even from meal to meal, but such variations are small compared to the early changes. This dynamism of the adult microbiome conceals a background of constancy.[21]

The exact pattern of succession will vary between different animals, because we turn out to be picky hosts. We are not just colonised by whatever microbes happen to land on us. We also have ways of selecting their microbial partners. We'll learn about these tricks, but for now let us simply note that the human microbiome is distinct from the chimpanzee microbiome, which looks different from the gorilla microbiome, just as the forests of Borneo (orang-utans, pygmy elephants, gibbons) are distinct from those in Madagascar (lemurs, fossas, chameleons) or New Guinea (birds of paradise, tree kangaroos, cassowaries). We know this because scientists have swabbed and sequenced their way around the entire animal kingdom. They have described the microbiomes of pandas, wallabies, Komodo dragons, dolphins, lorises, earthworms, leeches, bumblebees, cicadas, tube worms, aphids, polar bears, dugongs, pythons, alligators, tsetse flies, penguins, kakapos, oysters, capybaras, vampire bats, marine iguanas, cuckoos, turkeys, turkey vultures, baboons, stick insects, and so many more. They have sequenced the microbiomes of human infants, premature babies, children, adults, the elderly, pregnant women, twins, city dwellers from the USA or China, rural villagers from Burkina Faso or Malawi, hunter-gatherers from Cameroon or Tanzania, Amazonian people who had never been contacted before, lean and fat people, and those in perfect health versus those with disease.

These kinds of studies have blossomed. Even though the science of the microbiome is actually centuries old, it has picked up tremendous pace in the last few decades, thanks to technological improvements and the dawning realisation that microbes matter enormously to us – especially in a medical setting. They affect our bodies so extensively that they can determine how well we respond to vaccines, how much nourishment children can extract from their food, and how well cancer patients respond to their drugs. Many conditions, including obesity, asthma, colon cancer, diabetes, and autism, are accompanied by changes in the microbiome, suggesting that these microbes are at the very least a sign of illness, and at most a cause of it. If it's the latter, we might be able to substantially improve our health by tweaking our microbial communities: by adding and subtracting species, transplanting entire communities from one person to another, and engineering synthetic organisms. We can even manipulate the microbiomes of other animals, breaking partnerships that allow parasitic worms to afflict us with horrendous tropical diseases, while forging new symbioses that allow mosquitoes to fight off the virus behind dengue fever.

This is a rapidly changing field of science, and one still shrouded in uncertainty, inscrutability, and controversy. We cannot even identify many of the microbes in our bodies, let alone work out how they affect our lives or our health. But that is exciting! It is surely better to be on the crest of a wave, looking at the ride ahead, than to have already washed up on shore. Hundreds of scientists are now surfing that wave. Funds are flowing in. The number of relevant scientific papers has risen exponentially. Microbes have always ruled the planet but for the first time in history, they are *fashionable*. 'This was completely backwater science; now it's front-seat science,' says biologist Margaret McFall-Ngai. 'It's been fun to watch people realising that microbes are the centre of the universe, and to see the field blossom. We now know that they make up the vast diversity of the biosphere, that they live in intimate association with animals, and that animal biology was shaped by interacting with microbes. In my mind, this is the most significant revolution in biology since Darwin.'

Critics say that the popularity of the microbiome is undeserved, and that the majority of studies in the field amount to little more than fancy stamp collecting. So what if we know which microbes live on a pangolin's face, or in a person's gut? That tells us *what* and *where*, but not *why* or *how*. Why do some microbes live on some animals but not others, or on a few individuals but not everyone, or on certain body parts but not all of them? Why do we see the patterns that we see? How did those patterns arise? How do microbes first find their way into their hosts? How do they seal their partnerships? How do microbes and hosts change each other, once together? How do they cope if their alliances break down?

These are the deep questions that the field is trying to answer. In this book, I will show you how far we have come in addressing them, how much promise there lies in understanding and manipulating microbiomes, and how far we have to go to realise that promise. For now, let us note that these questions can be answered only by collecting small pieces of data, just as Darwin and Wallace did on their seminal voyages. The stamp collecting is important. 'Even Darwin's Journal was just a scientific travelogue, a pageant of colourful creatures and places, propounding no evolutionary theory,' wrote David Quammen.[22] 'The theory would come later.' Before that came a lot of hard graft. Classifying. Cataloguing. Collecting. 'If new continents are unexplored, before you find out why things are where they are, you need to find out where they are,' says Rob Knight.

It's in the spirit of exploration that Knight first approached San Diego Zoo. He wanted to swab the faces and skins of different mammals to characterise their microbiomes, as well as the chemicals – metabolites – that those microbes produce. These substances shape the environment in which the microbes live and evolve, and they also show what those microbes are doing, rather than just which ones are present. Surveying metabolites is like running an inventory of a city's art, food, inventions, and exports, rather than just doing a census of its citizens. Knight recently tried surveying the metabolites of human faces, but found that beauty products, like sunscreens and face creams, drowned out the natural microbial metabolites.[23] The solution: swab the faces of animals. After all, Baba the pangolin doesn't moisturise. 'We're hoping to get oral samples too,' says Knight. 'And maybe vaginal.' I raise an eyebrow. 'The breeding programmes here for the cheetahs and pandas have freezers and freezers full of vaginal swabs,' he assures me.

The zookeeper shows us a colony of naked mole rats skittering around a set of interconnected plastic tubes. They are distinctly unattractive animals, like wrinkled sausages with teeth. They are also incredibly weird: insensitive to pain, resistant to cancer, extremely long-lived, terrible at controlling their body temperature, and possessed of misshapen, incompetent sperm. They live in ant-like colonies with queens and workers. They also burrow, which makes them interesting to Knight. He has just secured a grant to study the microbiomes of animals that share specific traits or lifestyles: burrowing, flying, living in water, adaptations to hot and cold, and even intelligence. 'It's pretty speculative but the idea is that you might have microbial pre-adaptations to get the energy you need to do some of those more exotic things,' he says. Speculative, certainly, but not far-fetched. Microbes have opened many doors for animals, allowing them to take up all kinds of peculiar lifestyles that would normally be closed off to them. And when animals share habits, their microbiomes often converge. For example, Knight and his colleagues once showed that ant-eating mammals, including pangolins, armadillos, anteaters, aardvarks, and aardwolves (a type of hyena), all have similar gut microbes, even though they have been evolving independently for around 100 million years.[24]

We walk past a gang of meerkats, some upright and alert, others playing together. The lone female – the group's matriarch – is the only one Knight could potentially swab but she is old and has a heart condition. That's not uncommon. Meerkats will sometimes attack each other's pups or abandon their own, and when this happens, the zoo steps in to hand-raise the youngsters. They survive, but the keeper tells us that, for unknown reasons, they often develop heart problems when they get older. 'That's very interesting,' says Knight. 'Do you know anything about meerkat milk?' He asks because mammalian milk contains special sugars that infants cannot digest, but that certain microbes can. When a human mother breastfeeds her child, she isn't just feeding it; she is also feeding the child its first microbes, and ensuring that the right pioneers settle inside its gut. Knight wonders if the same applies to meerkats. Do the abandoned pups start their lives with the wrong microbes because they don't get mother's milk? Do those early changes affect their health in later life?

Knight is already working on other projects to improve the health of the zoo's animals. As we walk past a cage full of silvered langurs – beautiful, pewter-furred monkeys with electric facial fuzz – he tells me that he is trying to work out why some monkey species frequently suffer from inflammation of the colon (colitis) in captivity, while others do not. There's good reason to think that their microbes are involved. In people, cases of inflammatory bowel disease are usually accompanied by an overabundance of bacteria that provoke the immune system and a lack of those that restrain it. Several other conditions show similar patterns, including obesity, diabetes, asthma, allergies, and colon cancer. These are health problems re-envisioned as ecological ones, where no single microbe is at fault, yet an entire community has shifted into an unhealthy state. They are cases of symbiosis gone wrong. And if these distorted microbiomes actually cause the various conditions, it should be possible to restore good health by manipulating the microbes. Even if the microbial communities are changing as a *result* of a disease, they could still be useful in diagnosing a condition before symptoms become apparent. That's what Knight hopes to see in the monkeys; he is comparing animals with and without colitis, across different species, to see if there are signatures of disease that keepers could use to identify a symptomless animal at risk. Such studies might also help us to understand how the microbiome changes in people with inflammatory bowel disease.

Finally, we walk into a back room where several animals are being temporarily housed out of the public eye. One of the cages houses a giant shadow: a three-foot-long, black-furred creature that has the shape of a weasel but the countenance of a bear. It's a binturong: a large, shaggy civet which Gerald Durrell described as a 'badly made hearthrug'. The keeper reckons that we could easily swab its face and feet, but the real action lies further down. Binturongs have scent glands on either side of their anus, which produce a smell that's reminiscent of popcorn. Again, it seems likely that bacteria create the odours. Scientists have already characterised the microbial scents that drift from the scent glands of badgers, elephants, meerkats, and hyenas. The binturong awaits!

'Could we swab the anus?' I ask.

The keeper looks at the intimidating animal in the cage and then slowly back at us. He says, 'I . . . don't think so.'

When we look at the animal kingdom through a microbial lens, even the most familiar parts of our lives take on a wondrous new air. When a hyena rubs its scent glands on a blade of grass, its microbes write its autobiography for other hyenas to read. When a meerkat mother breastfeeds its pups, it builds worlds within their guts. When an armadillo slurps down a mouthful of ants, it feeds a community of trillions that, in turn, provide it with energy. When a langur or human gets sick, its problems are akin to a lake that's smothered by algae or a meadow that's overrun with weeds – ecosystems gone awry. Our lives are heavily influenced by external forces that are actually inside us, by trillions of things that are separate from us and yet very much a part of us. Scent, health, digestion, development, and dozens of other traits that are supposedly the province of *individuals* are really the result of a complex negotiation between host and microbes.

Knowing what we know, how would we even define an individual?[25] If you define an individual anatomically, as the owner of a particular body, then you must acknowledge that microbes share the same space. You could try for a developmental definition, in which an individual is everything that grows from a single fertilised egg. But that doesn't work either because several animals, from squids to mice to zebrafish, build their bodies using instructions encoded by both their genes *and* their microbes. In a sterile bubble, they wouldn't grow up normally. You could moot a physiological definition, in which the individual is composed of parts – tissues and organs – that cooperate for the good of the whole. Sure, but what about insects in which bacterial and host enzymes work together to manufacture essential nutrients? Those microbes are absolutely part of the whole, and an indispensable part at that. A genetic definition, in which an individual consists of cells that share the same genome, runs into the same problem.

Any single animal contains its own genome, but also many microbial ones that influence its life and development. In some cases, microbial genes can permanently infiltrate the genomes of their hosts. Does it really make sense to view them as separate entities? With your options running out, you could pass the buck to the immune system, since it supposedly exists to distinguish our own cells from those of intruders, to tell self from non-self. That's not quite true, either; as we will see, our resident microbes help to build our immune system, which in turn learns to tolerate them. No matter how we squint at the problem, it is clear that microbes subvert our notions of individuality. They shape it, too. Your genome is largely the same as mine, but our microbiomes can be very different (and our viromes even more so). Perhaps it is less that I *contain* multitudes and more that I *am* multitudes.

These concepts can be deeply disconcerting. Independence, free will, and identity are central to our lives. Microbiome pioneer David Relman once noted that 'loss of a sense of self-identity, delusions of self-identity and experiences of "alien control"' are all potential signs of mental illness.[26] 'Small wonder that recent studies of symbiosis have engendered substantial interest and attention'. But he also added that '[Such studies] highlight the beauty in biology. We are social creatures and seek to understand our connections to other living entities. Symbioses are the ultimate examples of success through collaboration and the powerful benefits of intimate relationships.'

I agree. Symbiosis hints at the threads that connect all life on Earth. Why can organisms as disparate as humans and bacteria live together and cooperate? Because we share a common ancestor. We store information in DNA using the same coding scheme. We use a molecule called ATP as a currency of energy. The same is true across all life. Picture a BLT sandwich: every component, from the lettuce and tomatoes to the pig that produced the bacon, to the yeast that baked the bread, to the microbes that surely sit on its surface, speaks the same molecular language. As Dutch biologist Albert Jan Kluyver once said, 'From the elephant to the butyric acid bacterium – it is all the same!'

Once we understand how similar we are, and how deeply the ties between animals and microbes extend, our view of the world will become immeasurably enriched. Mine certainly has. All my life, I have loved the natural world. My shelves are lined with wildlife documentaries and books bursting with meerkats, spiders, chameleons, jellyfish, and dinosaurs. But none of these talk about how microbes affect, enhance, and direct the lives of their hosts, and so they are incomplete – paintings without frames, cakes without icing, Lennon without McCartney. I now see how the lives of all these creatures depend upon unseen organisms that they live with but are unaware of, that contribute to and sometimes entirely account for their abilities, and that have existed on the planet for far longer than they have. It is a dizzying change in perspective, but a glorious one.

I have been visiting zoos ever since I was too small to remember (or to know that you shouldn't climb into the giant tortoise enclosure). But my visit to San Diego Zoo with Knight (and Baba) feels different.

Although the place is a riot of colour and noise, I realise that most of the life here is invisible and inaudible. At the main entrance, vessels full of microbes part with money so that they can file through gates and see differently shaped microbial vessels that loiter in cages and enclosures. Trillions of microbes, hidden within feather-coated bodies, fly through aviaries. Other hordes swing through branches or scuttle through tunnels. One bacterial throng, nestled within the backside of a black hearthrug fills the air with the redolent twang of popcorn. This is the living world as it actually is, and although it is still invisible to my eyes, I can finally see it.

2. THE PEOPLE WHO THOUGHT TO LOOK

Bacteria are everywhere, but as far as our eyes are concerned, they might as well be nowhere. There are a few extraordinary exceptions: *Epulopiscium fishelsoni,* a bacterium that lives only in the guts of the brown surgeonfish, is about the size of this full stop. But the rest cannot be seen without help, which means that for the longest time they weren't seen at all. In our imaginary calendar, which condenses Earth's history into a year, bacteria first appeared in mid-March. For virtually their entire reign, nothing was consciously aware of their existence. Their anonymous streak broke just a few seconds before the very end of the year, when a curious Dutchman had the whimsical notion of examining a drop of water through handmade lenses of world-beating quality.

In 1632, Antony van Leeuwenhoek was born in the city of Delft, a bustling hub of foreign trade permeated by canals, trees, and cobbled paths.[1] By day, he worked as a city official and ran a small haberdashery business. By night, he made lenses. It was a good time and place to do so: the Dutch had recently invented both the compound microscope and the telescope. Through small circles of glass, scientists were peering at objects too far or too small to see with the naked eye. The British polymath Robert Hooke was one. He gazed at all manner of minute things: fleas, lice clinging to hairs, the points of needles, peacock feathers, poppy seeds. In 1665 he published his observations in a book called *Micrographia,* complete with gorgeous and extraordinarily detailed illustrations. It became an instant bestseller in Britain. Small things had hit the big time.

Leeuwenhoek differed from Hooke in that he never went to university, was not a trained scientist, and spoke only Dutch rather than the more scholarly Latin. Even so, he taught himself to make lenses with a skill that no one else could match. The exact details of his technique are unknown but, broadly speaking, he would grind a bauble of glass into a smooth and perfectly symmetrical lens, less than two millimetres across. This he sandwiched between a pair of brass rectangles. He would then fix a specimen in front of the lens with a tiny pin, and adjust its position with a couple of screws. The resulting microscope looked like a glorified door hinge, and was little more than an adjustable magnifying glass. To use it, Leeuwenhoek had to hold it so that it was practically touching his face, while squinting through the tiny lens, preferably in bright sunlight. These single-lens models were much harder on the eye than the multi-lens compound microscopes that Hooke championed. But they produced clearer images at higher magnification. Hooke's instruments magnified objects by 20 to 50 times; Leeuwenhoek's did so by up to *270 times*. In their day they were easily the best microscopes on earth.

But Leeuwenhoek was 'more than a good microscope maker', observes Alma Smith Payne in *The Cleere Observer*. 'He was also an excellent microscopist – a user of microscopes.' He documented everything. He repeated observations. He conducted methodical experiments. Even though he was an amateur, the scientific method instinctively ran deep within him – as did a scientist's untrammelled curiosity about the world. Through his lenses, he gazed at animal hairs, fly heads, wood, seeds, whale muscle, skin flakes, and ox eyes. He saw marvels, and he showed them to friends, family, and scholars in Delft.

One such scholar, the physician Regnier de Graaf, was a member of the Royal Society, an esteemed and newly founded scientific guild based in London. He recommended Leeuwenhoek, whose microscopes 'far surpass those which we have hitherto seen', to his learned colleagues and implored them to make contact. Henry Oldenburg, the Society's secretary and the editor of its leading journal, did so, and eventually translated and published several of the outsider Leeuwenhoek's disarmingly rambling, informal letters that described red blood cells, plant tissues, and louse guts with matchless detail and care.

And then, Leeuwenhoek looked at some water – specifically, water of Berkelse Mere, a lake near Delft. Sucking some of the turbid liquid into a glass pipette and mounting it on his microscope, he saw that it was teeming with life: 'little green clouds' of algae, along with thousands of tiny, dancing creatures.[2] 'The motion of most of these animalcules in the water was so swift, and so various upwards, downwards and round about that 'twas wonderful to see,' he wrote, 'and I judged that some of these little creatures were above a thousand times smaller than the smallest ones I have ever yet seen upon the rind of cheese.'[3] They were protozoa – the diverse group of organisms that includes amoebas and other single-celled eukaryotes. Leeuwenhoek had become the first person ever to see them.[4]

In 1675, Leeuwenhoek used his lenses to look at rainwater, which had gathered in a blue pot outside his house. Again, a delightful menagerie appeared. He saw serpentine things that wound and unwound themselves, and ovals 'furnished with diverse tiny feet' – more protozoa. He also saw examples of an even tinier class of creature, a thousand times smaller than a louse's eye, which would 'turn themselves about with that swiftness as we see a top turn round' – bacteria! He looked at more water, from his study, his roof, Delft's canals, the nearby sea, and the well in his garden. The little 'animalcules' were everywhere. Life, it turned out, existed in untold numbers beyond the threshold of our perception, visible only to this one man and his superlative lenses. As historian Douglas Anderson later wrote, 'Almost everything he saw, he was the first human ever to see.' And more to the point, *why did he look at the water in the first place?* What on earth possessed this man to scrutinise rain that had collected in a pot? A similar question could be asked of many people throughout the entire history of microbiome research: they were the ones who thought to look.

In October 1676, Leeuwenhoek told the Royal Society about what he'd seen.[5] All of his missives were utterly unlike the stuffy scientific discourse of academic journals. They were full of local gossip and reports about Leeuwenhoek's health. ('The man needed a blog,' observed Anderson.) The October letter, for example, tells us about the weather in Delft that summer. But it also contains fascinatingly detailed accounts of the animalcules. They were 'incredibly small; nay, so small, in my sight, that I judged that even if 100 of these very wee animals lay stretched out one against another, they could not reach the length of a grain of coarse sand; and if this be true, then ten hundred thousand of these living creatures could scarce equal the bulk of a coarse grain of sand'. (He later noted that a sand grain is around 1/80th of an inch across, which would make these 'wee animals' 3 micrometres long. That is, more or less, the length of an average bacterium. The man was *astonishingly* accurate.)

If someone suddenly announced to you that they had seen a group of wondrous, invisible creatures that no one else had ever witnessed, would you believe them? Oldenburg certainly had his doubts, as he did about Leeuwenhoek's earlier descriptions of the 'animalcules'. Still, he published Leeuwenhoek's letter in 1677, in what Nick Lane calls 'an extraordinary monument to the open-minded scepticism of science'. Oldenburg did, however, add a cautionary note, saying that the Society wanted details of Leeuwenhoek's methods so that others could confirm his unexpected observations. Leeuwenhoek didn't exactly cooperate. His lens-making technique was a closely guarded secret. Instead of divulging it, he showed the animalcules to a notary, a barrister, a physician, and other gentlemen of repute, who assured the Royal Society that he could indeed see what he claimed to have seen. Meanwhile, other microscopists tried to duplicate his work – and failed. Even the mighty Hooke struggled at first, and succeeded only when he turned to the single-lens microscopes he so hated. His success vindicated Leeuwenhoek, and cemented the Dutchman's reputation. In 1680, this untrained draper was elected a Fellow of the Royal Society. And since he still couldn't read Latin or English, the Society agreed to write the diploma of membership in Dutch.

Having already become the first human to see microbes, Leeuwenhoek then became the first to see his own. In 1683 he noticed white, batter-thick plaque lodged between his teeth and, as was his wont, he looked at it through his lenses. More living things, 'very prettily a-moving'! There were long, torpedo-shaped rods that shot through the water 'like a pike', and smaller ones that spun around like a top. 'All the people living in our United Netherlands are not as many as the living animals that I carry in my own mouth this very day,' he reported. He drew these microbes, creating a simple image that has become the *Mona Lisa* of microbiology. He studied them in the mouths of local Delft citizens: two women, an eight-year-old child, and an old man who had reputedly never cleaned his teeth. He even added wine vinegar to his own scrapings and saw that the animalcules fell dead – the first account of antisepsis.

By the time he died in 1723, at the age of 90, Leeuwenhoek had become one of the Royal Society's most famous members. He bequeathed to them a black lacquered cabinet containing 26 of his amazing microscopes, complete with mounted specimens. Bizarrely, the cabinet disappeared and was never recovered; an especially tragic loss, since Leeuwenhoek never told anyone exactly how he made his instruments. In one letter, he complained that students were more interested in money or reputation than in 'discovering things hidden from our sight'. 'Not one man in a thousand is capable of such study, because it needs much time, and spending much money,' he lamented. 'And over and above all, most men are not curious to know: nay, some even make no bones about saying: What does it matter whether we know this or not?'[6]

His attitude almost killed his legacy. When others looked through their inferior microscopes they saw nothing, or imagined figments. Interest waned. In the 1730s, when Carl Linnaeus began classifying all life, he lumped all microbes into the genus *Chaos* (meaning formless) and the phylum Vermes (meaning worms). A century and a half would pass between the discovery of the microbial world and its earnest exploration.

* * *

Microbes are now so commonly associated with dirt and disease that if you show someone the multitudes that live in their mouth, they will probably recoil in disgust. Leeuwenhoek harboured no such revulsion. Thousands of tiny things? In his drinking water? In his *mouth*? In *everyone's mouth*? How exciting! If he suspected that they might cause disease, it didn't manifest itself in his writing, which was notable for its lack of speculation. Other scholars were not so restrained. In 1762, the Viennese doctor Marcus Plenciz claimed that microscopic organisms could cause sickness by multiplying in the body and spreading through the air. 'Every disease has its organism,' he said, presciently. Sadly, he had no evidence, and so no way of persuading others that these insignificant organisms were significant. 'I shall not waste time in efforts to refute these absurd hypotheses,' wrote one critic.[7]

Things started changing in the mid-nineteenth century, thanks to a cocky, confrontational French chemist named Louis Pasteur.[8] In short succession, he demonstrated that bacteria could sour liquor and putrefy flesh. And if they were responsible for both fermentation and decay, Pasteur contended, they might also cause disease. This 'germ theory' had been championed by Plenciz and others, but was still controversial. People more commonly thought that diseases were caused by bad air, or *miasma*, released from rotting matter. Pasteur showed otherwise in 1865, when he discovered that two conditions afflicting France's silkworms were caused by microbes. By isolating infected eggs, he stopped the illnesses from spreading and saved the silk industry.

Meanwhile, in Germany, physician Robert Koch was working on an epidemic of anthrax that was sweeping local farm animals. Other scientists had seen a bacterium, *Bacillus anthracis*, in the victims' tissues. In 1876, Koch injected this microbe into a mouse – which died. He recovered it from the dead rodent and injected it into another one – which also died. Doggedly he repeated this grim process for over 20 generations and the same thing happened every time. Koch had unequivocally shown that the bacterium caused anthrax. The germ theory of disease was right.

Microbes had effectively been rediscovered, and were immediately cast as avatars of death. They were germs, pathogens, bringers of pestilence. Within two decades of Koch's work on anthrax, he and many others had discovered the bacteria behind leprosy, gonorrhoea, typhoid, tuberculosis, cholera, diphtheria, tetanus, and plague. As with Leeuwenhoek, new tools led the way: better lenses; ways of growing pure cultures of microbes on plates of jelly-like agar; and new stains that helped microscopists to spot and identify bacteria. From identification, they skipped straight to elimination. Inspired by Pasteur, British surgeon Joseph Lister started using antiseptic techniques in his practice, forcing his staff to chemically sterilise their hands, instruments, and operating theatres and sparing countless patients from raging infections. Others searched for ways of blocking bacteria in the name of curing disease, improving sanitation, and preserving food. Bacteriology became an applied science, which studied microbes in order to repel or destroy them.

It didn't help that just before this wave of discoveries, in 1859, one Charles Darwin had published *On the Origin of Species*. 'Through this historical accident, the germ theory of disease developed during the gory phase of Darwinism, where the interplay between living things was regarded as a struggle for survival, when one had to be friend or foe, with no quarter given,' wrote microbiologist René Dubos.[9] 'This attitude moulded, from the beginning, all later attempts to control microbial diseases. It led to a kind of aggressive warfare against the microbes, aimed at their elimination from the sick individual and the community.'

This attitude persists. If I went to a library and lobbed a microbiology textbook out the window, I could easily concuss a passer-by. If I tore out all the pages that dealt with *beneficial* microbes, I could just about give someone a nasty paper cut. The narrative of disease and death still dominates our view of microbiology.

While the limelight-hogging germ theorists were identifying one deadly pathogen after another, other groups of biologists were toiling on the sidelines on work that would eventually cast microbes in a very different light.

Martinus Beijerinck, a Dutchman, was among the first to demonstrate their planetary importance. Reclusive, brusque, and unpopular, he had no love for people, except a few close colleagues, nor any love for medical microbiology.[10] Disease didn't interest him. He wanted to study microbes in their natural habitats: soil, water, plant roots. In 1888, he found bacteria that pulled nitrogen out of the air and turned it into ammonia for plants to use; later, he isolated species that contributed to the movement of sulphur through the soil and atmosphere. This work stimulated a rebirth of microbiology in Beijerinck's city of Delft – where Leeuwenhoek had first laid eyes on bacteria two centuries earlier. The members of this new-found Delft School, along with intellectual soulmates like the Russian Sergei Winogradsky, called themselves *microbial ecologists*. They revealed that microbes were not just threats to humanity but critical components of the world.

Newspapers of that era began talking about 'good germs', which nourished soil and helped to make booze and dairy products. According to a 1910 textbook, the 'bad germs' that everyone focused on were a 'small, specialised off-shoot of the realm of bacteria, and, broadly speaking, actually of minor importance'.[11] Most bacteria, it said, are decomposers that return nutrients from decaying organic matter back to the world. 'It is not an extravagant statement to say that without [them] . . . all life on earth would of necessity cease.'

Other turn-of-the-century microbiologists realised that many microbes shared the bodies of animals, plants, and other visible organisms. It became clear that lichens – those splotches of colour growing on walls, stones, barks, and logs – are composite organisms, consisting of microscopic algae that live alongside a fungus host, providing nutrients in exchange for minerals and water.[12] The cells of animals such as many sea anemones and flatworms also turned out to contain algae, while those of carpenter ants harboured living bacteria. The fungi that grow on tree roots, long thought to be parasites, were revealed as partners that provide nitrogen in exchange for carbohydrates.

This type of partnership gained a new term – *symbiosis*, from the Greek for 'together' and 'living'.[13] The word itself was a neutral one, implying any form of coexistence. If one partner benefited at the expense of the other, it was a parasite (or a pathogen if it caused disease). If it benefited without affecting its host, it was a commensal. If it benefited its host, it was a mutualist. All these styles of coexistence fell under the rubric of symbiosis.

These concepts emerged at an unfortunate time. In the shadow of Darwinism, biologists were talking of survival of the fittest. Nature was red in tooth and claw. Thomas Huxley, Darwin's bulldog, had compared the animal world to a 'gladiator's show'. Symbiosis, with its themes of cooperation and teamwork sat uneasily within this framework of conflict and competition. Nor did it fit with the idea of microbes as villains. Post-Pasteur, their presence had become a sign of sickness, and their absence a defining trait of healthy tissue. In 1884, when Friedrich Blochmann first saw the bacteria of carpenter ants, the idea of harmless resident microbes was so counter-intuitive that he performed linguistic somersaults to avoid describing them as what they were.[14] 'Plasma rodlets,' he called them, or 'very conspicuous fibrous differentiation of the egg plasma'. It took him years of rigorous investigation before he finally took a stand: 'One can scarcely do otherwise than declare these rodlets to be bacteria,' he finally wrote in 1887.

Meanwhile, other scientists had noticed that the guts of humans and other animals also contained legions of symbiotic bacteria. They caused no obvious disease or decay. They were just there – the 'normal flora'. 'With the advent of animals . . . it was inevitable that bacteria should from time to time be caught up on their bodies,' wrote Arthur Isaac Kendall, a pioneer in the study of gut bacteria.[15] The human body was just another habitat, and Kendall felt that its microbes deserved to be studied, rather than crushed or suppressed. That was easier said than done. Even then, it was clear that our microbes existed in dispiritingly large communities. Theodor Escherich, who discovered *E. coli*, the bacterium that has become a mainstay of laboratory science, once said, 'It would appear to be a pointless and doubtful exercise to examine and disentangle the apparently randomly appearing bacteria in normal faeces and the intestinal tract, a situation that seems controlled by a thousand coincidences.'[16]

Still, Escherich's contemporaries did their best. They characterised bacteria from cats, dogs, wolves, tigers, lions, horses, cattle, sheep, goats, elephants, camels, and humans, a century before *microbiome* became a buzzword.[17] They sketched out the basics of the human microbial ecosystem, several decades before the word *ecosystem* was even coined in 1935. They showed that microbes accumulate in our bodies from birth, and that the prevailing species vary between organs. They realised that the gut was especially rich, and that the microbes there change if animals eat different foods. In 1909, Kendall described the gut as a 'singularly perfect incubator' for bacteria whose activities were 'not in active opposition to those of the host'.[18] They might opportunistically cause disease when a host's resistance was lowered, but they were otherwise harmless.

Could they possibly be beneficial? Ironically, Pasteur, the man who cocked the gun in the long shoot-out with microbes, thought so. He argued that bacteria might be helpful – perhaps even essential – to life, those in cow stomachs were known to digest cellulose from plants and produce nutritious acids for their hosts to absorb. Kendall suggested that the microbes in the human gut might help their host by fighting foreign bacteria and preventing them from taking hold (although he doubted their digestive role).[19] The Russian Nobel laureate Elie Metchnikoff took these views to an extreme. Once described as a 'hysterical character out of one of Dostoevsky's novels',[20] he was a study in self-contradiction: a profound pessimist who tried to kill himself at least twice, yet wrote a book called *The Prolongation of Life: Optimistic Studies*. And in that book, published in 1908, he projected his contradictions onto the world of microbes.

On the one hand, Metchnikoff said that intestinal bacteria produce toxins that cause illness, senility, and ageing and were 'the principal cause of the short duration of human life'. On the other, he also believed that some microbes could *prolong* life. In this, he was inspired by Bulgarian peasants, who regularly drank soured milk and lived well past the age of 100. The two traits were connected, said Metchnikoff. The fermenting milk contained bacteria, including one that he called the Bulgarian bacillus. These made lactic acid, which killed the harmful life-shortening microbes in the peasants' intestines. Metchnikoff was so convinced by this idea that he started regularly quaffing sour milk himself. Others were so convinced by Metchnikoff – a respected scientist – that they did the same. (His claims even started a fashion for colostomy, and inspired Aldous Huxley to write *After Many a Summer*, in which a Hollywood tycoon injects himself with carp guts to alter his gut microbes and achieve immortality.) Humans had, of course, been drinking fermented dairy products for thousands of years, but they were now doing so with microbes in mind. This fad outlasted Metchnikoff himself, who died of heart failure at the age of 71.

Despite the efforts of Kendall, Metchnikoff and others, the study of the symbiotic bacteria, in both humans and other animals, was steamrollered by the increasing focus on pathogens. Public health messages started encouraging people to scour germs from their bodies and surroundings with antibacterial products and a regime of hyper-hygiene. Meanwhile, scientists discovered and mass-manufactured the first antibiotics – substances that overwhelmed both germs and the narrative around them. Finally, we had a chance of vanquishing these tiny foes. And with that chance, the study of symbiotic bacteria lapsed into a long drought, which continued well into the latter half of the twentieth century. A detailed history of bacteriology, published in 1938, failed to mention our resident microbes at all.[21] The leading textbook in the field gave them a lonely chapter, but mainly talked about how to distinguish them from pathogens. They were notable only because they had to be separated from their more interesting peers. If scientists studied bacteria, they mostly did so to understand other organisms better. It turned out that many aspects of biochemistry, like how genes are switched on or how energy is stored, were the same throughout the tree of life. By studying *E. coli*, scientists hoped to understand elephants. Bacteria became 'stand-ins for a universal, reductionist view of life', wrote historian Funke Sangodeyi. 'Microbiology became a kind of handmaiden science.'[22]

Its path to prominence was a slow one. New technologies helped, including ways of growing the oxygen-hating microbes that dominated animal guts, which allowed scientists to study huge groups of important microbes that had previously lain beyond their reach.[23] There were changes in attitude, too. Thanks to the microbial ecologists of the Delft School, scientists realised that bacteria should be studied as *communities* living in *habitats* – in this case, host animals – rather than as solitary organisms to be prodded in a test tube. People from peripheral branches of medicine, like dentistry and dermatology, studied the microbial ecology of their respective organs.[24] They 'set their work against the dominant microbiology of the day', wrote Sangodeyi. But they did so in isolation. Likewise, botanists studied plant microbes, and zoologists tackled the animal ones. Microbiology had splintered into several small fiefdoms, whose piecemeal efforts were easy to ignore. There was no coherent community of scientists who studied symbiotic microbes – no field to speak of. In the spirit of symbiosis, someone needed to assemble the parts into a greater whole.

Theodor Rosebury, an oral microbiologist, started doing that for the human microbiota in 1928. For more than thirty years, he collected every bit of research he could find, and in 1962 he wove those flimsy gossamer strands into a single sturdy tapestry: a groundbreaking tome called *Microorganisms Indigenous to Man*.[25] 'Nobody else, to my knowledge, has ever attempted such a book before,' he wrote. 'In fact, this seems to be the first time . . . that the subject has been treated as an organic unit.' He was right. His book was detailed, sweeping, and a forerunner of this one.[26] He described the common bacteria in every body part in considerable detail. He wrote about how these microbes colonise babies after birth. He suggested that they might produce vitamins and antibiotics, and prevent infections caused by pathogens. He said that the microbiome reverts to normal after bouts of antibiotics, but might be altered more permanently through chronic use. And he was right about most of it. 'Much of the neglect that came to be visited long ago on the normal flora has never been made good,' he wrote. 'It is part of the purpose of this book to suggest that it ought to be.'

It succeeded. Rosebury's synthesis galvanised a faltering field and spurred a lot of new research.[27] One of the scientists who added to that legacy was a charming French-born American named René Dubos. He had already made a name for himself. Emulating the ecological teachings of the Delft School, he studied soil microbes; from them, he had isolated drugs that helped to usher in the antibiotic age. But Dubos saw his drugs as tools for 'domesticating' microbes rather than as weapons for killing them. Even in his later work on tuberculosis and pneumonia he refrained from casting microbes as enemies, and he avoided militaristic metaphors. He was a consummate nature-lover at heart, and microbes were part of nature. 'It had been his lifelong credo that a living organism can be understood only through its relationships with everything else,' wrote his biographer Susan Moberg.[28]

He saw the value of our microbial symbionts, and was dismayed that their benefits had been overlooked. 'The knowledge that microorganisms can be helpful to man has never had much popular appeal, for men as a rule are more preoccupied with the danger that threatens their life than in the biological forces on which they depend,' he wrote. 'The history of warfare always proves more glamorous than accounts of co-operation. Plague, cholera, and yellow [fever] have found their way into the novel, the stage, and the screen, but no one has made a success story of the useful role played by microbes in the intestine or the stomach.'[29] Together with colleagues Dwayne Savage and Russell Schaedler, he helped to work out what they did. They showed that eliminating the indigenous species with antibiotics allowed poor colonisers to become dominant. They studied germ-free mice that had been raised in sterile incubators and showed that these rodents lived shorter lives, grew more slowly, developed abnormal guts and immune systems, and became susceptible to stress and infections. 'Several kinds of microbes play an essential role in the development and physiological activities of normal animals and man,' he wrote.[30]

But Dubos knew that he was just scratching the surface. 'It is certain that [the bacteria identified so far] present but a very small part of the total indigenous microbiota, and not the most important,' he wrote. The rest – perhaps as many as 99 per cent of them – simply refused to grow in a lab. This 'uncultured majority' was a daunting obstacle. Despite everything that had happened since Leeuwenhoek's day, microbiologists still knew nothing about most of the organisms they were meant to be studying. Powerful microscopes couldn't solve the problem. Techniques for culturing microbes couldn't solve the problem. A different approach was needed.

In the late 1960s a young American named Carl Woese began a weirdly niche project: he collected different species of bacteria and analysed a molecule called 16S rRNA, which was found in all of them. No other scientists saw the value of this work and Woese had no competitors: 'It was a one-horse race,' he would later say.[31] The race was expensive, slow, and dangerous, involving worrying amounts of radioactive liquids. But it was also revolutionary.

At the time, biologists relied solely on physical traits to deduce the relationships between species, comparing minutiae of size, shape, and anatomy to work out who was related to whom. Woese felt he could do a better job with the molecules of life: DNA, RNA, and proteins, which are universal to all living things. These molecules accumulate changes over time, so closely related species have more similar versions than distantly related ones. If Woese compared the right molecule across a diverse enough range of species, he believed, the branches and trunks of the tree of life would reveal themselves.[32]

He settled on 16S rRNA, which is produced by a gene of the same name. It forms part of the essential protein-making machinery that is found in all organisms, and so provided the unit of universal comparison that Woese craved. By 1976, he had profiled 16S rRNA from around

30 different microbes. And in June of that year he started work on the species that would change his life – and biology as we know it.

It came from Ralph Wolfe, who had become an authority on an obscure group of microbes called methanogens. These bugs could survive on little more than carbon dioxide and hydrogen, which they converted into methane. They lived in marshes, oceans, and human guts; the one Wolfe sent over – *Methanobacterium thermoautotrophicum* – was found in hot sewage sludge. Woese assumed, as did everyone else, that it would be just another bacterium, albeit one with strange proclivities. But when he looked at its 16S rRNA, he realised that it was decidedly un-bacterial. Accounts differ as to how fully he grasped what he saw, how exuberant or cautious he was, and whether he asked for the experiments to be repeated. But what is clear is that by December his team had sequenced several more methanogens and found the same pattern in all of them. Wolfe remembers Woese telling him, 'These things aren't even bacteria.'

Woese published his results in 1977, in a paper that rebranded the methanogens as the *archaebacteria,* later renamed simply as *archaea.*[33] They weren't weird bacteria, Woese insisted, but an entirely different form of life. It was an astonishing claim. Woese had lifted these obscure microbes out of muck and given them equal billing to the ubiquitous bacteria and the mighty eukaryotes. It was as if everyone was staring at a world map, only for Woese to quietly unfold a full third that had been hidden underneath.

As expected, his claims drew vociferous criticism, even from fellow iconoclasts. The journal *Science* would later dub him 'microbiology's scarred evolutionary', and he bore those scars right up to his death in 2012.[34] Today, his legacy is undeniable. His assertion that archaea are distinct from bacteria was correct. Perhaps more importantly, the approach he championed – comparing genes to work out how species are related to each other – is one of the most important in modern biology.[35] His methods also paved the way for other scientists, like his long-time friend Norman Pace, to *really* start exploring the microbial world.

In the 1980s, Pace started studying the rRNA of archaea that lived in extremely hot environments. He was especially excited by Octopus Spring, a deep blue cauldron in Yellowstone National Park whose water reached a scalding 91 degrees Celsius. The spring was full of unidentified heat-loving microbes, which grew in such huge swarms that they manifested as visible pink filaments. Pace remembers reading about the spring and rushing into his lab, shouting, 'Hey, guys, look at this! Kilogram quantities! Let's get a bucket and go up there.' One of his team said, 'But you don't even know what the organism is.'

And Pace replied: 'That's okay. We can sequence for it.'

He might as well have shouted, 'Eureka!' Pace had realised that, with Woese's methods, he no longer needed to *grow* microbes to study them. He didn't even need to *see* them. He could just pull DNA or RNA right out of the environment and sequence the lot. That would reveal what was living there *and* how they fitted into the microbial tree of life – biogeography and evolutionary biology, in one fell swoop. 'We took our bucket up to Yellowstone and did it,' he says. From the waters of that 'still, beautiful, and lethal place', Pace's team identified two bacteria and an archaeon. None of them had been cultured. All were new to science. The results, published in 1984,[36] marked the first time that anyone had discovered an organism from its genes alone. It would not be the last.

In 1991, Pace and his student Ed DeLong analysed samples of plankton, fished out of the Pacific Ocean. They found an even more complex community of microbes than in Yellowstone: 15 new species of bacteria, two of which were distinct from any known group. Slowly, the sparse bacterial tree of life sprouted new leaves, twigs, and sometimes entire trunks. In the 1980s, all known bacteria had fitted nicely into a dozen major groups, or phyla. By 1998, that number had blossomed to around 40. When I spoke to Pace, he told me that we now are up to 100, and around 80 of those have never been cultured at all. A month later, Jill Banfield announced the discovery of 35 new phyla from a single aquifer in Colorado.[37]

Freed from the yoke of cultures and microscopy, microbiologists could now carry out a more comprehensive census of the planet's microbes. 'That was always the goal,' says Pace. 'Microbial ecology had become a moribund science. People went out, overturned a rock, found a bacterium and thought it exemplary of what's out there. It was stupid. From the very first days of this, we just blew open the doors of the natural microbial world. I want that on my epitaph. It was a wonderful feeling and still is.'

They weren't restricted to 16S rRNA. Pace, DeLong and others soon developed ways of sequencing *every* microbial gene in a dollop of soil or a scoop of water.[38] They would extract the DNA from all the local microbes, cut it into small fragments, and sequence them together. 'We could get any damn gene we wanted,' says Pace. They could see who was there using 16S rRNA, but they could also work out what the local species were capable of by searching for vitamin synthesis genes or fibre-digesting genes or antibiotic resistance genes.

This technique promised to revolutionise microbiology; all it needed was a catchy name. Jo Handelsman provided one in 1998 – *metagenomics*, the genomics of *communities*.[39] 'Metagenomics may be the most important event in microbiology since the invention of the microscope,' she once said. Here, finally, was a way of understanding the full extent of life on Earth. Handelsman and others started studying the microbes that lived in Alaskan soils, Wisconsin grasslands, the acidic run-off from a Californian mine, the water from the Sargasso Sea, the bodies of deep-sea worms, and the guts of insects. And, of course, in the style of Leeuwenhoek, some microbiologists turned to themselves.

Like Dubos and many others who eventually fell in love with microbes, David Relman originally planned to kill them, having begun his career as a clinician working on infectious diseases. In the late 1980s, he used Pace's new technique to identify unknown microbes behind mysterious human diseases. At first he was deeply frustrated because every tissue sample that might harbour a new pathogen was always swamped by our normal microbiota. These residents were an annoying distraction – until Relman realised that they were interesting in their own right. Why not characterise *these* microbes, rather than the pathogenic minority?

So it was that Relman, starting a grand tradition of microbiologists sequencing their own microbiomes, asked his dentist to scrape some plaque from the crevices of his gums and dunk it in a sterile collection tube. He took the gunk back to his lab, and decoded its DNA. It could have led to nothing. The mouth was arguably the most well-studied microbial habitat in the human body. Leeuwenhoek had looked at it. Rosebury had examined it. Microbiologists had cultured nearly 500 strains of bacteria from its various niches. If any body part was immune to new discoveries, it would be the mouth. And yet Relman revealed a range of bacteria in his gums that vastly exceeded what he could grow from the same samples.[40] Even there, in the best-known of human habitats, a staggering number of unknown species awaited discovery. In 2005, Relman found the same pattern in the gut. Using three volunteers, he collected samples from various points along their intestines, and identified almost 400 species of bacteria and one archaeon – 80 per cent of which were new to science.[41] In other words, Dubos's hunch had been right: the microbiologists of his day had barely scratched the surface of the normal human flora.

That started to change in the early 2000s, when researchers conducted sequencing surveys all over the human body. Jeff Gordon, a pioneer we will meet in a later chapter, showed that our microbes control the storage of fat and the creation of new blood vessels, and that obese individuals have different gut microbes to lean ones.[42] Relman himself started describing the microbiota as an 'essential organ'. These trailblazers attracted collaborators from every corner of biology, as well as attention from the popular press, and millions of dollars in funding for large international projects.[43] For centuries, the human microbiome had lurked in the outfield of biology, championed by rebels and iconoclasts. Now it had become part of the establishment. Its story is the story of how ideas about the body and about science move from the periphery to the centre.

* * *

Next to the entrance of Amsterdam's Artis Royal Zoo there is a two-storey building with an image of a giant, striding figure on the side wall. He is made up of small, fluffy balls – orange, beige, yellow, and blue. He is a representation of the human microbiome, and with a friendly wave he invites passers-by into Micropia – the world's first museum devoted entirely to microbes.[44]

The museum opened in September 2014, after twelve years of development and cost 10 million euros. It is fitting that such a place should open in the Netherlands. In Delft, just 40 miles away, Leeuwenhoek first introduced the world to the hidden realm of bacteria. Today, a replica of one of his superlative microscopes is the first thing I see when I pass through Micropia's ticket barrier. It sits in a glass jar – humble, incongruously simple, and mounted upside-down. Around it are samples of things that Leeuwenhoek would have examined, including infusions of pepper, duckweed from a local pond, and dental plaque.

From there, I step into a lift with a friend and a small family. We look up to see ourselves reflected in a video feed on the ceiling. As the lift rises, the video dramatically zooms in to our faces, closer and closer, depicting eyelash mites, skin cells, bacteria, and eventually viruses. When the doors open on the second floor we see a sign made from little pinpricks of light, shimmering gently like a living colony. 'When you look from really close, a new world is revealed to you, more beautiful and spectacular than you would ever have imagined,' the words say. 'Welcome to Micropia.'

Immediately, we get a first-hand glimpse into that new world through a row of microscopes trained on mosquito larvae, water fleas, nematode worms, slime moulds, algae, and green pond bacteria. The latter are magnified 200 times, and it astonishes me to think that Leeuwenhoek's self-made microscope on the lower floor could do the same. He would have seen these wonders too, albeit in much less comfort. While he had to squint uncomfortably through a tiny lens, I can push my face against a comfortable padded eyepiece and look at a crisp digital display.

Beyond the microscopes, there's a full-size display that charts the biogeography of the human microbiome. Visitors stand in front of a camera, which scans their bodies and creates a microbial avatar on a full-size screen. The avatar, with skin outlined in white dots and organs represented in brighter colours, mimics their movements. They shuffle, it shuffles. They wave, it waves. By moving their hands they can select different organs and reveal information about the microbes in their skin, stomach, gut, scalp, mouth, nose, and more. They can learn who lives where, and what they do. Decades of discovery, from Kendall to Rosebury to Relman, are represented in that one exhibit. In fact, the entire museum is a tribute to history. There's a row of lichens, the composite organisms that alerted nineteenth-century scientists to the importance of symbiosis. Here a microscope reveals the lactic acid bacteria that Metchnikoff was so enamoured with – tiny spheres, magnified 630 times and very prettily a-moving.

I am struck by how unapologetic the information is, and how quickly the visitors accept the idea of a microbial world. No one recoils, or frowns, or wrinkles their nose. A couple stand on a red heart-shaped platform and lock lips in front of the 'Kiss-o-Meter', which tells them how many bacteria they just exchanged. A young woman gazes intently at a wall of stool samples from gorillas, capybaras, red pandas, wallabies, lions, anteaters, elephants, sloths, Sulawesi crested macaques and more, all collected from the nearby zoo and double-sealed in airtight jars and Perspex cases. A group of teens stares at a wall of backlit agar plates with mould and bacteria growing on them, some of which have come from everyday objects. They can make out the imprints of keys, phones, computer mice, remote controls, toothbrushes, doorknobs, and the rectangular outline of a euro note. They gawk at the orange dots of *Klebsiella*, blue mats of *Enterococcus*, and grey smudges of *Staphylococcus* that look like pencil shadings.

The family who rode up in the lift with me are staring at a beautiful rendering of Carl Woese's tree of life, which fans out over an entire wall. Animals and plants are relegated to a small circle in the corner, while bacteria and archaea dominate the trunks and branches. The dad was probably born before anyone even knew that archaea existed; now his children are learning about them in a major tourist attraction.

Micropia represents some 350 years of growing knowledge and changing attitudes to microbes. Here, they are not neglected B-listers or sinister villains. Here, they are fascinating, beautiful, and worthy of attention. Here, they are the stars. In *Middlemarch*, George Eliot wrote, 'Most of us, indeed, know little of the great originators until they have been lifted up among the constellations and already rule our fates.' She could have been talking both about the scientists who revealed the world of microbes to us, and about the microbes themselves.

3. BODY BUILDERS

'What you're looking for is something the size of a golf ball,' says Nell Bekiares.[1]

I'm in a lab in the University of Wisconsin-Madison, peering down into a small aquarium tank. It looks empty. I'm not seeing anything golf-ball-sized. I'm not seeing anything at all, except for a layer of sand. Then Bekiares wafts her hand through the water and something erupts outwards, releasing a cloud of viscous, black ink. It's a Hawaiian bobtail squid, a female, about the size of my thumb. Bekiares scoops the squid up in a bowl and it jets around, ghostly white in agitation, arms extended, fins beating furiously. As the squid calms down it tucks its arms under its body and mooches on them, changing shape from a dart to a large jelly bean. Its skin changes too. Tiny pinpricks of colour quickly expand into flat discs, coloured in dark brown, red, and yellow and dotted with iridescent flecks. The squid is not white any more. Now, it looks like an autumnal scene painted by Seurat.

'When they're brown like that, they're happy,' says Bekiares. 'Brown is pretty good. Often, the males are more pissed off. They'll be inking, inking, jetting around. When they shoot water at your face or chest, it certainly seems intentional.'

I'm rather taken. The squid oozes personality. And it is spectacularly beautiful.

There are no other animals in the bowl but the squid is not alone. Two chambers in its undersides – its light organs – are full of luminous bacteria called *Vibrio fischeri*, which cast a downward glow. This glow is too faint to see under the lab's fluorescent lights but would be clearer in the shallow reef flats around Hawaii, where the squid lives. At night, the light from the bacteria supposedly matches the moonlight that wells down from above, cancelling out the squid's silhouette and hiding it from predators. This animal casts no shadow.

The squid may be invisible from below but it is easy to spot from above. All you have to do is fly to Hawaii, wait till nightfall, and wade through knee-deep water with a headlamp and net. With good reflexes, you can snag half a dozen before sunrise. And once caught, they're just as easy to keep, feed, and breed. 'If they can live in the middle of Wisconsin, they can live anywhere,' says Margaret McFall-Ngai, the zoologist who runs this particular lab. Poised, elegant, and effusive, McFall-Ngai has been studying the squid and its luminous bacteria for almost three decades. She has elevated it into an icon of symbiosis and, in the process, has become iconic herself. Her colleagues bill her variously as an outspoken iconoclast, an enthusiastic if unexpected skateboarder, and a tireless advocate for microbes since well before 'microbiome' became a fashionable buzzword. 'She talks about "the New Biology", and that's all caps when Margaret says it,' one biologist told me. She didn't always think like this. It was the squid that changed her mind.[2]

When McFall-Ngai was a graduate student, she studied a fish that also carried a glowing bacterium. She was captivated by it, but frustrated. The fish proved impossible to breed in the lab, so every individual that she worked with had already been colonised by bacteria. So she couldn't answer any of the questions that really intrigued her. What happens when the partners first meet? How do they establish a connection? What stops other microbes from colonising the host? Then a colleague said to her: 'Hey, have you heard about this squid?'

The Hawaiian bobtail squid was familiar to embryologists and its glowing bacterium was known to microbiologists but the *partnership* between them had been wholly neglected – and the partnership was what mattered to McFall-Ngai. To study it, she needed a partner of her own, someone whose understanding of the bacteria could complement her zoological expertise. That person was Ned Ruby. 'I think I was the third microbiologist she came to and the first who said yes,' he says. The two of them formed a professional bond and, shortly after, a romantic one. Ruby's laid-back surfer-guy yin complemented McFall-Ngai's intense stateswoman yang. They have, as one of their friends told me, 'a real symbiosis'. Today, they run adjacent labs and share the same squid.

The animals live in tanks that line a narrow corridor. There's room for 24 at any one time. Whenever a new batch arrives, Bekiares, the lab manager, chooses a letter and all the students christen the animals accordingly. The female whom I met is Yoshi. Yahoo, Ysolde, Yardley, Yara, Yves, Yusuf, Yokel, and Yuk (Mr) sit in nearby tanks. The females have 'date-night' every two weeks. After they mate, they are left in a nursery room with tanks full of PVC piping, into which they lay hundreds of eggs. These take a few weeks to hatch. When we visit the nursery, there's a plastic cup on a shelf with a few dozen baby squid bobbing inside, each just a few millimetres long. Ten female squid can produce 60,000 juveniles in a year, which is one reason why they are such phenomenal lab animals. Here's another: the hatchlings are born sterile. In the wild, they would be colonised by *V. fischeri* within a few hours. In the lab, McFall-Ngai and Ruby can control the hatchlings' introduction to any symbionts. They can label *V. fischeri* cells with glowing proteins and track them as they make their way into the squid's light organs. They can watch the partnership begin.

It begins with physics. The surface of the light organ is covered in mucus and fields of beating hairs called cilia. The hairs create a turbulent current that draws in particles of bacterial size but no bigger. These microbes amass in the mucus, *V. fischeri* among them. Physics now gives way to chemistry. When one *V. fischeri* cell touches the squid, nothing happens. Two cells: still nothing. But if just *five* cells make contact, they switch on scores of squid genes. Some of these genes produce a cocktail of antimicrobial chemicals that leave *V. fischeri* unharmed while creating an inhospitable environment for other microbes. Others release enzymes that break down the squid's mucus, producing a substance that attracts even more *V. fischeri*. These changes explain why *V. fischeri* soon dominates the mucus layer, even though other bacteria initially outnumber it by a thousand to one. It, and it alone, has the ability to transform the surface of the squid into a landscape that attracts more of its kind and deters competitors. It's like the protagonists of science-fiction stories, who terraform inhospitable planets into comfortable homes – except it terraforms an animal.

Once it changes the squid on the outside, *V. fischeri* begins to move inwards. It slips through one of a few pores, travels down a long duct, squeezes through a bottleneck, and finally reaches several blind-ended crypts. Its arrival changes the squid even further. The crypts are lined with pillar-like cells that now become bigger and denser, enveloping the arriving microbes in a tight embrace. As the bacteria accommodate to the remodelled interiors, the door shuts behind them. The entrance to the crypts narrows. The ducts constrict. The fields of cilia waste away. The light organ reaches its mature form. Having been colonised by the right bacteria – and again, *V. fischeri* is the only microbe that ever makes this journey – it won't be colonised again.

Well, so what? This seems an arcane amount of detail to know about the life of one obscure animal. But the squid's particulars hide a profound implication; one that McFall-Ngai immediately picked up on. In 1994, after her first wave of squid studies was complete, she wrote, 'The results of these studies are the first experimental data demonstrating that a specific bacterial symbiont can play an inductive role in animal development.'

In other words, microbes sculpt animal bodies.

How? In 2004, McFall-Ngai's team showed that two molecules on *V. fischeri*'s surface underlie its transformative powers: peptidoglycan (PGN) and lipopolysaccharide (LPS). That was a surprise. At the time, these chemicals were known only in the context of disease. They were described as *pathogen-associated molecular patterns*, or PAMPs, telltale substances that alert animal immune systems to burgeoning infections. But *V. fischeri* is no pathogen. It is related to the bacterium that causes cholera in humans, but it doesn't harm the squid at all. So McFall-Ngai took the acronym, swapped the pathogenic P for a more inclusive microbial M, and rebranded these molecules as MAMPs: *microbe-associated molecular patterns*. The new term is symbolic of microbiome science as a whole. It tells the world that these molecules aren't just signs of disease. They can trigger debilitating inflammation but they can also start a beautiful friendship between an animal and a bacterium. Without them, the light organ never reaches its normal, final form. Without them, the squid survives but never quite completes its journey to full maturity.

It is now clear that many animals, from fish to mice, grow up under the influence of bacterial partners, often under the auspices of the same MAMPs that shape the squid's light organ.[3] Thanks to these discoveries, we can start to see development – the process where an animal transforms from a single cell into a fully functioning adult – in a new light.

If you carefully isolate a fertilised egg – human, squid, any will do – and watch it under the microscope, you will eventually see it divide into two, then four, then eight. The ball of cells gets bigger. It folds, bulges, and contorts. The cells exchange molecular signals that tell each other which tissues and organs to create. Body parts start to form. An embryo grows, and will continue to do so as long as it gets enough nutrients. The whole sequence seems self-contained, barrelling along like an immensely complicated computer program that runs itself. But the squid and other animals tell us that development is more than this. It progresses using instructions in an animal's genes, but also in the genes of its microbes. It is the result of an ongoing negotiation – a conversation between several species, only one of which is doing the actual developing. It is the unfolding of an entire ecosystem.

The easiest way of checking if an animal needs microbes to develop properly is to deprive it of them. Some just die: the dengue-carrying mosquito *Aedes aegypti* makes it to larva-hood but fails to progress beyond that.[4] Others tolerate sterility better. The bobtail squid merely loses its luminescence; that might not matter in McFall-Ngai's lab but it would make the uncamouflaged animal an easy target in the wild.

Scientists have also raised germ-free versions of almost all the most common lab animals, including zebrafish, flies, and mice. These animals also survive but are, however, changed. 'The germ-free animal is, by and large, a miserable creature, seeming at nearly every point to require an artificial substitute for the germs he lacks,' wrote Theodor Rosebury. 'He is as a child might be if we could keep him under glass, entirely protected against the buffets of the outside world.'[5]

The weird biology of germ-free animals is most obvious in the gut. A well-functioning gut needs a big surface area for absorbing nutrients, which is why its walls are densely lined with long, finger-like pillars. It needs to constantly regenerate the cells at its surface, which get sloughed off by the passing tide of food. It needs a rich network of underlying blood vessels to carry nutrients to and fro. And it needs to be sealed – its cells must stick tightly to each other to prevent foreign molecules (and microbes) from leaking into those blood vessels. All of these essential properties are compromised without microbes. If zebrafish or mice grow up in the absence of bacteria their guts don't develop fully, their pillars are shorter, their walls leakier, their blood vessels look more like sparse country lanes than a dense urban grid, and their cycle of regeneration pedals in a lower gear. Many of these glitches can be rectified simply by giving the animals a normal complement of microbes or even isolated microbial molecules.[6]

The bacteria don't physically reshape the gut themselves. Instead, they work *via their hosts*. They are more management than labour. Lora Hooper demonstrated this by infusing into germ-free mice a common gut bacterium called *Bacteroides thetaiotaomicron* – or B-theta to its friends.[7] She found that the microbe activated a wide range of mouse genes that are involved in absorbing nutrients, building an impermeable barrier, breaking down toxins, creating blood vessels, and creating mature cells. In other words, the microbe told the mice how to use *their own genes* to make a healthy gut.[8] Scott Gilbert, a developmental biologist, calls this idea *co-development*. It's as far as you can get from the still-lingering idea that microbes are just threats. Instead, they actually help us become who we are.[9]

Sceptics might argue that mice, zebrafish, and bobtail squid don't *need* microbes to develop: a germ-free mouse still looks like a mouse, walks like a mouse, and squeaks like a mouse. It's not as if you remove the bacteria and suddenly get a totally different animal. But germ-free animals live in undemanding environments: climate-controlled bubbles with plentiful food and water, zero predators, and no infections of any kind. In the brutal wild, they wouldn't last long. They could exist but probably wouldn't persist. They can develop alone, but they're better off with their microbial partners.

Why? Why have animals effectively outsourced parts of their development to other species? Why not just do everything in-house? 'I think it's unavoidable,' says John Rawls, who has worked with germ-free mice and squid. 'Microbes are a necessary part of animal life. There's no getting rid of them.' Remember that animals emerged in a world that had already been teeming with microbes for billions of years. They were the rulers of the planet long before we arrived. And when we *did* arrive, *of course* we evolved ways of interacting with the microbes around us. It would be absurd not to, like moving into a new city wearing a blindfold, earplugs, and a muzzle. Besides, microbes weren't just unavoidable: they were *useful.* They fed the pioneering animals. Their presence also provided valuable cues to areas rich in nutrients, to temperatures conducive to life, or flat surfaces upon which to settle. By sensing these cues, pioneering animals gained valuable information about the world around them. And as we shall see, hints of those ancient interactions still abound today.

Nicole King is far from home. She normally runs a lab at the University of California at Berkeley, but she's currently on vacation in London. She is about to take her eight-year-old son Nate to a matinee of the musical *Billy Elliot*, on condition that he sit patiently on a park bench next to us for half an hour while we talk about a little-known group of creatures called choanoflagellates. King is one of the few scientists who studies them intently, and since she affectionately calls them 'choanos', I will too.

They are found in water all around the world, from tropical rivers to the seas beneath Antarctic ice. As we talk, Nate, who has been quietly doodling on a pad, pipes up excitedly and draws one. He pens an oval with a sinuous tail and a collar of rigid filaments, like a sperm wearing a skirt. The beating tail drives bacteria and other detritus towards the collar, where they are trapped, engulfed, and digested; choanos are active predators. Nate's drawing captures their essence beautifully. In particular, it nails the fact that choanos are single-celled creatures. They're eukaryotes like you and me, complete with deluxe features like mitochondria and nuclei, which bacteria don't have. But, like bacteria, they consist of just one free-swimming cell.[10]

Sometimes, those cells show a social side. King's favoured species, *Salpingoeca rosetta,* often forms colonies or rosettes. Her son can draw these too – dozens of choanos with their heads facing inwards and their tails flailing outwards, like some kind of hairy raspberry. It looks like a group of choanos that have swum towards each other, but it is actually the result of division rather than collision. Choanos reproduce by dividing in two, but sometimes the two daughter cells fail to split completely and end up connected by a short bridge. This happens again and again, until there's a sphere of linked cells, enveloped in a single sheath. That's the rosette. It would be an obscure piece of biological trivia were it not for the fact that choanos are the closest living relatives of all animals.[11] They are the distant cousins of every frog, scorpion, earthworm, wren, and starfish. For King, who wants to understand how the animal kingdom first evolved, choanos are fascinating. And the process that creates the rosette, where a single cell becomes a multicellular cluster, is especially so.

We know very little about what the first animals looked like because their soft bodies didn't fossilise. They came and went like a winter breath, leaving no imprint upon the world. But we can make some educated guesses about them. All modern animals are multicellular creatures that begin life as a hollow ball of cells and eat other things for sustenance, so it's reasonable to think that our common ancestor shared the same traits.[12] These rosettes, then, are modern representations of what the first animals may have looked like. And the process that creates them, where a single cell divides into a cohesive colony, recapitulates the kind of evolutionary transition that gave rise to those proto-animals, and eventually to squirrels, pigeons, ducks, children and every other beast in the park where King and I are talking. Studying these innocuous, obscure, single-celled creatures is as close as King can get to filming the shrouded origins of our entire kingdom.

Her relationship with *S. rosetta* has been a rocky one. She knew it formed colonies in the wild, but couldn't persuade it to do so in her lab. In her hands, and those of other scientists, these social creatures mysteriously turned into loners. She changed their temperature, nutrient levels, acidity . . . nothing worked. King only solved the problem by giving up. Frustrated, she turned to a different goal: sequencing *S. rosetta*'s genome. That brought its own troubles. King had been feeding *S. rosetta* on bacteria, but she now had to get rid of these cells so that their genes wouldn't contaminate the sequencing results. So, she fed the choanos with a battery of antibiotics and, to her surprise, disrupted their ability to form colonies entirely. If they were reluctant to form colonies before, they were now utterly set against it. Something about *the bacteria* had been making them sociable.

Graduate student Rosie Alegado took the original water samples, isolated the microbes within, and fed them to the choanos one by one. Out of 64 species, just one bacterium restored the rosettes. That explained why King's original experiments never worked: *S. rosetta* forms colonies only when it meets the right microbe. Alegado identified the culprit and named it *Algoriphagus machipongonensis* – a new species, but part of the Bacteroidetes lineage that dominates our guts.[13] She also identified how the bacteria induce the rosettes: by releasing a fat-like molecule called RIF-1. 'I called it RIF for rosette-inducing factor, and I numbered it 1 because I'm sure there are others,' she says. She was right. The team have since identified several other molecules from many other microbes that can shove the choanos towards colonial life.

Alegado suspects that all these substances are a sign that food is near. The choanos are better at catching bacteria as a group than they are on their own, so if they sense bacteria nearby, they unite. 'I think the choanos are eavesdropping,' says Alegado. 'They're slow swimmers, and the Bacteroidetes are good indicators that they have entered an area with great resources and food. Then, they can commit to making a rosette.'

What to make of all this? Did bacteria drive the origin of animals, by providing cues that prompted our single-celled ancestors to form multicellular colonies? King advises caution. Today's choanos are our *cousins*, not our ancestors. It would be a big leap to deduce from their behaviour what ancient choanos did, let alone how they reacted to ancient microbes. King isn't prepared to do that yet. She now wants to check if modern animals respond to bacteria in the same way. If that's the case – if the same bacteria direct the development of choanos *and* animals via the same molecules – it would substantially strengthen the idea that this is an ancient phenomenon that played out at our origins. 'In the oceans in which the first animals evolved, I think there's no controversy that there was a ton of bacteria,' says King. 'They were diverse. They dominated the world, and animals had to accommodate to them. It's not a stretch to think that some molecules produced by bacteria may have influenced the development of the first animals.' No, it's not a stretch – especially given what still happens in Pearl Harbor.

On the morning of 7 December, 1941, a large squadron of Japanese fighter planes launched a surprise attack on the US naval base at Pearl Harbor in Hawaii. The USS *Arizona* was an early casualty; when she sank, she took more than 1,000 officers and crew with her. The other seven battleships in the harbour were either destroyed or heavily damaged, along with 18 more ships and 300 aircraft. Today, the harbour is a more tranquil place. Though it is still an important naval base and still home to several mighty ships, its greatest threat comes not from the sky, but from the sea.

You can see what happens to the ships by throwing a random scrap of metal into the water. Within hours, bacteria start growing on it. Algae might follow. There may be clams or barnacles. But eventually, within days, white tubes appear. They're tiny – each just a few centimetres long and a few millimetres wide. But soon there are hundreds of them. Then thousands. Millions. Eventually, the entire surface looks like a frozen shag pile rug. These tubes get everywhere: on rocks, pilings, fishing cages, and ships. If an aircraft carrier sits in the harbour for a few months, the tubes will amass on its hull in layers several centimetres deep. The technical term is *biofouling*. The lay version is 'a pain in the ass'. The Navy sometimes sends divers down to the ships to cover propellers and other sensitive structures in plastic bags so that the tubes can't clog them up.[14]

Each of these white cylinders houses, and is made by, an animal. The Navy folks call it 'the squiggly worm'. Michael Hadfield, a marine biologist at the University of Hawaii, knows it as *Hydroides elegans*. It was first described in Sydney Harbour and has since shown up in the Mediterranean, the Caribbean, the coast of Japan, Hawaii – any bay with warm water and ships. By clinging to man-made hulls, this master stowaway has colonised the whole world.

Hadfield started studying the squiggly worms in 1990, at the Navy's behest. He was already an expert on marine larvae, and the Navy wanted him to test a range of anti-fouling paints to see if any could repel the worms. But the real trick, he thought, would be to work out why the worms decide to settle down at all. What makes them suddenly appear on hulls?

This is an ancient question. In his wonderful biography of Aristotle, Armand Marie Leroi writes: 'A naval squadron, [Aristotle] says, once anchored off Rhodos and a lot of earthenware was thrown overboard. The pots collected mud and then living oysters. Since oysters can't move on to pots, or indeed anywhere, they must have arisen from the mud.'[15] This idea of spontaneous generation remained fashionable for centuries, but is hopelessly wrong. The truth behind the abrupt appearance of oysters and tube worms is more banal. These animals, like corals, sea urchins, mussels and lobsters, have larval stages that drift through the open ocean until they find somewhere to land. The larvae are microscopic, extraordinarily abundant (there might be 100 in a drop of seawater), and utterly unlike their adult counterparts. A baby sea urchin looks more like a shuttlecock than the pincushion it will become. A larval *H. elegans* looks like a wall-plug with eyes, not a long, tube-covered worm. It's hard to believe that it's the same animal.

At some point, the larvae settle down. They abandon their youthful wanderlust and remodel their bodies into sedentary adult shapes. This process – metamorphosis – is the most important moment in their lives. Scientists once suspected that it happened randomly, with the larvae settling in arbitrary places and surviving if they were lucky enough to hit a good location. In fact, they are purposeful and selective. They follow clues like chemical trails, temperature gradients, and even sounds, to find the best spots for metamorphosis.

Hadfield soon learned that *H. elegans* was drawn to bacteria and specifically to biofilms – the slimy mats of densely packed bacteria that quickly grow on submerged surfaces. When a larva finds a biofilm, it swims along the bacteria, pressing its face against them. After a few minutes, it anchors itself by extruding a thread of mucus from its tail, and secretes a transparent sock around its body. Firmly fastened, it begins to change. It loses the small beating hairs that once propelled it through the water. It gets longer. It grows a ring of tentacles around its head for snagging morsels of food. It starts laying down its hard tube. It is now an adult and it will never move again. This transformation utterly depends upon bacteria. To *H. elegans*, a clean, sterile beaker is like Neverland – a place of eternal immaturity.

The worms don't respond to any old microbe. Of the many strains in Hawaiian waters, Hadfield found that only a few could induce metamorphosis, and only one did so strongly. Its gargled mouthful of a name is *Pseudoalteromonas luteoviolacea*. Mercifully, Hadfield just calls it P-luteo. More than any other microbe, this one excels at turning larval worms into adults. Without the bacteria, the worms would never reach adulthood.[16]

They wouldn't be the only ones. Some sponge larvae also alight on surfaces and transform when they encounter bacteria. So do mussels, barnacles, sea squirts, and corals. Oysters belong on the list; sorry, Aristotle. *Hydractinia,* a tentacled relative of jellyfish and sea anemones, reaches adulthood when it touches bacteria that live on the shells of hermit crabs. The oceans are swarming with baby animals that only complete their life cycles upon contact with bacteria – and often P-luteo in particular.[17]

If these microbes suddenly disappeared, what would happen? Would these animals all become extinct, unable to mature and reproduce? Would coral reefs – the richest ecosystem in the oceans – fail to form without bacterial surveyors to scout out the right surfaces first? 'I don't think I've ever said anything that grand,' says Hadfield with the characteristic caution of a scientist. Then, surprising me, he adds, 'But it's a fair thing to say. Certainly, not every larva in the sea needs a bacterial stimulus, and there are so many larvae out there that haven't been tested. But, between tube worms and corals and sea anemones and barnacles and bryozoans and sponges . . . I could go on and on. There are examples in all of those groups where bacteria are the key.'

Again, one might ask: Why rely on bacterial cues? It's possible that the microbes improve a larva's grip on a surface or provide molecules that keep pathogens at bay. But Hadfield thinks that their value is simpler. The presence of a biofilm provides a larval animal with important information. It means that: (a) there's a solid surface, (b) which has been around for a while, (c) isn't too toxic, and (d) has enough nutrients to sustain microbes. Those reasons are as good as any to settle down. The better question would be: Why *wouldn't* you rely on bacterial cues? Or better still: what choice do you have? 'When the larvae of the first marine animals were ready to come down, there wasn't a clean surface,' says Hadfield, echoing Rawls and King. 'They were all covered in bacteria. It's not surprising that differences in those bacterial communities would be the original cue for settlement.'

* * *

King's choanos and Hadfield's worms are both exquisitely tuned to the presence of microbes, and dramatically transformed by them. Without bacteria, the sociable choanos would forever be solitary, and the larval worms would forever be immature. These are beautiful examples of how thoroughly microbes can shape the bodies of animals (or animal cousins). And yet, they aren't symbioses in the classical sense. The worms don't actually harbour P-luteo in their bodies, and they don't seem to interact with the bacterium after they become adults. Their relationship is transient. They are like tourists asking passers-by for directions and then moving on. But other animals form more lasting and co-dependent relationships with microbes.

The flatworm *Paracatenula* is one such creature. This tiny animal, which lives in warm ocean sediments all over the world, takes symbiosis to an extreme. Up to half of its centimetre-long body consists of bacterial symbionts, packed into a compartment called the trophosome that fills up 90 per cent of the worm. Pretty much everything behind the brain is either microbe or living quarters for microbes. Harald Gruber-Vodicka, who studies the flatworm, describes the bacteria as both its motor and its battery – they provide it with energy, and store that energy in the form of fats and sulphur compounds. These stores give the flatworm its bright white colour. They also fuel its most extraordinary ability.[18] *Paracatenula* is a master of regeneration. Cut it in two, and both ends become fully functional animals. The back half will even re-grow a head and brain. 'Chop them up and you can get ten,' says Gruber-Vodicka. 'That's probably what they do in nature. They get longer and longer, and then one end breaks off and there are two.' This skill depends entirely on the trophosome, the bacteria inside it, and the energy they lock away. As long as a fragment of flatworm contains enough symbionts, it can produce an entire animal. If the symbionts are too scarce, the fragment dies. Counter-intuitively, this means that the only bit of the flatworm that can't regenerate is the bacteria-free head. The tail will re-grow a brain but the brain alone will not produce a tail.

Paracatenula's partnership with microbes is typical of the entire animal kingdom, including you and me. We might not have the flatworm's wondrous healing powers, but we do host microbes *inside our bodies* and interact with them throughout our lives. Unlike Hadfield's tube worms, whose bodies are transformed by environmental bacteria at a single point in time, *our* bodies are continuously built and reshaped by the bacteria inside us. Our relationship with them isn't a one-off exchange but a continuous negotiation.

We have already seen that microbes influence the development of the gut and other organs, but they can't rest after the job is done. It takes work to keep an animal's body going. In the words of Oliver Sacks, 'Nothing is more crucial to the survival and independence of organisms – be they elephants or protozoa – than the maintenance of a constant internal environment.'[19] And in maintaining such constancy, microbes are crucial. They affect the storage of fat. They help to replenish the linings of the gut and skin, replacing damaged and dying cells with new ones. They ensure the sanctity of the blood–brain barrier – a web of tightly packed cells that lets nutrients and small molecules pass from blood to brain, but bars the way to larger substances and living cells. They even influence the relentless remodelling of skeletons, in which fresh bone is deposited and old stuff is reabsorbed.[20]

Nowhere is this steady influence more clear than in the immune system: the cells and molecules that collectively protect our bodies from infection and other threats. It's obscenely complicated. Picture an immense Rube Goldberg-esque machine, consisting of a seemingly endless array of components that spawn, trigger, and signal to one another. Now picture the same machine as a creaky, half-finished mess, where every part is either half-formed, low in number, or wired incorrectly. That's what the immune system looks like in a germ-free rodent. That's why these animals are, as Theodor Rosebury put it, 'susceptible to infection in general, retaining an infantile immaturity towards the perils of the world'.[21]

This tells us that an animal's genome doesn't provide everything it needs to create a mature immune system. It also needs input from a microbiome.[22] Hundreds of scientific papers, on species as diverse as mice, tsetse flies and zebrafish, have shown that microbes help to shape the immune system in some way. They influence the creation of entire classes of immune cells, and the development of organs that make and store those cells. They are especially important early in life, when the immunity machine is first constructed and tunes itself to the big, bad world. And once the machine is chugging away, microbes continue to calibrate its reactions to threats.[23]

Take inflammation: a defensive response, where immune cells rush to the site of an injury or infection, leading to swelling, redness, and heat. It's important for protecting the body against threats; without it, we'd be riddled with infections. But it becomes a problem if it spreads throughout the body, lasts too long, or launches at the slightest provocation: that leads to asthma, arthritis, and other inflammatory and autoimmune diseases. So, inflammation must be triggered at the right time, and controlled appropriately. Suppressing it is as important as activating it. Microbes do both. Some species stimulate the production of hawkish pro-inflammatory immune cells, while others induce dove-like anti-inflammatory cells.[24] Between them, they allow us to react to threats without overreacting. Without them, this balance disappears, which is why germ-free mice are prone to both infections *and* autoimmune diseases: they can neither mount an appropriate immune response when one is needed, nor fend off an inappropriate one during quieter times.

Let's pause to note how peculiar this all is. The traditional view of the immune system is full of military metaphors and antagonistic lingo. We see it as a defence force that discriminates self (our own cells) from non-self (microbes and everything else), and eradicates the latter. But now we see that microbes craft and tune our immune system in the first place!

Consider just one example: a common gut bacterium called *Bacteroides fragilis* or 'B-frag'. In 2002, Sarkis Mazmanian showed that this particular microbe can fix some of the immune problems in germ-free mice. Specifically, its presence restores normal levels of 'helper T cells', a crucial class of immune cell that rallies and coordinates the rest of the ensemble.[25] Mazmanian didn't even need the entire microbe.

He showed that a single sugar molecule in its coat, polysaccharide A (PSA), could boost the numbers of helper T cells on its own. This result was the first time anyone had shown that a single microbe – no, a single microbial *molecule* – could correct a specific immune problem. Mazmanian's team later showed that PSA can prevent and cure inflammatory diseases like colitis (which affects the gut) and multiple sclerosis (which affects nerve cells), at least in mice.[26] These are diseases of overreaction; PSA offers health through tranquillity.

But remember that PSA is *a bacterial molecule*: exactly the type of substance that, according to common wisdom, the immune system should see as a threat. PSA ought to trigger inflammation. Instead, it does the opposite: it *quells* inflammation and *calms* the immune system. Mazmanian calls it a 'symbiosis factor' – a chemical message from microbe to host that says: *I come in peace*.[27] This clearly shows that the immune system isn't innately hard-wired to tell the difference between a harmless symbiont and a threatening pathogen. In this case, it's the microbe that makes that distinction clear.

How, then, can we possibly view the immune system as an armada of destructive troops, belligerently bent on destroying microbes? It is evidently more subtle than that. It can come to a disastrous boil in one's own body, as in the case of autoimmune diseases like type 1 diabetes or multiple sclerosis. It also simmers gently in the presence of countless native microbes, like B-frag. I think it's more accurate to see the immune system as a team of rangers in charge of a national park – as ecosystem managers. They must carefully control the numbers of resident species, and expel problematic invaders.

But here's the twist: the creatures of the park hired the rangers in the first place. They taught their guardians which species to care for and which to evict. And they're constantly producing chemicals like PSA that affect how alert and responsive the rangers are. The immune system isn't just a means of controlling microbes. It is at least partly controlled *by microbes*. It's yet another route through which our multitudes preserve our bodies.

* * *

If you list all the species in a particular microbiome, you can tell who's there. If you list all the genes in those microbes, you can tell what they are capable of.[28] But if you list all the chemicals the microbes produce – their metabolites – you can tell what those species *are actually doing*. We have already met many of these chemicals, such as the symbiosis factor PSA, and the two squid-manipulating MAMPs that McFall-Ngai identified. There are hundreds of thousands more, and we're only just starting to understand what they all do.[29] These substances are the means by which animals converse with their symbionts. Many scientists are now trying to eavesdrop on these exchanges – and they're not the only ones. The molecules that microbes make can also extend beyond the bodies of their hosts, drifting through the air to convey messages at a distance. You can smell some of these communiqués if you head to the savannahs of Africa.

Of all Africa's large predators, spotted hyenas are the most sociable. A lion pride might comprise a dozen individuals, but a hyena clan has between 40 and 80. They won't all be in the same place together; small subgroups repeatedly form and dissolve over the course of the day. These dynamics make hyenas wonderful subjects for budding field biologists. 'You can observe lions in the field, but they'll just lie there, and you can work with wolves for years and just see scats or hear howls,' says hyena aficionado Kevin Theis. 'But hyenas . . . there are greetings, reintroductions, dominance and submissive signalling. You'll have young cubs trying to learn their place within the clan, immigrant males doing a run-through to see who's there. Their social lives are incredibly more complex.'

They deal with that complexity using a wide repertoire of signals – including chemical ones. A spotted hyena will straddle a long grass stalk and extrude a scent gland from its backside. It drags the gland across the stalk, leaving behind a thin paste. The colour can vary from black to orange, and the consistency from chalky to runny. And the smell? 'To me, it smells like fermenting mulch, but other people think it smells like Cheddar cheese or cheap soap,' says Theis.

He had been studying the pastes for years when a colleague asked him whether bacteria were involved in their odours. Theis was stumped. Then he found that other scientists had proposed that very idea in the 1970s, arguing that many mammals have bacteria in their scent glands, which ferment fats and proteins to produce smelly airborne molecules. Variations in these microbes could explain why different species have their own distinctive aromas – remember the popcorn-scented binturong from San Diego zoo?[30] They might also provide a badge of identity, revealing information about their host's health or status. And when individuals play, jostle, and mate they might share microbes that give them a characteristic group fragrance.

The hypothesis made sense but people struggled to validate it. Several decades on, with genetic tools at his disposal, Theis had no such problem. Working in Kenya, he collected paste samples from the glands of 73 anaesthetised hyenas. By sequencing the DNA of the resident microbes, he found more types of bacteria than all the previous surveys put together. He also showed that these bacteria, and the chemicals they produce, vary between spotted and striped hyenas, between spotted hyenas from different clans, between males and females, and between fertile and infertile ones.[31] Based on these differences, the paste could act as chemical graffiti that reveal who their makers are, which species they're from, how old they are, and whether they're ready to mate. By impregnating grass stalks with their smelly microbes, hyenas spray their personal tags all over the savannah.

This is still a hypothesis. 'We need to manipulate the scent microbiome and see if the odour profiles change,' says Theis. 'Then, we need to show that when the odours change, the hyenas pay attention and respond.' In the meantime, other scientists have found similar patterns in the scent glands and urine of other mammals, including elephants, meerkats, badgers, mice, and bats. The whiff of an old meerkat is distinct from Eau de Youngster. The stink of a male elephant differs from that of a female.

Then, there's us. The human armpit is not unlike a hyena's scent gland – warm, moist, and rich in bacteria. Each species creates its own aromas. *Corynebacterium* will convert sweat into something that smells like onions, and testosterone into something that smells either like vanilla, urine, or nothing, depending on the sniffer's genes. Do these scents make useful signals? Apparently so! This armpit microbiome is surprisingly stable – and so are our armpit odours. Every person has their own distinctive pong, and in several experiments, volunteers have been able to distinguish people from the smell of their T-shirts. They've even managed to match the smells of identical twins. Maybe, like hyenas, we can also glean information about each other by sniffing the messages sent out by our microbes. It's not just mammals, either. The gut bacteria of the desert locust produces parts of the aggregation pheromone that encourages these solitary insects to form sky-canvassing swarms. The gut bacteria of German cockroaches account for their revolting tendency to congregate around each other's faeces. And the giant mesquite bugs rely on their symbionts to make an alarm pheromone that they use to warn each other of danger.[32]

Why should animals rely on microbes to make these chemical signals? Theis offers the same reason that Rawls, King, and Hadfield did: it's inevitable. Every surface is populated by microbes, which release volatile chemicals. If those chemical cues reflect a trait that's useful to know about – say, gender, strength, or fertility – the host animal might evolve scent organs to nourish and harbour those specific microbes. Eventually, the inadvertent cues turn into full-blown signals. So, by creating airborne messages, microbes could affect the behaviour of animals far outside their original hosts. And if that's the case, it shouldn't be surprising to learn that they can affect animal behaviour in more local ways.

In 2001, neuroscientist Paul Patterson injected pregnant mice with a substance that mimics a viral infection and triggers an immune response. The mice gave birth to healthy pups but as the babies grew into adults, Patterson started noticing interesting quirks in their behaviour. Mice are naturally reluctant to enter open spaces but these mice were especially so. They were easily startled by loud noises. They would groom themselves over and over, or repeatedly try to bury a marble. They were less communicative than their peers, and they shied away from social contact. Anxiety, repetitive movements, social problems: in his mice, Patterson saw reflections of two human conditions – autism and schizophrenia. Those similarities weren't entirely unexpected. Patterson had read that pregnant women who incur serious infections, like flu or measles, are more likely to have kids with autism and schizophrenia. He thought that a mother's immune responses might somehow affect the development of her baby's brain. He just didn't know how.[33]

The penny dropped several years later when Patterson was having lunch with his colleague Sarkis Mazmanian, who discovered the anti-inflammatory effects of the gut bacterium B-frag. Together, the scientists realised that they had been looking at two halves of the same problem. Mazmanian had shown that gut microbes affect the immune system, and Patterson had found that the immune system affects the developing brain. And they realised that Patterson's mice had gut problems in common with actual autistic children: both were more likely to have diarrhoea and other gastrointestinal disorders, and both harboured unusual communities of gut microbes. Perhaps, the duo reasoned, those microbes were somehow affecting behavioural symptoms in both mice and kids? And perhaps, they reasoned, fixing those gut problems might also lead to changes in behaviour?

To test this idea, the duo fed B-frag to Patterson's mice. The results were remarkable. The rodents became keener to explore, harder to startle, less prone to repetitive movements, and more communicative. They were still reluctant to approach other mice, but in every other respect B-frag had reversed the changes caused by their mothers' immune responses.

How? And why? Here's the best guess: By mimicking a viral infection in the pregnant mothers, the team triggered an immune response that landed their offspring with an excessively permeable gut, and one with an unusual collection of microbes. Those microbes produced chemicals that entered the bloodstream and travelled to the brain, where they triggered atypical behaviours. The top culprit is a toxin called 4-ethylphenylsulfate (4EPS), which can trigger anxiety in otherwise healthy animals. When the mice swallowed B-frag, this microbe sealed up their guts and stemmed the flow of 4EPS (and other substances) to their brain, reversing their atypical symptoms.

Patterson died in 2014 but Mazmanian is now carrying on his friend's work. His long-term goal is to develop a bacterium that people can swallow to control some of the more difficult symptoms of autism. That might be B-frag: it certainly worked well in the mice, and happens to be the most heavily depleted microbe in the guts of people with autism. Parents with autistic children, who read about his work, regularly email him about where to get the bacterium. Many such parents are already giving probiotics to their kids to help with their gut problems, and some claim to have seen improvements in behaviour. Mazmanian now wants hard clinical evidence to accompany these anecdotes. He is optimistic.

Others are more sceptical. The most obvious critique, as science writer Emily Willingham puts it, is that 'mice don't have autism, which is a human neurobiological construct shaped in part by social and cultural perceptions of what is considered *normal*'.[34] Is a mouse repeatedly burying a marble really like a child rocking back and forth? Is a lower frequency of squeaks the same as being unable to talk to other people? If you squint just so, the similarities jump out. Look again, and you might see parallels to other conditions; indeed, Patterson's mice were originally bred to model schizophrenia rather than autism. Then again, Mazmanian's team recently did an experiment which hints that the two sets of behaviour are related. They transferred gut microbes from children with autism into mice, and found that the rodents developed the same quirks that Patterson saw, such as repetitive behaviour and social aversion.[35] This suggests that the microbes are at least partly responsible for these behaviours. 'I don't think anyone would ever claim that you can reproduce autism in a mouse model,' says Mazmanian, sanguinely. 'It's inherently limited, but it is what it is.'

At the very least, Patterson and Mazmanian showed that tweaking a mouse's gut microbes – or even a single microbial molecule, 4EPS – could change its behaviour. So far, we have seen that microbes can influence the development of guts and bones, blood vessels and T cells. Now we've seen that they can sway the brain too – the organ that, more than any other, makes us who we are. It is a disquieting thought. We put such a premium on our free will that the prospect of losing independence to unseen forces informs many of our deepest societal fears. Our darkest fiction is full of Orwellian dystopias, shadowy cabals, and mind-controlling supervillains. But it turns out that the brainless, microscopic, single-celled organisms that live inside us have been pulling on our strings all along.

On 6 June, 1822, on an island in the Great Lakes, a 20-year-old fur trader named Alexis St Martin accidentally took a musket shot to his side. The only doctor on the island was an army surgeon named William Beaumont. When Beaumont arrived on the scene, St Martin had been bleeding for half an hour. His ribs were cracked, his muscles shredded. A bit of burnt lung was poking out of his side. His stomach had a finger-wide hole in it, with food leaking out. 'In this dilemma I considered my attempt to save his life entirely useless,' Beaumont later wrote.[36]

He tried, though. He took St Martin into his home and, against all odds, after many surgeries and months of care, managed to stabilise him. But St Martin never completely healed. His stomach attached itself to the corresponding hole in his skin, creating a permanent port-hole into the outside world – an 'accidental orifice', in Beaumont's words. With fur-trapping out of the question, St Martin joined Beaumont as a handyman-cum-servant. Beaumont treated the man as a guinea pig. At the time, people knew next to nothing about how digestion worked. In St Martin's wound, Beaumont saw a literal window of opportunity. He collected many samples of stomach acid, and he sometimes dangled food through the open hole to watch it being digested in real time. The experiments continued until 1833, after which the men finally parted ways. St Martin returned to Quebec, where he died as a farmer at the age of 78. Beaumont became known as the Father of Gastric Physiology.[37]

Among Beaumont's many observations, he noticed that St Martin's mood affected his stomach. When the man became angry or irritable – and it's hard to imagine not getting irascible when a surgeon is dangling food through the hole in your side – his rate of digestion changed. That was the first clear sign that the brain affects the gut. Almost two centuries later, this maxim seems all too familiar. We lose our appetite when our mood changes, and our mood changes when we feel hungry. Psychiatric problems and digestive problems often go hand in hand. Biologists speak of a 'gut–brain axis' – a two-way line of communication between the gut and the brain.

We now know that gut microbes are part of this axis, in both directions. Since the 1970s, a trickle of studies have shown that any kind of stress – starvation, sleeplessness, being separated from one's mother, the sudden arrival of an aggressive individual, uncomfortable temperatures, overcrowding, even loud noises – can change a mouse's gut microbiome. The opposite is also true: the microbiome can affect a host's behaviour, including its social attitudes and its ability to deal with stress.[38]

In 2011, this trickle of studies became a flood. Within months of each other, several scientists published fascinating papers showing that microbes can affect brain and behaviour.[39] At Sweden's Karolinska Institute, Sven Petterson found that germ-free mice were less anxious and took more risks than their microbe-laden cousins. But if these mice were colonised by microbes as pups, they grew up into adults that behaved in the usual cautious ways. On the other side of the Atlantic, Stephen Collins from McMaster University made a similar discovery almost by accident. A gastroenterologist by training, he was looking at how probiotics affect the guts of germ-free mice. 'One of my technicians said to me: There's something wrong with this probiotic because it's making the mice jumpy,' he recalls. 'They seem different.' Collins then worked with two common strains of lab mice, one of which is naturally more timid and anxious than the other. If he colonised germ-free versions of the bolder strain with microbes from the timid strain, they became more timid themselves. The opposite was also true: germ-free versions of the timid mice were emboldened by the microbes of their more intrepid cousins. It was as dramatic a result as Collins could have hoped for: by swapping the bacteria in the animals' guts, he had also swapped part of their personalities.

As we have seen, germ-free mice are odd creatures with many physiological changes that could have impinged on their behaviour. So it was important that John Cryan and Ted Dinan from the University of Cork in Ireland found similar results, but in normal mice with complete microbiomes. They worked with the same strain of timid mice that Collins studied, and managed to change the animals' behaviour by feeding them with a single strain of *Lactobacillus rhamnosus* – a bacterium commonly used in yoghurts and dairy products. After the mice ingested this strain, known as JB-1, they were better able to overcome anxiety: they spent more time in the exposed parts of a maze, or the centre of an open field. They were also better at resisting negative moods: when dropped into a bottle of water, they spent more time paddling away than floating aimlessly.[40] These kinds of test are commonly used to test the effectiveness of psychiatric drugs, and JB-1 was behaving rather like substances with anti-anxiety *and* antidepressant properties. 'It was like the mice were on low doses of Prozac or Valium,' says Cryan.

To find out what the bacterium was doing, the team looked in the brains of the mice. They saw that JB-1 changed how different parts of the brain – those involved in learning, memory, and emotional control – responded to GABA, a pacifying chemical that quiets the buzz of excitable neurons. Again, there were striking parallels to human mental disorders: problems with GABA responses have been implicated in both anxiety and depression, and a group of anti-anxiety drugs called benzodiazepines work by enhancing GABA's effects. The team also worked out *how* the microbes were affecting the brain. Their main suspect was the vagus nerve. It's a long, branching nerve that carries signals between the brain and visceral organs like the gut – a physical embodiment of the gut–brain axis. The team severed it, and found that the mind-altering JB-1 lost all its influence.[41]

These studies, and others that followed, all showed that changing a mouse's microbiome can change its behaviour, the chemicals in its brain, and its susceptibility to the mouse versions of anxiety and depression. But they also have many inconsistencies. Some studies found that microbes only affect the brains of young mice; others that adolescents and adults are also affected. Some found that bacteria make rodents less anxious; others, more so. Some show that the vagus nerve is vital; others emphasise that microbes can produce neurotransmitters like dopamine and serotonin, which carry messages from one neuron to another.[42] These contradictions aren't unexpected – when two things as fiendishly complex as the microbiome and the brain collide, it would be naïve to expect clean results.

The big question now is whether any of this matters in real life. Are these subtle microbial influences, which show up in the controlled environments of laboratory rodents, actually important in the real world? Cryan understands that scepticism is justified, and that there is only one way to quell it: they need to go beyond rodent experiments. 'We have to go into humans,' he says.

There is a smattering of research looking at whether people behave differently after doses of antibiotics or probiotics, but they are plagued by methodological problems and ambiguous results. In one of the more promising studies (albeit, still a small one), Kirsten Tillisch found that women who ate twice-daily servings of a microbe-rich yoghurt showed less activity in parts of the brain involved in processing emotions, compared to women who ate microbe-free milk products. The meaning of these differences is open to debate, but they do at least show that bacteria can affect human brain activity.[43]

The real test will be to see if bacteria can help people to cope with stress, anxiety, depression, and other mental health issues. Already, there are some signs of success. Stephen Collins has just completed a small clinical trial in which a probiotic bacterium – a proprietary

Bifidobacterium strain owned by a food company – reduced symptoms of depression in people with irritable bowel syndrome.[44] 'It's the first demonstration, I think, of the ability of a probiotic to reduce abnormal behaviour in a patient group,' he says. Meanwhile, John Cryan and Ted Dinan are close to finishing their own trial, to see if probiotics – or, in their words, psychobiotics – can help people to cope with stress. Dinan, a psychiatrist who runs a clinic for people with depression, is measured about his hopes. 'I must say that I was profoundly sceptical that giving an animal a microbe could change their behaviour,' he says. He's now convinced, but he still believes 'it's highly unlikely that we'll come up with a cocktail of probiotics that will treat severe depression. But there's potential at the milder end of the spectrum. There are lots of people who don't want to take antidepressants or find therapy too expensive, and if we can give them an effective probiotic, that would be a major advance in psychiatry.'

These studies are already forcing scientists to view different aspects of human behaviour through a microbial lens. Drinking lots of alcohol makes the gut leakier, allowing microbes to more readily influence the brain – could that help to explain why alcoholics often experience depression or anxiety? Our diet reshapes the microbes in our gut – could those changes ripple out to affect our minds?[45] The gut microbiome becomes less stable in old age – could that contribute to the rise of brain diseases in the elderly? And could our microbes manipulate our food cravings in the first place? If you reach for a burger or a chocolate bar, what exactly is pushing that hand forward?

From your perspective, choosing the right item on a menu is the difference between a good meal and a bad one. But for your gut bacteria, the choice is more important. Different microbes fare better on certain diets. Some are peerless at digesting plant fibres. Others thrive on fats. When you choose your meals, you are also choosing which bacteria get fed, and which get an advantage over their peers. But they don't have to sit there and graciously await your decision. As we have seen, bacteria have ways of hacking into the nervous system. If they released dopamine, a chemical involved in feelings of pleasure and reward, when you ate the 'right' things, could they potentially train you to choose certain foods over others? Do they get a say in your menu picks?[46]

For now, it's just a hypothesis – but not a far-fetched one. Nature is full of parasites that control the minds of their hosts.[47] The rabies virus infects the nervous system and makes its carriers violent and aggressive; if they lash out at their peers, and inflict bites and scratches, they pass the virus on to new hosts. The brain parasite *Toxoplasma gondii* is another puppetmaster. It can only sexually reproduce in a cat; if it gets into a rat, it suppresses the rodent's natural fear of cat odours and replaces it with something more like sexual attraction. The rodent scurries *towards* nearby cats, with fatal results, and *T. gondii* gets to complete its life cycle.[48]

The rabies virus and *T. gondii* are outright parasites, selfishly reproducing at the expense of their hosts, with detrimental and often fatal results. Our gut microbes are different. They are natural parts of our lives. They help to construct our bodies – our gut, our immune system, our nervous system. They benefit us. But we shouldn't let that lure us into a false sense of security. Symbiotic microbes are still their own entities, with their own interests to further and their own evolutionary battles to wage. They can be our partners, but they are not our friends. Even in the most harmonious of symbioses, there is always room for conflict, selfishness, and betrayal.

4. TERMS AND CONDITIONS APPLY

In 1924, Marshall Hertig and Simeon Burt Wolbach found a new microbe inside common brown mosquitoes, *Culex pipens*, which they had collected near Boston and Minneapolis.[1] It looked a bit like the *Rickettsia* bacteria that Wolbach had previously identified as the cause of Rocky Mountain spotted fever and typhus. But this new microbe didn't seem responsible for any disease – and so was largely ignored. It took twelve years for Hertig to formally name it *Wolbachia pipientis*, in honour of his friend who found it and the mosquito that carried it. And it took many more decades for biologists to realise just how special this bacterium really is.

It is not unusual for science writers who regularly write about microbiology to pick a favourite bacterium, much as people would choose a favourite film or band. *Wolbachia* is mine. It is breathtaking in its behaviour and majestic in its spread. It is also the perfect example of the dual nature of microbes – all microbes – as partners or parasites.

In the 1980s and 1990s, after Carl Woese showed the world how to identify microbes by sequencing their genes, biologists started finding *Wolbachia* everywhere. People who were independently studying bacteria that could manipulate the sex lives of their hosts realised that they were all working on the same thing. Richard Stouthamer discovered a group of asexual, all-female wasps, which only reproduced by cloning themselves. This trait was the work of a bacterium, *Wolbachia*: when Stouthamer treated the wasps with antibiotics, the males suddenly reappeared and both sexes started mating again. Thierry Rigaud found bacteria in woodlice that transformed males into females by interfering with the production of male hormones; it was *Wolbachia,* too. In Fiji and Samoa, Greg Hurst found that a bacterium was killing the male embryos of the magnificent blue-moon butterfly, so that the females outnumbered the males by a hundred to one. Again: *Wolbachia.* Maybe not exactly the same strain, but all were different versions of the microbe from Hertig and Wolbach's mosquito.[2]

There's a reason why all of these strategies are bad news for males. *Wolbachia* can only pass to the next generation of hosts in eggs; sperm are too small to contain it. Females are its ticket to the future; males are an evolutionary dead end. So it has evolved many ways of screwing over male hosts to expand its pool of female ones. It kills them, as in Hurst's butterflies. It feminises them, as in Rigaud's woodlice. It eliminates the need for them entirely by allowing females to reproduce asexually, as in Stouthamer's wasps. None of these manipulations is unique to *Wolbachia,* but it is the only bacterium to use them all.

Where *Wolbachia* does allow males to survive, it still manipulates them. It often changes their sperm so that they cannot successfully fertilise eggs unless the eggs are infected with the same strain of *Wolbachia.* From the females' perspective, this incompatibility means that infected females (which can mate with whomever they like) gain a competitive advantage over uninfected females (which can only mate with uninfected males). With every passing generation, the infected females become more common, as do the *Wolbachia* they carry. This is called cytoplasmic incompatibility, and it's *Wolbachia*'s most common and most successful strategy – the strains that use it spread so quickly through a population that they typically infect 100 per cent of their potential hosts.

Aside from these misandrist tricks, *Wolbachia* also excels at invading ovaries and entering egg cells, so it quickly becomes a hand-me-down that insects pass on to their offspring. It is also unusually good at jumping into new hosts, so even if it breaks up with any one species, it has dozens of new ones to inhabit. 'I might find the same *Wolbachia* strain in a beetle in Australia and a fly in Europe,' says Jack Werren, who studies the bacterium. For these reasons, *Wolbachia* has become exceptionally common. One recent study estimated that it infects at least four in every ten species of arthropods – the animal group that includes insects, spiders, scorpions, mites, woodlice, and more. That is a preposterous proportion! The majority of the 7.8 million or so living animal species are arthropods. If *Wolbachia* infects 40 per cent of them,[3] it is arguably the most successful bacterium in the world, at least on land.[4] And, rather tragically, Wolbach never knew. He died in 1954, unaware that his name had been grafted onto one of the greatest pandemics in the history of life.

In many animals, *Wolbachia* is a reproductive parasite: an organism that manipulates the sex lives of its hosts to further its own ends. The hosts suffer. Some die, others become sterile, and even unaffected individuals must live in a skewed world with few potential mates. *Wolbachia* might seem like the archetypal 'bad microbe', but it has a beneficent side, too. It provides some unknown benefit to certain nematode worms, which cannot survive without it. It protects some flies and mosquitoes from viruses and other pathogens. The wasp *Asobara tabida* cannot make eggs without it. In bed bugs, *Wolbachia* is a nutritional supplement: it makes B-vitamins that are lacking in the blood that the bugs drink. Without it, the bugs are stunted and infertile.[5]

The most striking use for *Wolbachia* becomes apparent if you walk through a European apple orchard in the autumn. Among the yellow and orange leaves, you might find some with small green islands, defiantly resisting the seasonal decay. These are the work of the spotted tentiform leaf miner, a moth whose caterpillars live inside the leaves of apple trees. Almost all of them carry *Wolbachia*. In these insects, the microbe releases hormones that stop the leaves from yellowing and dying. They are the means by which the caterpillar holds back the autumn, to give itself enough time to become an adult. If you cure *Wolbachia* the leaves will die and fall, as will the caterpillars inside.

Wolbachia, then, is a microbe of many guises. Some strains act as ur-parasites, selfish manipulators of such skill that they have spread throughout the world on the wings and legs of legions of hosts; they kill animals, warp their biology, and lay restrictions upon their choices. Other strains are mutualists, boons, indispensable allies. Some are both. And in this multifarious nature, *Wolbachia* is not alone.

Here is a strange but critical sentiment to introduce in a book about the benefits of living with microbes: there is no such thing as a 'good microbe' or a 'bad microbe'. These terms belong in children's stories. They are ill-suited for describing the messy, fractious, contextual relationships of the natural world.[6]

In reality, bacteria exist along a continuum of lifestyles, between 'bad' parasites and 'good' mutualists. Some microbes, like *Wolbachia,* slide from one end of the parasite-mutualist spectrum to the other, depending on the strain, and on the host they find themselves in. But many exist at both ends of the continuum at once: the stomach bacterium *Helicobacter pylori* causes ulcers and stomach cancer, but also protects against oesophageal cancer – and it's the same strains that account for both these pros and cons.[7] Others can change roles in the same hosts, depending on the context. All of this means that labels like mutualist, commensal, pathogen, or parasite don't quite work as badges of fixed identity. These terms are more like states of being, like hungry or awake or alive, or behaviours like cooperating or fighting. They're adjectives and verbs rather than nouns: they describe how two partners relate to one another at a given time and place.

Nichole Broderick found a great example of this when she was studying a soil-dwelling microbe called *Bacillus thuringiensis,* or Bt. It produces toxins that can kill insects by punching holes in their guts. Farmers have exploited this ability since the 1920s, by spraying Bt onto crops as a living pesticide. Even organic farmers do this. The bacterium's effectiveness is undeniable, but for decades scientists had the wrong idea about *how* it kills. They assumed that its toxins inflict so much damage that their victims starve to death. But this couldn't be the whole story. It takes more than a week for a caterpillar to starve, and Bt kills in half that time.

Broderick found out what was really going on – and almost by accident.[8] She suspected that caterpillars would have gut microbes that protect them from Bt, so she treated them with antibiotics and exposed them to the pesticide. With the microbes gone, she expected them to die even faster. In fact, they all survived. It turned out that the gut bacteria, rather than protecting the caterpillars, are the means through which Bt kills. They are harmless if they stay in the gut, but they can pass through the holes created by Bt toxins and invade the bloodstream. When the caterpillar's immune system senses them, it goes berserk. A wave of inflammation spreads through the caterpillar's body, damaging its organs and interfering with its blood flow. This is sepsis. It's what kills the insect so quickly.

The same thing probably happens to millions of people every year. We humans are also infected by pathogens that create holes in our guts; and we also get sepsis when our usual gut microbes cross into our bloodstream. As in the caterpillars, the same microbes can be good in the gut, but dangerous in the blood. They're only mutualists by virtue of where they live. The same principles apply to so-called 'opportunistic bacteria' that live in our bodies – they are normally harmless but they can cause life-threatening infections in people whose immune system is weakened.[9] It is all down to context. Even symbionts as essential and long-standing as mitochondria, the energy-providing power plants that exist in all animals' cells, can wreak havoc if they end up in the wrong place. A cut or a bruise can split some of your cells apart and spill fragments of mitochondria into your blood – fragments that still keep some of their ancient bacterial character. When your immune system spots them, it mistakenly assumes that an infection is under way and mounts a strong defence. If the injury is severe, and enough mitochondria are released, the resulting body-wide inflammation can build into a lethal condition called systemic inflammatory response syndrome (SIRS).[10] SIRS can be worse than the original injury. Absurdly, it's simply the result of a human body mistakenly overreacting to microbes that have been domesticated for over two billion years. Just as a weed is a flower in the wrong place, our microbes might be invaluable in one organ but dangerous in another, or essential inside our cells but lethal outside them. 'If you go immunosuppressed for a little bit, they'll kill you. When you die, they'll eat you,' says coral biologist Forest Rohwer. 'They don't care. It's not a nice relationship. It's just biology.'

So, the world of symbiosis is one in which our allies can disappoint us and our enemies can rally to our side. It's a world where mutualisms shatter for the matter of a few millimetres.

Why are these relationships so tenuous? Why do microbes so easily slide between pathogen and mutualist? For a start, these roles are not as contradictory as you might imagine. Think about what a 'friendly' gut microbe needs to set up a stable relationship with its host. It must survive in the gut, anchor itself so it doesn't get swept away, and interact with its host's cells. These are all things that pathogens must do, too. So both characters – mutualists and pathogens, heroes and villains – often use the same molecules for the same purposes. Some of these molecules get saddled with negative names, like 'virulence factors', because they were first discovered in the context of disease, but they are inherently neutral. They are just tools, like computers, pens, and knives: they can be used to do wonderful things and terrible things.

Even helpful microbes can indirectly harm us, by creating vulnerabilities that other parasites and pathogens can exploit. Their very presence creates openings. An aphid's microbes, though essential, release airborne molecules that attract the marmalade hoverfly. This black-and-white insect, which looks like a wasp, is death to aphids. Its larvae can eat hundreds of them over a lifetime, and the adults find prey for their offspring by sniffing out Eau de Microbiome – a scent that the aphids can't avoid giving off. The natural world is full of these inadvertent lures. You are giving some off right now. Certain bacteria can even turn their owners into magnets for malarial mosquitoes, whilst others put off the little bloodsuckers. Ever wonder why two people can walk through a midge-filled forest and one emerge with dozens of welts while the other just has a smile? Your microbes are part of the answer.[11]

Pathogens can also use our microbes to launch their invasions, as is the case with the virus that causes polio. It grabs molecules on the surface of gut bacteria as if they were reins, using them to ride the bacteria towards a host's cells. The virus has a better grip on mammalian cells and becomes more stable at our warm body temperatures *after* touching our gut microbes. These microbes inadvertently turn it into a more effective virus.[12]

So, symbionts don't come for free. Even when they help their hosts, they create vulnerabilities. They need to be fed, housed, and transmitted, all of which costs energy. And most importantly, like every other organism, they have their own interests – which often clash with those of their hosts. If a maternally inherited symbiont like *Wolbachia* did away with males it would get more hosts in the short term, at the risk of driving those hosts extinct in the long term. If a few of a bobtail squid's bacteria stopped glowing they'd save on energy, but if enough of them went dark, the squid would lose its protective glow and the whole alliance would be swallowed by a watchful predator. If my gut microbes suppressed my immune system they would grow more readily, but I would get sick.

Almost every major partnership in the natural world is like this. Cheats are always a problem. Betrayal lurks perpetually on the horizon. Couples might work well together, but if one partner can get the same benefits without spending as much energy or effort, it will do so unless punished or policed. H. G. Wells wrote about this in 1930: 'Every symbiosis is, in its degree, underlain with hostility, and only by proper regulation and often elaborate adjustment can the state of mutual benefit be maintained. Even in human affairs, the partnerships for mutual benefit are not so easily kept up, in spite of me being endowed with intelligence and so being able to grasp the meaning of such a relation. But in lower organisms, there is no such comprehension to help keep the relationship going. Mutual partnerships are adaptations as blindly entered into and as unconsciously brought about as any others.'[13]

These principles are easy to forget. We like our black-and-white narratives, with clear heroes and villains. In the last few years, I've seen the viewpoint that 'all bacteria must be killed' slowly give ground to 'bacteria are our friends and want to help us', even though the latter is just as wrong as the former. We cannot simply assume that a particular microbe is 'good' just because it lives inside us. Even scientists forget this. The very term *symbiosis* has been twisted so that its original neutral meaning – 'living together' – has been infused with positive spin, and almost flaky connotations of cooperation and harmony. But evolution doesn't work that way. It doesn't necessarily favour cooperation, even if that's in everyone's interests. And it saddles even the most harmonious relationships with conflict.

We can see this clearly if we temporarily leave the world of microbes and think a little bigger. Take oxpeckers. These brown birds can be found in Africa, clinging to the flanks of giraffes and antelope. They're classically viewed as cleaners that pick ticks and blood-sucking parasites off their hosts. But they also peck at open wounds – a less helpful habit that stymies the healing process and increases the risk of infection. These birds crave blood, and they satisfy that craving in ways that either profit their hosts, or punish them. A similar dynamic goes on in coral reefs, where a small fish called the cleaner wrasse runs a health spa. Big fish arrive, and the wrasse picks parasites from their jaws, gills, and other hard-to-reach places. The cleaners get meals, and the clients get healthcare. But the cleaners sometimes cheat by nipping bits of mucus and healthy tissue. The clients punish them by taking their business elsewhere, and the cleaners themselves will castigate any colleagues that annoy potential customers. Meanwhile, in South America, acacia trees rely on ants to defend them from weeds, pests, and grazers. In return, they give their bodyguards sugary snacks to eat and hollow thorns to live in. It looks like an equitable relationship, until you realise that the tree laces its food with an enzyme that stops the ants from digesting other sources of sugar. The ants are indentured servants. All of these are iconic examples of cooperation, found in textbooks and wildlife documentaries. And each of them is tinged with conflict, manipulation, and deceit.[14]

'We need to separate *important* from *harmonious*. The microbiome is incredibly important but it doesn't mean that it's harmonious,' says evolutionary biologist Toby Kiers.[15] A well-functioning partnership could easily be seen as a case of reciprocal exploitation. 'Both partners may benefit but there's this inherent tension. Symbiosis *is* conflict – conflict that can never be totally resolved.'

It can, however, be managed and stabilised. The waters around Hawaii aren't full of dark squid.[16] Many *Wolbachia*-infected insects still have males. My immune system is working reasonably well. All of us have found ways of stabilising our relationship with our microbes, of promoting fealty rather than defection. We have evolved ways of selecting which species live with us, restricting where they sit in our bodies, and controlling their behaviour so they are more likely to be mutualistic than pathogenic. Like all the best relationships, these ones take work. Every major transition in the history of life – from single-celled to multi-celled, from individuals to symbiotic collectives – has had to solve the same problem: how can the selfish interests of individuals be overcome to form cooperative groups?

How, in other words, do I contain my multitudes?

Containing our multitudes is not unlike a bit of agriculture. We use fences and barriers to mark the boundaries of our gardens. We use fertiliser to feed the plants. We uproot and poison incipient weeds. And we set the garden in a place with the right temperature, soil, and levels of sunlight to nourish whatever we want to grow. Animals use equivalents of all of these measures to set the terms and conditions for their microbial partnerships.[17] We will meet each one in turn.

To begin with, every body part on every species has its own zoological terroir – its unique combination of temperature, acidity, oxygen levels, and other factors that dictate what kinds of microbe can grow there. The human gut might seem like nirvana for microbes, with its regular baths of food and fluid. But it is a challenging environment, too. That food supply comes in a fast-flowing torrent, so microbes need to grow quickly or carry molecular anchors to maintain a foothold. The gut is a dark world, so microbes that depend on sunlight to make their food cannot thrive. It lacks oxygen, which explains why the overwhelming majority of gut microbes are anaerobes – organisms that ferment their food, and grow without this supposedly essential gas. Some of them are so reliant on the absence of oxygen that they die in its presence.

The skin is different: it varies from cool, dry deserts like the forearm to warm, humid jungles like the groin or armpits. Sunlight is abundant, but is also a problem because of the ultraviolet radiation it contains. Oxygen matters here too, and since most of the skin is exposed to fresh air, aerobes thrive. However, concealed niches, like sweat glands, can support the growth of oxygen-hating anaerobes like *Propionibacterium acnes*, the microbe that causes acne. All over our bodies, the laws of physics and chemistry sculpt bundles of biology.

Animals can also actively manipulate the conditions within themselves to lay out welcome mats and cordon off forbidden zones. Our stomachs secrete powerful acids that keep most bacteria at bay, except for a few tolerant specialists like *H. pylori*. Carpenter ants don't have acid-secreting stomachs, but they do produce formic acid from a gland at their back-ends. Normally, they spray the stuff as a defensive weapon, but by sucking the acid from their own bums they can acidify their digestive tracts to keep out unwanted microbes.[18]

These conditions set out the main admission requirements for life in our bodies. They are crude filters, which roughly dictate the types of microbes that can share our lives, while marking out the places where they can live. But we also need more specific ways of fine-tuning our microbial communities, and firmer blockades for keeping them in place. Remember that location is important: microbes can easily switch from beneficial allies to fatal threats depending on where they are. So, many animals set up actual barriers for walling off their microbial gardens. We have evolved good fences for making good neighbours. The bobtail squid has crypts for housing its luminous partners. The regenerating flatworm *Paracatenula* devotes most of its body to housing microbes. Stinkbugs have an extremely narrow corridor midway down their digestive tract, which stops the flow of food and fluid and turns the back half of the gut into a roomy apartment for microbes. And up to a fifth of insect species enclose their symbionts within special cells called bacteriocytes.[19]

Bacteriocytes have repeatedly evolved in different lineages. Some insects slot them between other cells; others bundle them together into organs called bacteriomes, which branch off from the gut like clusters of grapes. Whatever their origin, their functions are the same: contain and control bacterial symbionts; stop them from spreading into other tissues; and hide them from the immune system. Bacteriocytes are not luxury accommodation. A single one can contain tens of thousands of bacteria, packed so tightly that they make sardine cans look roomy. They are cells in more ways than one.

They are also tools of control. Despite the old and mutually dependent relationships that many insects have with their symbionts, there is still plenty of room for conflict. If that sounds strange, think about the millions of people diagnosed with cancer every year. Cancer is a disease of cellular rebellion, where a cell strikes out against the regulations of its own body. It grows and divides uncontrollably, producing tumours that can jeopardise the life of its host. If human cells can do this when they are actually part of the same animal, it is easy to imagine that a bacterium like *Blochmannia*, which is still a separate organism from its ant host, might do the same. It could turn into a kind of symbiotic cancer that replicates unchecked, soaks up energy that the ant needs for itself, and invades cells it shouldn't.[20]

With bacteriocytes, insects can stop this from happening. Insects can control the movement of nutrients across the bacteriocytes, withholding them from any cheating symbionts that violate the terms of their tenancy and fail to provide the requisite benefits. They can bombard the captive microbes with damaging enzymes and antibacterial chemicals to keep their populations under tight control. The cereal weevil – a long-snouted beetle that devours rice and other grains – does this to the *Sodalis* bacteria in its bacteriocytes, which produce chemicals that make up the weevil's hard protective shells. When the insect first makes that shell in adulthood, it relaxes its control of the bacteria, which quadruple in number. But once the shell is set, the weevil no longer needs its microbial companions – and kills them. It recycles the contents of its bacteriocytes, *Sodalis* and all, into raw materials, and makes the cells self-destruct. With its cellular prisons, the weevil can expand its population of domesticated bacteria when the situation demands it, and do away with them when their partnership no longer bears fruit.[21]

Containment is tougher for backboned animals like ourselves. We have to control a far larger consortium of microbes than any insect, and we have to do it without bacteriocytes. Most of our microbes live *around* our cells, not inside them. Just think about your gut. It's a long and heavily folded tube that, if spread out fully, would cover the surface of a football field. Swarming within that tube are trillions of bacteria. There's just one layer of epithelial cells – the ones that line our organs – stopping them from penetrating the walls of the gut and reaching the blood vessels that could carry them to other parts of the body. The gut epithelium is our main point of contact with our fellow microbes, but also our greatest point of vulnerability. Simple aquatic animals like corals and sponges have it even worse. Their entire bodies are little more than layers of epithelium immersed in a bath of microbes. And yet they too can control their symbionts. How?

For a start, they use mucus, the same slimy goo that clogs your nose when you have a cold. 'You can't go wrong with mucus, because mucus is cool,' says Forest Rohwer.[22] He should know – he has been collecting samples of the stuff from across the animal kingdom for years. Nearly all animals use mucus to cover tissues that are exposed to the outside world. For us, that means guts, lungs, noses, and genitals. For corals, it means everything. In each case, the goo acts as a physical barrier. Mucus is made from giant molecules called mucins, each consisting of a central protein backbone with thousands of sugar molecules branching off it. These sugars allow individual mucins to become entangled, forming a dense, nearly impenetrable thicket – a Great Wall of Mucus that stops wayward microbes from penetrating deeper into the body. And if that wasn't deterrent enough, the wall is manned by viruses.

When you think of viruses, you probably think of Ebola, HIV, or influenza: well-known villains that make us sick. But most viruses infect and kill microbes instead. These are called bacteriophages – literally, 'eaters of bacteria' – or phages, for short. They all have angular heads on top of spindly legs, rather like the Lunar Lander that delivered Neil Armstrong to the Moon. When they touch down on a bacterium, they inject their DNA and turn the microbe into a factory for making more phages. These eventually burst out of their host in fatal fashion. Phages don't infect animals, and they far outnumber the viruses that do. The trillions of microbes in your gut can support *quadrillions* of phages.

A few years ago, Rohwer's team member Jeremy Barr noticed that phages love mucus. In a typical environment, there will be 10 phages for every bacterial cell.[23] In mucus, there will be 40. The same fourfold spike in phage concentrations exists in human gums, mouse guts, fish skins, marine worms, sea anemones, and corals. Imagine hordes upon hordes of them, stuck head-first, their legs outstretched and waiting to embrace passing microbes in a lethal hug. And these mucus-bound phages might be more than just crude tools for killing microbes. Rohwer suspects that animals, by changing the chemical composition of their mucus, could potentially recruit specific phages, which kill some bacteria while providing safe passage to others. Perhaps this is one way in which we select for our favoured microbial partners.

This concept has profound implications. It suggests that phages – which, remember, are *viruses* – have a mutually beneficial relationship with animals, including us. They keep our microbes in check and we, in return, help them to reproduce by offering them a world full of bacterial hosts. Phages are 15 times more likely to find a victim if they stick to mucus. And since mucus is universal in animals, and phages are universal in mucus, this partnership probably started at the dawn of the animal kingdom. In fact, Rohwer suspects that phages were the original immune system – the means through which the simplest animals controlled the microbes at their door.[24] These viruses were already plentiful in the environment. It was a simple matter of concentrating them by giving them a layer of mucus in which to anchor themselves. From this basic beginning, more complex means of control emerged.

Take the mammalian gut. The mucus that covers it comes in two layers: a dense inner one that sits directly on top of the epithelial cells, and a loose outer one beyond that. The outer layer is full of phages, but it's also a place where microbes can anchor themselves and build thriving communities. They abound here. By comparison, very few of them exist in the dense inner layer. That's because the epithelial cells liberally spray this zone with antimicrobial peptides (AMPs) – small molecular bullets that take out any encroaching microbes. They create what Lora Hooper calls a demilitarised zone: a region immediately in front of the lining of the gut, where microbes cannot settle.[25]

If any microbes successfully weave their way through the mucus, run the gauntlet of phages and AMPs, and sneak through the epithelium, there's a battalion of immune cells on the other side to swallow and destroy them. These cells aren't just sitting around, waiting for the worst to happen. They are surprisingly proactive. Some reach through the epithelium to check for microbes on the other side, as if feeling around through the slats of a fence. If they find bacteria in the demilitarised zone, they capture them and bring them back across. By taking these prisoners, the immune system gets regular intel about the species that dominate the mucus, and can prepare antibodies and other appropriate countermeasures.[26]

These measures – the mucus, the AMPs, and the antibodies – also determine the species that get to stay in the gut.[27] We know this because scientists have bred many lines of mutant mice that lack one or more of these components. They all end up with irregular collections of microbes, and usually some kind of inflammatory disease. So the gut's immune system isn't an undiscriminating barrier; it isn't haphazardly mowing down any microbe that gets close. It is selective in its control. It's reactive, too. For example, many bacterial molecules stimulate gut cells to produce more mucus; the more bacteria there are, the more heavily fortified the gut becomes. Likewise, gut cells release certain AMPs upon receiving bacterial signals; they aren't constantly shooting into the demilitarised zone, but firing when their targets get too close.[28]

You could view this as the immune system calibrating the microbiome: the more microbes there are, the more strongly the immune system pushes back against them. Alternatively, you could say that *the microbes* are calibrating the immune system, triggering responses that create a suitable niche for themselves while pushing out their competitors. This latter view makes sense when you consider that many of our most common gut microbes have adaptations for coexisting with the immune system. All of which makes for a very different view of immunity than the classical portrait, which is all about destroying microbes that threaten to make us sick. As I write this, Wikipedia still defines the immune system as 'a system of biological structures and processes within an organism that protects against disease'. If the system activates, it is because it has sensed a pathogen – a threat that it then wipes out. To many scientists, however, warding off pathogens is just a bonus trick. The immune system's main function is to manage our relationships with our resident microbes. It's more about balance and good management than defence and destruction.

Backboned animals or vertebrates, like ourselves, own especially complex immune systems, which can create bespoke and long-lasting defences against specific threats; that's why we stay immune to childhood infections like measles, or to those we've been vaccinated against. It's not that we are more vulnerable to infections than other animals. Rather, squid expert Margaret McFall-Ngai thinks, this more intricate immune system evolved to control a more complex microbiome, allowing vertebrates to more precisely select which species live in their bodies, and to maintain those finely tuned relationships over time. Rather than limiting microbes, our immune system evolved to support *even more* of them.[29]

Think back to the previous chapter, in which I portrayed the immune system as a team of rangers carefully managing a national park. If microbes breach the park's fences – the mucus – the rangers push them back and fortify the barrier. They cull any species that becomes too dominant in the park, and they chuck out any pathogens that invade from the outside world. They keep equilibrium within the community, and constantly defend this balance from threats both foreign and domestic.

The rangers only get time off at the very start of our lives, when in microbiological terms we are blank slates. To allow our first microbes to colonise our newborn bodies, a special class of immune cells suppresses the rest of the body's defensive ensemble, which is why babies are vulnerable to infections for their first six months of life.[30] It's not because their immune system is immature, as is commonly believed: it's because it is deliberately stifled to give microbes a free-for-all window during which they can establish themselves. But without the immune system's full selective powers, how can a mammalian baby ensure that it gets the right communities?

Its mother helps. Mother's milk is full of antibodies which control the microbial populations of adults – and babies take up these antibodies during breastfeeding. When immunologist Charlotte Kaetzel engineered mutant mice that could not produce one of these antibodies in their milk, she found that their pups grew up with bizarre gut microbes.[31] They were full of species that are typically found in people with inflammatory bowel diseases, and many of these bacteria wormed their way through the gut walls to inflame the lymph nodes lying underneath. As we saw earlier, many harmless bacteria are harmless only by virtue of where they are. Milk keeps them restrained. And it does much more than that. Milk is one of the most astounding ways in which mammals control their microbes.

At the University of California, Davis, there is a block of terracotta-walled buildings overlooking a large vineyard and a garden bursting with summer vegetables. It resembles a Tuscan villa that has somehow been teleported into the western US. It is, in fact, a research institute, and one whose residents are obsessed with the science of milk. They are led by a short bundle of nervous energy named Bruce German. If there was a world title in extolling the virtues of milk, German would surely hold it. I meet him in his office, shake his hand, and ask, 'Why are you interested in milk?' Half an hour later, he is still monologuing his answer while bouncing on a gym ball and kneading a tattered shred of bubble wrap.

Milk is the perfect source of nutrition, he says – a 'superfood' that is actually worthy of the title. This isn't a common view. To date, the number of scientific publications about milk is tiny compared to the number devoted to other bodily fluids like blood, saliva, or even urine. The dairy industry has spent an unimaginable fortune on extracting more and more milk from cows, but very little on understanding just what this white liquid is or how it works. Medical funding agencies saw it as irrelevant; as German puts it, 'it doesn't have anything to do with the diseases of middle-aged white men'. And nutritionists saw it as a simple cocktail of fats and sugars that could be easily duplicated and replaced by formulas. 'People said it's just a bag of chemicals,' says German. 'It's anything but that.'

Milk is a mammalian innovation. Every mammal mother, whether platypus or pangolin, human or hippo, feeds her baby by literally dissolving her own body to make a white fluid that she secretes through her nipples. The ingredients of that fluid have been tweaked and perfected through 200 million years of evolution to provide all the nutrition that infants need. Those ingredients include complex sugars called oligosaccharides. Every mammal makes them but human mothers, for some reason, churn out an exceptional variety – scientists have identified over 200 human milk oligosaccharides, or HMOs, so far.[32] They are the third-biggest part of human milk, after lactose and fats, and they should be a rich source of energy for growing babies.

But babies cannot digest them.

When German first learned about HMOs, he was gobsmacked. Why would a mother spend so much energy manufacturing these complicated chemicals if they were indigestible and therefore useless to her child? Why hasn't natural selection put its foot down on such a wasteful practice? Here's a clue: these sugars pass through the stomach and the small intestine unharmed, and land in the large intestine where most of our bacteria live. So, what if they aren't food for babies at all? What if they are food for microbes?

This idea dates back to the early twentieth century, when two very different groups of scientists made discoveries that, unbeknownst to them, were closely connected.[33] In one camp, paediatricians found that microbes called *Bifidobacteria* (or Bifs to their friends) were more common in the stools of breast-fed infants than bottle-fed ones. They argued that human milk must contain some substance that nourishes these bacteria – something that later scientists would call the 'bifidus factor'. Meanwhile, chemists had discovered that human milk contains carbohydrates that cow milk does not, and were gradually whittling this enigmatic mixture down to its individual components – including several oligosaccharides. These parallel tracks met in 1954, thanks to a partnership between Richard Kuhn (chemist, Austrian, Nobel laureate) and Paul Gyorgy (paediatrician, Hungarian-born American, breast-milk advocate). Together, they confirmed that the mysterious bifidus factor and the milk oligosaccharides were one and the same – and that they nourished gut microbes. (It often takes partnerships between different branches of science to understand the partnerships between different kingdoms of life.)

By the 1990s, scientists knew that there were more than 100 HMOs in milk, but had only characterised a few. No one knew what most of them looked like or which species of bacteria they fed. The common wisdom was that they nourished all Bifs equally. German wasn't satisfied. He wanted to know exactly who the diners were and what dishes they were ordering. To do that, he took a cue from history and assembled a diverse team of chemists, microbiologists, and food scientists.[34] Together, they identified all the HMOs, pulled them out of milk, and fed them to bacteria. And, to their chagrin, nothing grew.

The problem soon became clear: HMOs are not an all-purpose food for Bifs. In 2006, the team found that the sugars selectively nourish one particular subspecies called *Bifidobacterium longum infantis*, or *B. infantis* for short. As long as you provide it with HMOs, it will outcompete any other gut bacterium. A closely related subspecies – *B. longum longum* – grows weakly on the same sugars. The ironically named *B. lactis*, a common fixture of probiotic yoghurts, doesn't grow at all. Another probiotic mainstay, *B. bifidum*, does slightly better but is a fussy, messy eater. It breaks down a few HMOs and takes in the pieces it likes. By contrast, *B. infantis* devours every last crumb with a cluster of 30 genes – a comprehensive cutlery set for eating HMOs.[35] No other Bif has this genetic cluster; it is unique to *B. infantis*. Human milk has evolved to nourish this microbe and it, in turn, has evolved into a consummate HMOvore. Unsurprisingly, it is often the dominant microbe in the guts of breast-fed infants.

It earns its keep. As it digests HMOs, *B. infantis* releases short-chain fatty acids (SCFAs) that feed an infant's gut cells – so while mothers nourish this microbe, the microbe in turn nourishes the baby. Through direct contact, *B. infantis* also encourages gut cells to make adhesive proteins that seal the gaps between them, and anti-inflammatory molecules that calibrate the immune system. These changes only happen when *B. infantis* grows on HMOs; if it gets lactose instead, it survives but doesn't engage in any repartee with the baby's cells. It unlocks its full beneficial potential only when it feeds on breast milk. Likewise, for a child to reap the full benefits that milk can provide, *B. infantis* must be present.[36] For that reason, David Mills, a microbiologist who works with German, actually sees *B. infantis* as part of milk, albeit a part that is not made in the breast.[37]

Human breast milk stands out among that of other mammals: it has five times as many types of HMO as cow's milk, and several hundred times the quantity. Even chimp milk is impoverished compared to ours. No one knows why this difference exists, but Mills offers a couple of good guesses. One involves our brains, which are famously large for a primate of our size, and which grow incredibly quickly in our first year of life. This fast growth partly depends on a nutrient called sialic acid, which also happens to be one of the chemicals that *B. infantis* releases while it eats HMOs. It is possible that by keeping this bacterium well fed, mothers can raise brainier babies. This might explain why, among monkeys and apes, social species have more milk oligosaccharides than solitary ones, and a greater range of them to boot. Larger groups mean more social ties to remember, more friendships to manage, and more rivals to manipulate. Many scientists believe that these demands drove the evolution of primate intelligence; perhaps they also fuelled the diversity of HMOs.

An alternative idea involves diseases. Pathogens can easily bounce from one host to another, so group-living animals need ways of protecting themselves against rampant infections. HMOs provide one such defence. When pathogens infect our guts, they almost always begin by latching onto glycans – sugar molecules – on the surface of our intestinal cells. But HMOs bear a striking resemblance to these intestinal glycans, so pathogens sometimes stick to them instead. They act as decoys to draw fire away from a baby's own cells. They can block a roll call of gut villains including *Salmonella*; *Listeria*; *Vibrio cholerae*, the culprit behind cholera; *Campylobacter jejuni*, the most common cause of bacterial diarrhoea; *Entamoeba histolytica*, a voracious amoeba that causes dysentery and kills 100,000 people every year; and many virulent strains of *E. coli*. They may even be able to obstruct HIV, which might explain why most infants who suckle from infected mothers don't get infected despite drinking virus-loaded milk for months. Every time scientists have pitted a pathogen against cultured cells in the presence of HMOs, the cells have come out smiling. This helps to explain both why breast-fed babies have fewer gut infections than bottle-fed ones and why there are so many HMOs. 'It makes sense that they would need to be diverse enough to handle a range of pathogens, from viruses to bacteria,' says Mills. 'I think it's the amazing diversity that provides a constellation of protections.'[38]

The team is just getting started. They have set up an impressive milk-processing facility in their mock-Tuscan institute to discover the many unfamiliar secrets of this most familiar of fluids. In the main lab, which Mills runs with food scientist Daniela Barile, there are two huge steel drums in which milk is stored, a pasteuriser that looks like a cappuccino machine, and a riot of other equipment for filtering the liquid and breaking it down into its components. Hundreds of empty white buckets are stacked on a nearby rack. 'They're normally full,' Barile tells me.

The full buckets are kept in a huge walk-in freezer that's chilled to an intensely uncomfortable −32 degrees Celsius. On a nearby bench, there's a row of wellies ('When we process, there's milk all over,' says Barile), a hammer for chipping ice ('The door's not closing properly'), and, inexplicably, a ham slicer (I don't ask). We pop our heads inside. White buckets are arrayed on pallets and shelves, containing some 600 gallons of milk between them. A lot of this is cow's milk, donated by dairies, but a surprising amount came from human breasts. 'Lots of women pump milk and store it, and once their kid weans, they think: Now what do we do with it? People then hear about us and we get donations,' says Mills. 'We got 80 litres, collected over two years, from someone random at Stanford University, who said: I have all this milk, do you guys want it?' Yes, they did. They need all the milk they can get.

Their plan is to study the components of milk – HMOs and beyond. There are fats and proteins with glycans stuck to them too: how do these affect *B. infantis* and other Bifs? And there are phages, as well – lots of them. German has teamed up with Jeremy Barr to see if mothers use breast milk to provide babies with a starter pack of symbiotic viruses. They have already found something weird: phages are great at sticking to mucus, but they do so ten times more efficiently if there's breast milk around. Something in the milk helps them anchor in place. The culprits seem to be little spheres of fat, encased in proteins that resemble those in mucus. If you let a glass of milk sit in the open, the layer of fat that forms on the surface is full of these globules. They provide nutrition to a baby, but they might also give baby's first viruses a foothold in the gut.

When Barr tells me about this, I'm astounded. It means that the measures by which we shape and control our microbiome – the phages, the mucus, the various arms of the immune system, and the ingredients in milk – are *all connected*. I've discussed them as if they were separate tools, but they are all part of a huge interwoven system for stabilising our relationships with our microbes. In this counter-intuitive reality, viruses can be allies, immune systems can support microbes, and a breastfeeding mother isn't just feeding a baby but also setting up an entire world. And breast milk? German was right: it's far more than a bag of chemicals. It nourishes baby and bacteria, infant and *infantis* alike. It's a preliminary immune system that thwarts more malevolent microbes. It is the means by which a mother ensures that her children have the right companions, from their first days of life.[39] And it prepares the baby for life ahead.

Once we are weaned, it falls to us to nourish our own microbes. We do this partly through our diet, which provides a diverse flood of branching sugar molecules – glycans – to replace the lost HMOs. But we also make our own glycans; the mucus in our gut is full of these, which provide rich pastures for our gut microbes. By continuing to offer the right foods we nurture bacteria that are likely to be beneficial, and exclude the ones that pose more danger. This imperative to feed our microbes is so strong that we do it even when we ourselves stop eating. When animals get sick, we frequently lose our appetite – a sensible tactic that diverts energy from foraging and towards getting better. It also means that our gut microbes experience a temporary famine. Sick mice deal with this problem by releasing emergency rations: a simple sugar called fucose. Gut microbes can snip off this sugar and feed on it, staying alive while they wait for their hosts to resume normal service.[40]

The *Bacteroides* group, which excels at eating these glycans, soon become the most common microbes in the gut. But crucially, glycans are so diverse that no single species of bacterium has the right tools for eating all of them. This means that by swallowing or making a wide range of glycans we can support an abundance of different bacteria. Some are unfussy generalists like pigeons or raccoons; others are choosy specialists like pandas or anteaters. They form food webs where some microbes break down the biggest and hardiest molecules and release smaller fragments that others mop up. They make pacts, where two species feed each other, each digesting a different food while creating waste chemicals that its partner can use. They form truces by adjusting their metabolic antics to avoid competing with their neighbours.[41]

These interactions matter, because they foster stability. If a single bacterium was too efficient at harvesting glycans, it might eat away the mucus barrier itself, creating openings through which other microbes could enter. But if there are hundreds of competing species, they can all keep each other from gluttonously monopolising the food supply. By offering a wide array of nutrients we feed a wide range of microbes and stabilise our enormous, diverse communities. And those communities, in turn, make it harder for pathogens to invade. By setting the table correctly, we ensure that the right guests turn up to dinner, while gatecrashers are locked out. Our mothers started that trend by feeding us at the start of our lives, and we carry on their work thereafter.

There is another way for hosts to reduce their conflict with their microbes, and it's an extreme one: they can become so co-dependent that they effectively act as a single entity.[42] This happens when bacteria find their way *inside* the cells of their hosts and are faithfully transmitted from parents to offspring. The two parties' fates are now entwined. They still have their own interests, but these overlap to such an extent that any remaining disagreements become negligible.

Such arrangements, which are especially common among insects, tend to trap microbes in a predictable spiral of simplification. In the cells of their hosts, they are restricted to small populations and kept apart from other bacteria. Their isolation allows harmful mutations to build up in their DNA. Any gene that's not essential becomes faulty and useless, before disappearing altogether.[43] If you stuck a new symbiont into an insect and played the evolutionary tape on fast-forward, you'd see violent turmoil as its genome contorted and crunched, warped and shrank. Eventually, they end up with shrivelled genomes, close to the minimum necessary for life. A typical free-living microbe like *E. coli* has a genome consisting of about 4,600,000 DNA letters. The smallest known symbiont, *Nasuia*, has just 112,000. If *E. coli*'s genome were the size of this book, you'd have to rip out everything past the Prologue to get something like *Nasuia*. These symbionts are fully domesticated – unable to survive on their own, and corralled forever within the cosseted environments of their insect hosts.[44] And the hosts often become dependent on their shrunken symbionts for nutrients or other vital benefits. This is the same process that transformed an ancient bacterium into mitochondria, the essential structures that we cannot live without.

These fusions are powerful ways of mitigating conflict between hosts and microbes, but they still come with a dark side. John McCutcheon, a tall, bald biologist with glasses and a wide smile, realised this after studying the 13-year periodical cicada. This black-bodied, red-eyed bug spends most of its life as a nymph, living underground and drinking from plant roots. After thirteen years of this indolent existence, the cicadas all emerge at the same time, filling the air with their cacophonous song. And after a lot of frenzied sex, they all die at the same time, covering the ground with their decaying bodies. Since these bugs have such a weird lifestyle, McCutcheon suspected that they might have equally weird symbionts. He was right – but he had no idea *how* weird they'd be.

The DNA sequences from the cicada's symbionts were a mess. They looked as if they should all belong to the same genome, but it was as if someone had given McCutcheon the jumbled pieces from several incomplete copies of the same jigsaw puzzle. Confused, he moved on to another cicada: a shorter-lived and fuzzier species from South America. He found the same problem: fragments of DNA that just wouldn't assemble into a single genome. They would, however, assemble into *two*.

The two genomes belonged to bacteria that descended from a symbiont called *Hodgkinia*. Once this microbe got inside the fuzzy cicada it somehow split into two separate 'species' – *within* the insect.[45] These daughter species have both lost genes that the original

Hodgkinia had, but they each jettisoned different ones. Their current genomes, though pale shadows of their former selves, are perfectly complementary. They are like two halves of a former whole: there's nothing that the original *Hodgkinia* could do that the two daughters can't do together.

It took McCutcheon almost a year to work out what was going on, but once he had, the mystery of the 13-year cicada's jumbled symbionts became a lot clearer. That insect also contains *Hodgkinia*, but rather than dividing into two species, the bacterium had split into who knows how many. Its DNA eventually assembled into at least 17 distinct rings, and maybe as many as 50. Is each one a different species? Or are there lineages whose genomes are split between different rings? No one knows. Regardless, the team have now looked at many other cicadas and often found the same pattern. In one Chilean cicada, *Hodgkinia* has split into six complementary genomes.[46]

In all these cases, the genes for making vital vitamins are scattered across the genomes of the cicadas and their many *Hodgkinia* symbionts, so the entire ensemble can survive only if every member is present. In the short term, they'll be fine. In the long term . . . who knows? If *Hodgkinia* continues to break up into smaller and smaller pieces, all of which are important, the entire community becomes incredibly precarious. The loss of one might doom them all. 'It's like watching a train wreck or a slow-motion extinction event,' says McCutcheon. 'It makes me think differently about symbiosis.' He always saw it as a positive force, one that provides both partners with benefits and opportunities. But it can also be a trap, where the partners become increasingly vulnerable in their dependency. Nancy Moran, McCutcheon's former advisor, calls this an 'evolutionary rabbit hole' – a metaphor that implies a 'generally irreversible journey into a very odd world where the usual rules do not apply'.[47] Once both partners tumble down the rabbit hole, it can be hard for them to escape. And at the bottom, there is no wonderland – only extinction.

This is the price of symbiosis. Even when microbes aren't as crucial to their hosts as a cicada's symbionts are, they still exert a powerful influence on our lives and our health. When they go rogue, the consequences can be disastrous. That's why humans and other animals have evolved so many ways of stabilising their multitudes. We restrict them by relying on the chemistry of our bodies. We corral them with physical barriers. We can go for the carrot, by nourishing them with dedicated foods. We can beat them with the stick, by using phages, antibodies, and other parts of our immune system. We have many solutions to the ever-present conflicts that exist with our microbes, and many ways of enforcing our contracts with them.

Unfortunately, we humans have inadvertently developed just as many ways of breaking those contracts.

5. IN SICKNESS AND IN HEALTH

Take a globe and spin it until the side that faces you is largely blue. You are now staring into the Pacific Ocean, in all its daunting immensity. Now stab your finger into its heart. Down a bit. Right a bit. You are now prodding the Line Islands, a linear constellation of eleven tiny land masses, slashing their way through the middle of nowhere. Around 3,500 miles from California, 3,800 miles from Australia, and 4,900 miles from Japan, the Line Islands epitomise isolation. They are about as far away from anything else as you can get without leaving the planet. That is how far Forest Rohwer had to travel to find the most beautiful coral reefs he had ever seen.

In August 2005, Rohwer dove off the deck of the *White Holly* and plunged into the waters of Kingman Reef, the northernmost of the Line Islands, the tip of the slash.[1] Through ethereally clear water, he saw a huge wall of coral rising up from the depths and carpeting the seafloor. It was a Hollywood reef, the reef of Pixar's *Finding Nemo*, a beautifully lit ecosystem with an A-list cast: manta rays, dolphins, walls of big-eye jacks, schools of fang-toothed Cubera snappers, and sharks galore. At least fifty grey reef sharks circled the divers, each one roughly human-sized. But Rohwer and his fellow scientists were unconcerned; they knew that sharks were a sign of a healthy reef, and were thrilled to see them in such numbers. Besides, they mostly feed at night, so as long as the researchers made it back to the ship before sunset, they would be fine. They cut it close. By the time the last scientist climbed on board, the sun was tickling the horizon and, as Rohwer later wrote, 'the "lots of sharks" had turned into "my God, there are lots of sharks".'

Seven hundred kilometres to the south-east, at Christmas Island (now known as Kiritimati), everything was different. There, Rohwer saw 'some of the deadest reefs' he'd ever seen. The vibrant, layered, bountiful world of Kingman was replaced by fields of ghostly slime-covered coral skeletons, as if some force had swept through the reef and drained it of both life and colour. The water was turbid and flecked with particles. The fish were scarce. The sharks were gone. In a hundred hours of diving, the scientists didn't see a single one.

It wasn't always like this. When James Cook arrived at Christmas Island in 1777, his navigator documented 'sharks innumerable'. Even in the late nineteenth century, the big predators were still around and the reefs were still healthy. That changed in 1888 when people started colonising the island in earnest. Today, it houses around 5,500 residents – a tiny number, but enough to have wiped out the sharks and ruined the reefs. Kingman, by contrast, has always been uninhabited. With just three football pitches' worth of dry land, there's nothing for settlers to settle on. Its above-ground inhospitability made it an underwater sanctuary. To Rohwer, it is a window into the past, to the glorious reefs that greeted Captain Cook. Christmas Island, however, is a glimpse into our desolate coral-free future and, as we shall see, into many common human diseases.

Corals are animals with soft tubular bodies crowned by stinging tentacles. You rarely see them like this because they hide in limestone, which they secrete themselves. It is these rocky skeletons that combine to form mighty reefs – undersea landscapes of branches, shelves, and boulders, which play home to countless marine animals. Corals have been building reefs for hundreds of millions of years but their days of underwater architecture may be drawing to a close. Caribbean populations have largely collapsed. Australia's mighty Great Barrier Reef has lost most of its coral. A full third of reef-building coral species face extinction, imperilled by many threats. The carbon dioxide that humans unleash into the atmosphere warms the oceans by trapping the sun's heat. In these warmer seas, corals expel the algae that live inside their cells and provide them with nutrients. Bereft of these partners, they become weak and ghostly. The carbon dioxide also dissolves in the oceans directly, acidifying them. This depletes the minerals that corals need to build their reefs, which start to wear away. Hurricanes, ships, and voracious starfish erode them even further. Starving, pallid, homeless, and deprived of mortar, the poor corals get sick. They fall victim to a colour chart of pestilence: white pox, black band disease, pink line disease, red band disease. There are dozens of these syndromes, and in recent decades they have become more common.

This trend is unusual. Infections typically become more common when hosts live in *high* densities that facilitate transmission, but coral diseases seem to have risen as their host populations have declined. That is because only some of these illnesses are caused by specific pathogens. The others have more complicated origins: they seem to be caused by large groups of microbes working together, or by bacteria that are normal parts of a coral's microbial world. It was that world that caught Rohwer's attention.

Rohwer has scraggly black hair, a laid-back demeanour, and a high voice. He dresses almost entirely in shades of black and charcoal, and wears silver jewellery. He is a pioneer of metagenomics: the game-changing method that we met in the second chapter, in which scientists survey microbes by sequencing all their genes. Rohwer first used this technique to catalogue the viruses in the open ocean. Then, he moved on to corals. Other scientists had already shown that corals are smothered in microscopic life. Every square centimetre of their surface contains 100 million microbes, more than ten times as many as on a similar patch of human skin or forest soil. Coral reefs may have a reputation as wonderlands of diversity but that diversity is largely invisible. Forget rays, turtles, and eels: bacteria and viruses comprise most of a reef's biology, and most of them have never been studied.

What do these microbes do? 'First and foremost,' says Rohwer, 'they occupy space.' A coral's body has only so many places in which microbes can live, and only so many food sources. If benign species fill those niches, dangerous ones can't invade, so a diverse microbiome, through its mere presence, creates a blockade against disease. This effect is called colonisation resistance. Disrupt it, and infections become more common. This, Rohwer suspected, was the underlying explanation for the loss of so many reefs. All the stressors that weaken the corals – the warming seas, acidic waters, and nutrient overloads – disrupt their partnerships with their microbes, leaving them with distorted and impoverished communities that are vulnerable to disease, or that might even *cause* disease.[2]

To test this idea, Rohwer needed to study a variety of reefs, from the pristine to the despoiled. Hence the *White Holly*. Over two months, the ship sailed down the four northern Line Islands and up a gradient of human activity, from Kingman Reef (uninhabited) to Palmyra Atoll (a few dozen people) to Fanning Island (2,500 residents) to Christmas Island (5,500 residents). While other scientists counted fish and scooped up coral, Rohwer and his colleague Liz Dinsdale studied the microbes. They captured seawater from each site and filtered it through glass wafers with holes so small that even viruses couldn't squeeze through. They scraped the microbes off these ubersieves and stained them with fluorescent dyes. Under a microscope, they glowed. 'The fortunes of the coral – good health or decline – were written in these little pinpoints of light,' Rohwer later wrote.

Dinsdale and Rohwer found that as humans become more common, so do microbes. From Kingman to Christmas Island, top predators such as sharks went from dominant parts of the reefs to bit-players, coral cover fell from 45 per cent to 15 per cent, and the water contained 10 times as many microbes and viruses. All of these trends are connected in a complicated web of cause and effect that revolves around a turf war between corals and their ancient rivals: the so-called 'fleshy algae'.

Some algae are coral allies: they live in their cells and provide them with food, or form tough pink crusts that link separate colonies into a sturdy whole. But the fleshy algae are antagonists that compete with corals for space. If the algae rise, the corals fall, and vice versa. In most reefs, fleshy algae are kept in check by grazers like surgeonfish and parrotfish, which nibble them down to well-trimmed lawns. But humans kill the grazers with spears, hooks, and nets. We also kill top predators like sharks, leading to population explosions of medium-sized predators, which then take out the grazers. Either way, we give the algae an advantage. The well-trimmed lawns become overgrown fields, and the neighbouring corals start to die. Jennifer Smith, who was also on the Line Islands expedition, demonstrated this through a simple experiment. She placed nubbins of coral and scraps of algae in adjacent aquaria, connected by the same water but separated by one of those extremely fine filters. Microbes could not pass through but chemicals in the water could. Within two days, all the corals were dead. Something in the water, released by the algae, was killing them. A toxin? Possibly, but when Smith treated the corals with antibiotics, they survived. Not a toxin, then. Not spreading microbes, either – the filters would have blocked their path. No, the algae were making *something* that killed the corals via *their own* microbes.

That something turned out to be dissolved organic carbon (DOC); essentially, sugars and carbohydrates in the water. When algae get too numerous on a reef they make huge amounts of DOC and create a banquet for coral microbes. These algal sugars would normally flow up the food chain to be locked away in the bodies of grazers and, ultimately, sharks; a single shark represents the stored energy of several tons of algae. But if all the sharks die, those sugars remain at the bottom of the food web where, instead of fuelling the flesh of fish, they build the cells of microbes. Nourished by this feast, the microbes bloom so explosively that they consume all the surrounding oxygen, choking the corals.

But DOC doesn't nourish all microbes equally. Being high in energy and easy to digest – Rohwer compares it to hamburgers – it preferentially enriches fast-growing species, especially pathogens. Around Kingman Reef, just 10 per cent of the local microbes belonged to families that could cause coral diseases. But around Christmas Island, half the microbes belonged to such families. 'You wouldn't want to swim there,' Rohwer wrote. 'Unfortunately, the corals have no choice.' No wonder, then, that Christmas Island has about twice as much diseased coral as Kingman, despite having only a quarter as much coral. (A later survey would show that Christmas Island does still have a few healthy reefs: former nuclear testing sites, where fear of radiation has repelled fishermen and saved both fish and corals.) Those waters are like a dirty hospital ward, full of immunocompromised patients. And as with such patients, the corals are only rarely killed by exotic pathogens travelling in from afar. Mostly, the things that infect them are opportunistic parts of their own microbiome, which exploit the rich supply of DOC at the expense of their host.

The sequence of events that Rohwer describes is a loop. As corals die, they create more space for algae, which release even more DOC, which nourishes even more pathogens, which kill even more corals. Eventually, this cycle spins so quickly that the entire reef shifts, dramatically and perhaps irreversibly, from a fish-and-coral state to an algal state. 'It's horrible and so fast,' says Rohwer. 'A coral reef will go down in a year. You have a beautiful reef and then it's dead.'

All the main stressors that weaken reefs can kick off this cycle. In 2009, Rohwer's team subjected pieces of coral to either higher temperatures, acidified water, an increase in nutrients, or more DOC. In response, the coral microbiomes shifted from those found in healthy reefs to the pathogenic communities that flourish on diseased corals. There was also more evidence of virulence genes, which bacteria use to infect their hosts, and more viruses, related to those that cause herpes in humans. Herpes viruses can hide out in the genomes of their hosts, lying dormant until some kind of stress reactivates them. When they re-emerge, these latent viruses can cause cold sores in humans. It's unclear what they inflict upon corals but disease seems likely.[3]

Humans can set off this vicious cycle in other unexpected ways. In 2007, an 85-foot fishing vessel ran aground on Kingman Reef, possibly because of an engine fire. Its origins, its name, and the fate of its crew are unknown. Its effects, however, have been appallingly clear. As the ship fell apart, its pieces rained down on the underlying reef, creating a kilometre-long dead zone quite unlike the usual fields of bleached rubble. Instead, these corals are covered in dark algae and shrouded in especially turbid water. They are called black reefs. They are a marine vision of Tolkien's Mordor, and they happen when a boatload of iron lands in an ecosystem that is generally poor in nutrients. The iron acts as fertiliser for fleshy algae, which grow so vigorously that even grazing fish can't trim them back fast enough. The algae then trigger Rohwer's cycle: more DOC, more microbes, more pathogens, more disease, more dead corals.

Rohwer's team saw black reefs in other parts of the Line Islands, always associated with shipwrecks, and always down-current of the debris. Unlike places such as Christmas Island, where corals are almost uniformly degraded, black reefs can show up in pristine waters. 'You can literally imagine that this is all nice reef,' says Rohwer, gesturing at a table, 'and *this* part is dead.' He slams his hands in the middle. 'Any place where there's a piece of iron, even if it's just a bolt, will have a little black reef around it.'

In 2013, the US Fish and Wildlife Service removed the offending ship from Kingman. A crew of workers lifted thousands of pounds of debris by hand, sliced it with plasma-cutters and chainsaws, and rafted the fragments out. Only the main engine, all 5,000 iron-rich pounds of it, remains. With most of the debris gone, the corals may recover.

Other reefs are not so lucky. Their woes stem not from a one-off influx of iron but from the unrelenting pressure of human activity. Rohwer's team also measured that activity for 99 sites across the Pacific, coming up with a single unified score that reflected the combined influence of fishing, industry, pollution, shipping and more. For the same sites, they calculated a 'microbialisation score', a measure of the proportion of energy in the ecosystem that went into microbes rather than fish. The two measures increased in clear and direct proportion with each other. As humans make our presence felt, we disturb the ancient relationships between corals and their microbes, converting the vivid splendour of fish-filled reefs into bleak algal barrens submerged in a pathogenic soup.

This, according to Rohwer, is how a coral reef dies – weakened by a cabal of threats and eventually overwhelmed by its own microbes. It's not the only explanation for failing reefs, but it's certainly a compelling and sweeping one – a Grand Unified Theory of Coral Death. It shows how the largest sharks are connected to the smallest viruses. It tells us that the invisible part of the reef is what ultimately decides its fate. Rohwer puts it plainly: 'Even though coral reefs are incredibly complex, microbes are the main determinants of [their] health and decline.'

Think about microbial diseases. Think about influenza, AIDS, measles, Ebola, mumps, rabies, smallpox, tuberculosis, plague, cholera, and syphilis. All of these maladies, though different from each other, fit a similar pattern. They are caused by a single microbe: a virus or bacterium that infects our cells, reproduces at our expense, and triggers a predictable panoply of symptoms. This causal agent can be identified, isolated, and studied. With luck, it can be removed, ending the affliction.

Rohwer's work with corals hints at a different type of microbial disease, one without a single obvious culprit.[4] These illnesses are caused by *communities* of microbes, which have shifted into configurations that harm their hosts. None is a pathogen in its own right; instead, the entire community has shifted to a *pathogenic state*. There's a word for such a state: dysbiosis.[5] It is a term that evokes imbalance and discord in place of harmony and cooperation. It is the dark reflection of symbiosis, the antithesis of all the themes we have seen so far.

Recall that every individual animal, whether human or coral, is an ecosystem in itself. It grew up under the influence of its microbes and continues to engage them in a lively negotiation. Remember also that these partners often have competing interests and that hosts need to control their microbes, keeping them in line by offering the right food, confining them to specific tissues, or placing them under immune surveillance. Now imagine that something disrupts that control. It jostles the microbiome, changing the proportions of species within it, the genes they activate, and the chemicals they produce. This altered community still communicates with its host, but the tenor of their conversation changes. Sometimes it becomes, quite literally, inflammatory, as microbes over-stimulate the immune system or wheedle their way into tissues where they don't belong. In other cases, microbes might start to opportunistically infect their hosts.

That's dysbiosis. It's not about individuals failing to repel pathogens, but about breakdowns in communication between different species – host and symbiont – that live together. It is disease, recast as an *ecological* problem. Healthy individuals are like virgin rainforests or lush grasslands or Kingman Reef. Sick individuals are like fallow fields or scum-covered lakes or the bleached reefs of Christmas Island – ecosystems in disarray. This is a more complicated view of health, and one that raises important questions. Foremost among them: are such changes the cause of disease, or merely its consequence?

'So, what's in the thermos?' I ask.

I'm standing in a lift at Washington University in St Louis, with Jeff Gordon and two of his students, one of whom is holding a metal canister.

'Just some faecal pellets in tubes,' she says.

'They're microbes from healthy children, and also from some who are malnourished. We transplanted them into mice,' explains Gordon, as if this was the most normal thing in the world.

Jeff Gordon is arguably the most influential human microbiome scientist working today. He is also one of the hardest to get in touch with. It took me six years of writing about his work to get him to answer my emails, so visiting his lab is a hard-won privilege. I arrive expecting someone gruff and remote. Instead, I find an endearing and affable man with crinkly eyes, a kindly smile, and a whimsical demeanour. As he walks around the lab, he calls people 'professor' – including his students. His aversion to the media comes not from aloofness, but from a distaste for self-promotion. He even refrains from scientific conferences, preferring to stay out of the limelight and in his laboratory. Ensconced there, Gordon has done more than most to address how microbes affect our health, and which connections are, in his words, 'causal not casual'. But when asked about his influence, he tends to deflect credit onto students and collaborators, past and present.[6]

Gordon's figurehead status is all the more remarkable because long before the microbiome crossed his mind, he was already a well-established scientist who had published hundreds of studies on the development of the human gut. In the 1990s, he started to suspect that bacteria influence this process, but he was also struck by how difficult it would be to test that idea. At the time, Margaret McFall-Ngai was showing that microbes can influence a squid's development, but she was working with just one species of bacteria. The human gut contained thousands. Gordon needed to isolate parts of this daunting whole and examine it under controlled conditions. He needed that critical resource which scientists demand but biology withholds: control. In short, he needed germ-free mice – and lots of them.

The lift doors open, and I follow Gordon, his students, and the thermos of frozen pellets into a large room. It is filled with rows of sealed chambers made of transparent plastic. These isolators are some of the strangest environments in the world: habitats that are genuinely free of bacteria. The only living things inside them are mice. The isolators contain everything they need: drinking water, brown nuggets of chow, straw chips for bedding, and a white styrofoam hutch for mating in privacy. The team irradiates all of these items to sterilise them before piling them into loading cylinders. They sterilise the cylinders by steaming them at a high temperature and pressure, before hooking them to portholes in the back of the isolators, using connecting sleeves that they also sterilise. It is laborious work, but it ensures that the mice are born into a world without microbes, and grow up without microbial contact. They exemplify the concept of 'gnotobiosis', from the Greek for 'known life'. We know exactly what lives in these animals – which is nothing. Unlike every other mouse on the planet, each of these rodents is a mouse and nothing more. An empty vessel. A silhouette, unfilled. An ecosystem of one. They do not contain multitudes.[7]

Each isolator has a pair of black rubber gloves affixed to two portholes, through which the researchers can manipulate what's inside. The gloves are thick. When I stick my hands in, I quickly start sweating. Awkwardly, I pick up one of the mice by its tail. It sits snugly on my palm, white-furred and pink-eyed. It is a strange feeling: I'm holding this animal but only via two black protrusions into its hermetically sealed world. It is sitting on me and yet completely separated from me. When I stroked Baba the pangolin, we exchanged microbes. When I stroke this mouse, we exchange nothing.

There are now dozens of similar germ-free facilities around the world, and they are among our most powerful tools for understanding how the microbiome works. But when isolator technology was developed in the 1940s, and refined a decade later, it proved unpopular.[8] No one had a use for the germ-free animals. But Gordon realised that they were perfect for his needs. He could load germ-free mice with specific microbes, feed them with pre-defined diets, and do so again and again in controlled and repeatable conditions. He could treat them as living bioreactors, in which he could strip down the baffling complexity of the microbiome into manageable components that he could systematically study.

In 2004, Gordon's team used the sterile rodents to run an experiment that would set the entire lab on a focused path.[9] They inoculated the mice with microbes harvested from the guts of conventionally raised rodents. Normally, germ-free rodents can eat as much as they like without putting on weight, but this enviable ability disappeared once their guts were colonised. They didn't start eating any more food – if anything, they ate slightly less – but they converted more of that food into fat and so piled on the pounds. Mice are clearly different to humans, but their biology is similar enough for scientists to use them as stand-ins in everything from drug testing to brain research; the same applies to their microbes. Gordon reasoned that if those early results apply to humans, *our* microbes must surely influence the nutrients that we extract from our food, and thus our body weight. Here was a meaty, fascinating, and medically relevant area that his team could sink its teeth into.

Next, the team showed that obese people (and mice) have different communities of microbes in their guts.[10] The most obvious difference lay in the ratio of the two major groups of gut bacteria: obese people had more Firmicutes and fewer Bacteroidetes than their leaner counterparts. This raised an obvious question: does extra body fat tilt the Bacteroidetes/Firmicutes see-saw or, more tantalisingly, does the tilt make individuals fatter? The team couldn't answer that question by relying on simple comparisons. They needed experiments.

That's where Peter Turnbaugh came in. Then a graduate student in the lab, he harvested microbes from fat and lean mice, and then fed them to germ-free rodents. Those that got microbes from lean donors put on 27 per cent more fat, while those with obese donors packed on 47 per cent more fat. It was a stunning result: Turnbaugh had effectively transferred obesity from one animal to another, simply by moving their microbes across. 'It was an "*Oh my god*" moment,' says Gordon. 'We were thrilled and inspired.' These results showed that the guts of obese individuals contain altered microbiomes that can indeed cause obesity, at least in some contexts. The microbes were perhaps harvesting more calories from the rodents' food, or affecting how they stored fat. Either way, it was clear that microbes don't just go along for a ride; sometimes, they grab the wheel.

They can also turn it in both directions. While Turnbaugh showed that gut microbes can lead to weight gain, others have found that they can trigger weight loss. *Akkermansia muciniphila*, one of the more common species of gut bacteria, is over 3,000 times more common in normal mice than in those genetically predisposed to obesity. If obese mice eat it, they lose weight and show fewer signs of type 2 diabetes. Gut microbes also partly explain the remarkable success of gastric bypass surgery: a radical operation that reduces the stomach to an egg-sized pouch and connects it directly to the small intestine. After this procedure, people tend to lose dozens of kilograms, a fact typically accredited to their shrunken stomachs. But the operation also restructures the gut microbiome, increasing the numbers of various species, including *Akkermansia*. And if you transplant these restructured communities into germ-free mice, those rodents will *also* lose weight.[11]

The world's media treated these discoveries as both salvation and absolution for anyone who struggles with their weight. Why bother adhering to strict dietary guidelines when a quick microbial fix is seemingly around the corner? Why flagellate yourself over excessive calories when it turns out that bacteria have rigged the scales? 'Fat? Blame the bugs in your guts,' wrote one newspaper. 'Overweight? Microbes might be to blame,' echoed another. These headlines are wrong. The microbiome does not replace or contradict other long-understood causes of obesity; it is thoroughly entangled with them. Another of Gordon's students, Vanessa Ridaura, demonstrated this by using mice to stage battles between the gut microbes of lean and obese people.[12] First, she loaded these human communities into germ-free rodents. Next, she housed the mice in the same cages. Remember that mice readily eat each other's droppings and so constantly fill their guts with their neighbours' microbes. When this happened, Ridaura saw that the 'lean' microbes invaded guts that were already colonised by 'obese' communities, and stopped their new hosts from putting on weight. The opposite invasions never worked: the obese communities could never gain a foothold if the lean ones were around.

It's not that the lean communities were inherently superior. Instead, Ridaura had tipped the battles in their favour by feeding her mice with plant-heavy chow. The complex fibres in these meals created many opportunities for microbes with the right digestive enzymes – 'job openings for them to fill', in Gordon's words. The obese communities had few species that could fill those positions but the lean communities were brimming with qualified candidates, including fibre-busting specialists like B-theta. So, when obese communities colonised lean guts, they found that every morsel of food was already being devoured and every niche had been filled. By contrast, when the lean communities entered obese guts, they found a glut of uneaten fibre – and flourished. Their success only evaporated when Ridaura fed the mice with fatty, low-fibre chow, designed to represent the worst extremes of the Western diet. Without fibre, the lean communities couldn't establish themselves or stop the mice from putting on weight. They could only infiltrate the guts of mice that *ate healthily.* The old dietary advice still stands, overenthusiastic headlines be damned.

An important lesson emerged: microbes matter but so do we, their hosts. Our guts, like all ecosystems, aren't defined just by the species within them but also by the nutrients that flow through them. A rainforest isn't just a rainforest because of the birds, insects, monkeys, and plants within it, but also because ample rain and sunlight fall from above, and bountiful nutrients lurk in the soil. If you threw the forest's inhabitants into a desert, they would fare badly. Gordon's team have learned that lesson several times over in the lab – and also in Malawi.

Malawi has one of the highest rates of child mortality in the world, and half of the deaths are due to malnourishment. But malnourishment comes in different forms. There's marasmus, where kids end up emaciated and skeletal. There's also kwashiorkor, where fluids leak from blood vessels, leading to puffy swollen limbs, distended stomachs, and damaged skin. The latter has long been shrouded in mystery. It is said to be caused by protein-poor diets, but how can that be when children with kwashiorkor often don't eat any less protein than those with marasmus? For that matter, why do these children often fail to get better despite eating protein-rich food delivered by aid organisations? And why is it that one child might get kwashiorkor while their identical twin – who shares all the same genes, lives in the same village, and eats the same food – gets marasmus instead?

Jeff Gordon thinks that gut microbes are involved and might explain the differences in health between children who, on paper, look identical. After his team carried out their groundbreaking obesity experiments, he started to wonder: if bacteria can influence obesity, could they also be involved in its polar opposite – malnutrition? Many of his colleagues thought it unlikely but, undeterred, Gordon launched an ambitious study. His team went to Malawi and collected regular stool samples from a group of infants as they grew from one-year-olds into three-year-olds. They found that babies with kwashiorkor don't go through the normal progression of gut microbes of their healthy counterparts. Instead of diversifying and maturing with age, their inner ecosystem became stagnant. Their microbiological age soon lagged behind their biological age.[13]

When the team transplanted these immature communities into germ-free mice, the rodents lost weight – but only if they also ate food that mirrored the nutrient-poor Malawian diet. If the mice ate standard rodent chow, they didn't lose much weight no matter whose bacteria they were carrying. As in Ridaura's work, it was the combination of poor food and the wrong microbes that mattered. The kwashiorkor microbes seemed to interfere with chemical chain reactions that fuel our cells, making it harder for children to harvest energy from their food – food that contains very little energy to begin with.

The standard treatment for malnutrition is an energy-rich, fortified blend of peanut paste, sugar, vegetable oil, and milk. But Gordon's team found that the paste only has a brief effect on the bacteria of children with kwashiorkor (which perhaps explains why it doesn't always work). As soon as they revert to their normal Malawian diet, their microbes also boomerang back to their earlier impoverished state. Why?

Imagine a ball, sitting in a valley and surrounded by steep slopes. If you shove the ball, it will roll up a slope, slow down, and eventually fall back to its original starting position. To get the ball all the way up the slope, over the top, and into a neighbouring valley, you need one really big push, or several small sequential ones. This is how ecosystems work: they have a certain resilience to change, which must be overcome if they're to be pushed into a different state. Picture a healthy coral reef as the ball. Rising temperatures give it a gentle nudge. An algal incursion pushes it further up the slope. A smattering of iron propels it even further. Finally, the loss of sharks takes it over the summit and into the next valley, where it falls to the bottom and settles into a new algal-dominated state. It's unhealthier – dysbiotic, even – but, as before, it has resilience. Pushing it back from an algal dominion to a healthy, fish-filled reef will take a lot of effort.[14]

The same kinds of change happen in our bodies. Now, the ball is a child's gut. A poor diet changes the microbes within. It also impairs the child's immune system, changing its ability to control the gut microbiome and opening the door to harmful infections that disrupt the communities even further. And once these communities start wrecking the gut, they stop it from absorbing nutrients efficiently, leading to even worse malnutrition, more severe immune problems, more distorted microbiomes, and so on. Up and up the ball goes, until it crests the summit and slips into the next dysbiotic valley. Once microbiomes end up there, it can be hard to pull them back.

Next to my desk, mounted on the wall, is a thermostat. It's an old one, and thus a dial rather than a digital display. If I turn it down, it sets the temperature of the house at a cool simmer; if I turn it up, it allows a fiery heat to build. Somewhere in the middle, always one tiny adjustment away, is the ideal setting, a point of perfect comfort. The immune system, for all its intricacy, is a lot like that dial. It works like an 'immunostat', which, rather than stabilising temperature, stabilises our relationships with our microbes.[15] It manages the benign trillions that live with us, while thwarting invasions by an infectious minority. If it is set too low, it becomes relaxed, missing threats and leaving us open to infections. If it is set too high, it becomes jumpy, falsely attacking our own microbes and triggering chronic inflammation. It must tread a fine line between these extremes, balancing the cells and molecules that induce inflammation with those that repress it. It must react without overreacting. But over the last half-century, we have gradually pushed our immunostats to higher settings through a combination of sanitation, antibiotics, and modern diets. We've ended up with immune systems that go berserk at harmless things like dust, molecules in our food, our resident microbes, and even our own cells.

Such is the case with inflammatory bowel disease, or IBD.[16] The condition involves severe inflammation of the gut, which manifests as chronic pain, diarrhoea, weight loss, and fatigue. It usually starts in teenagers and young adults, hitting them at the prime of their lives, saddling them with social stigma, and forcing them to undergo tough treatments. Even if medicines and surgeries can bring the symptoms under control, people live with the lifelong spectre of relapse. Both major types of IBD – ulcerative colitis and Crohn's disease – have been around for centuries, but rates have soared since World War II, especially in developed countries.

The causes of IBD are still unclear. Scientists have identified over 160 genetic variants that are tied to the disease, but since these variants are common in the general population and relatively stable in their prevalence, they cannot possibly explain the disease's precipitous rise. They do, however, point to a different culprit. Most of them are involved in producing mucus, solidifying the lining of the gut, or regulating the immune system – all things that keep microbes in line. And although human genes don't change fast enough to account for the sudden rise of IBD, microbes do.

Scientists have long suspected a microbial culprit behind IBD, but despite extensive investigations, they haven't successfully accused any particular pathogen. It's more likely that the problem, as with Rohwer's corals and Gordon's malnourished kids, lies in a community of normal microbes gone rogue. The gut microbiomes of IBD patients certainly differ from those of their healthy peers, but the list of potential suspects seems to change with every new study – unsurprisingly, perhaps, since IBD is so diverse. Nonetheless, some broad patterns have consistently emerged. The IBD microbiome tends to be less diverse and less stable than its healthier counterparts. It lacks anti-inflammatory microbes, including fibre-fermenters like *Faecalibacterium prausnitzii* and *B. fragilis*. In their place are blooms of inflammatory species like *Fusobacterium nucleatum* and invasive strains of *E. coli*.

These microbes clearly play a crucial role yet no single species makes or breaks the ecosystem. The condition looks like a disease of dysbiosis. The entire community becomes more inflammatory, dialling the host's immunostat up to the twitchiest of settings. How did these communities come about? Was it something dietary that nourished inflammatory species? Antibiotics that killed off the anti-inflammatory ones? Genetic variants that altered the host's immune system, disrupting its ability to manage its microbes? The last of these seems possible: Wendy Garrett has shown that mutant mice which lack important immune genes end up with unusual communities of gut microbes, and those communities can trigger signs of IBD when transplanted into healthy mice. This also suggests that the microbiome can contribute to the disease, rather than simply reacting to its presence. But do these microbes actually instigate inflammation, or do they simply perpetuate it once it takes hold? If they are perpetuators, what initially inflamed the gut? An infection? An environmental toxin? Some foodstuff that disrupted the lining of the gut? Genetic variants that had already made the host's immune system prone to overreacting?

All of these possibilities could be true. Untangling them is tricky, not least because no one knows who is going to get IBD beforehand. Without that foresight, it becomes nigh-impossible to watch how the microbiome changes as the disease first manifests, and thus to truly discern the direction of cause and effect. The best that anyone has been able to do is to show that microbes are already dysbiotic in people who have only recently been diagnosed.[17] There is almost certainly no single trigger, microbial or otherwise, that causes IBD. It probably takes several hits to knock the ecosystem into an inflammatory state.

Herbert 'Skip' Virgin published a case study that beautifully supports this idea.[18] He worked with mice that had a genetic mutation common in people with Crohn's disease. Those rodents developed inflamed guts, but only if they were infected by a virus that knocked out part of their immune system, *and* were exposed to an inflammatory toxin, *and* had a normal set of gut bacteria. If any of these triggers was missing, the mice stayed healthy. It was the combination of genetic susceptibility, viral infection, immune problems, environmental toxin, *and their microbiome* that gave them IBD. This complexity helps to explain why the disease is so variable. Every case has its own convoluted history of hits.

These principles apply to other inflammatory diseases too, including type 1 diabetes, multiple sclerosis, allergies, asthma, rheumatoid arthritis and more.[19] All of them involve gung-ho immune systems that launch misdirected assaults at imagined threats. 'One of the common denominators is a simmering level of inflammation in the host. That lies at the heart of all of these problems,' says Justin Sonnenburg, formerly part of Gordon's team. 'Something has happened to give more weight to the pro-inflammatory side and less weight to the anti-inflammatory side. Why do Westerners live in such a hyper-inflammatory state?' And why, as in IBD, have we moved into that state in the last half-century – a period when these once-rare diseases all became much more common? 'For these modern plagues, all the lines are going in the same direction,' Sonnenburg adds. 'All the trends are the same. There have got to be a few major factors in our modern lifestyle that explain a large proportion of this. There aren't going to be 30 different things that we're doing that cause 30 different diseases. My guess is that there are five, or three, or maybe even one thing that explains 90 per cent [of cases] of 90 per cent of these diseases. It seems that there's got to be a single unifying cause.'

In 1976, a paediatrician named John Gerrard noticed a peculiar pattern of diseases among the people of Saskatoon, the Canadian city that he had called home for twenty years. The city's white population was more likely to get allergic diseases like asthma, eczema, and hives than the indigenous Metis communities, while the latter were more often infected by tapeworms, bacteria, and viruses. Gerrard wondered if those trends were connected, if allergic disease 'is the price paid by some members of the white community for their relative freedom from diseases due to viruses, bacteria and [worms]'. In 1989, on the other side of the Atlantic, epidemiologist David Strachan came to a similar conclusion after studying 17,000 British children. The more older siblings they had, the less likely they were to get hay fever.

'These observations . . . could be explained if allergic diseases were prevented by infection in early childhood, transmitted by unhygienic contact with older siblings,' Strachan wrote, in a paper alliteratively entitled 'Hay fever, hygiene, and household size'. The middle 'h' was crucial. It eventually gave the idea its name: the hygiene hypothesis.[20]

The hypothesis, as it now stands, contends that children in developed countries no longer run the gauntlet of infectious diseases that they used to, and so grow up with inexperienced, jumpy immune systems.[21] They are healthier in the short term, but they launch panicked immune responses to harmless triggers, like pollen. This concept delineated an unenviable trade-off between infectious and allergic disease, as if we were destined to suffer one or the other. Later versions of the hygiene hypothesis shifted the emphasis away from pathogens and more towards benevolent microbes that educate our immune systems, or environmental species that lurk in mud and dust, and even parasites that establish long-lasting but tolerable infections. They have been christened 'old friends'.[22] They have been part of our lives throughout our evolutionary history, but their tenure has become shakier of late.

Their disappearance isn't just due to stricter personal cleanliness, as the word 'hygiene' unhelpfully implies. It's also due to the various trappings of urbanisation: smaller families; a move from muddy countryside to concrete cities; a preference for chlorinated water and sanitised food; and a growing distance from livestock, pets and other animals. All of these changes have been consistently linked to a higher risk of allergic and inflammatory diseases, and all of them reduce the range of microbes that we are exposed to. A single dog can have a huge effect. When Susan Lynch hoovered up the dust of 16 homes, she found that those without furry pets were 'microbial deserts'. Those with cats were far richer in microbes, and those with dogs were richer still.[23] It turned out that man's best friend is a chauffeur for man's old friends.

Dogs carry microbes from the outdoors to the indoors, offering us a bigger library of species with which to populate our developing microbiomes. When Lynch fed these dog-associated dust microbes to mice, she found that the rodents became less sensitive to various allergens. The dusty meals also increased the numbers of over 100 bacterial species in the rodents' guts, at least one of which could protect the mice from allergens. This is the essence of the hygiene hypothesis and its various spin-offs: exposure to a broader range of microbes can change the microbiome and suppress allergic inflammation – at least in mice.

But pets are not our most important sources of old microbial friends. That honour goes to our mothers. When babies emerge from the womb they are colonised by mum's vaginal microbes – an endowment that creates chains of transmission which cascade through generations. This, too, is changing. Around a quarter of babies in the UK and a third of those in the USA are now born by Caesarean section, many of which are elective. Maria Gloria Dominguez-Bello found that if babies are born through a cut in their mother's abdomen, their starter microbes come from her skin and the hospital environment, instead of her vagina.[24] It's not clear what these differences mean in the long term, but just as an island's first colonists influence the species that eventually settle upon it, the effects of a child's first microbes could ripple through future communities. This might explain why C-section babies are more likely to develop allergies, asthma, coeliac disease, and obesity later in life. 'The baby's immune system is naïve at birth and whatever it sees first will start its education,' says Dominguez-Bello. 'Their immune system might be compromised if they start recognising the wrong guys instead of the normal good ones. It could make a difference for the rest of their life.'

Bottle-feeding might exacerbate these problems. As we saw, breast milk engineers a baby's ecosystem. It provides more microbe colonists for a baby's gut, and HMOs – those microbe-feeding sugars in breast milk – that nourish co-adapted companions like *B. infantis*. These abilities might overwrite any initial differences caused by a C-section birth, but 'if you go for a C-section and bottle-feeding, I'd certainly say that [your baby] is on a different trajectory,' says milk expert David Mills. Once we are weaned onto solids, that trajectory can veer even further astray if we fail to feed our microbial friends with the right foods. Saturated fats can nourish inflammatory microbes. So can two common food additives, CMC and P80, used to lengthen the shelf life of ice cream, frozen desserts, and other processed foods; they also suppress anti-inflammatory bugs.[25]

Dietary fibre has the opposite effects. This is a catch-all term for various complex plant carbohydrates that our microbes can digest. Fibre has been a mainstay of health advice ever since Denis Burkitt, an Irish missionary surgeon, noticed that rural villagers in Uganda eat up to seven times more fibre than Westerners. Their stools are five times heavier, but pass through the intestine twice as quickly. In the 1970s, Burkitt evangelically promoted the idea that this fibre-rich diet explained why Ugandans rarely suffer from diabetes, heart disease, colon cancer, and other diseases that are more common in the developed world. Some of this difference undoubtedly arises because these chronic diseases are more common in old age, and life expectancy is higher in the West. Nonetheless, Burkitt was on the right track. 'America is a constipated nation,' he said, indelicately. 'If you pass small stools, you have big hospitals.'[26]

He didn't quite know why, though. He imagined fibre as a 'colonic broom', which swept the intestines free of carcinogens and other toxins. He wasn't thinking about microbes. We now know that when bacteria break down fibre, they produce chemicals called short chain fatty acids (SCFAs); these trigger an influx of anti-inflammatory cells that bring a boiling immune system back down to a calm simmer. Without fibre, we dial our immunostats to higher settings, predisposing us to inflammatory disease. To make matters worse, when fibre is absent, our starving bacteria react by devouring whatever else they can find – including the mucus layer that covers the gut. As the layer disappears, bacteria get closer to the gut lining itself, where they can trigger responses from the immune cells underneath. And without the restraining influence of the SCFAs, those responses can easily build to extreme proportions.[27]

Lack of fibre also reshapes the gut microbiome. As we have seen, fibre is so complex that it creates openings for a wide range of microbes with the right digestive enzymes. If those openings close for long enough, the pool of applicants shrinks. Erica Sonnenburg, Justin's wife and colleague, demonstrated this by putting mice on a low-fibre diet for a few months.[28] The diversity in their gut microbiome crashed. It rebounded when the mice ate fibre again, but not fully; many species had gone AWOL and never returned. When these mice bred, they gave birth to pups that started off with a slightly impoverished microbiome. And if those pups ate more low-fibre food too, even more microbes fell off the radar. As the generations ticked by, more and more old friends broke contact. This could explain why Westerners carry a much lower diversity of gut microbes than rural villagers from Burkina Faso, Malawi, and Venezuela.[29] We not only eat fewer plants, we also heavily process the ones we do eat. For example, the milling process that converts wheat into flour removes most of the fibre in the kernels. We are, in the words of the Sonnenburgs, 'starving our microbial self'.

If it isn't bad enough to cut off the routes by which microbes reach us, and then starve those that make it, we also assault the remaining survivors with the greatest disruptors of all – antibiotics. Microbes have been using these substances to fight each other for as long as they have existed. Humans first tapped into this ancient arsenal in 1928 – by accident. On returning to his lab from a holiday in the country, British chemist Alexander Fleming noticed that a mould had landed in one of his bacterial cultures and had carved out a kill-zone of slaughtered microbes all around it. From that mould, Fleming isolated a chemical that he named penicillin. A dozen years later, Howard Florey and Ernst Chain worked out a way of mass-producing the substance, turning this obscure fungal chemical into the saviour of countless Allied troops during World War II. Thus began the modern antibiotic era. In quick succession, scientists developed one new class of antibiotic after another, grinding many deadly diseases under a pharmaceutical boot heel.[30]

But antibiotics are shock-and-awe weapons. They kill the bacteria we want as well as those we don't – an approach that's like nuking a city to deal with a rat. We don't even need to see the rat to begin the massacre: many antibiotics are prescribed needlessly to treat viral infections they have no hope of countering. The drugs are so wantonly used that, on a given day, between one and three per cent of the developed world takes an antibiotic of some kind. One estimate suggests that the average American child gets nearly three courses of antibiotics before her second birthday, and ten before her tenth.[31] Meanwhile, other studies have shown that even short courses of antibiotics can change the human microbiome. Some species temporarily disappear. The overall diversity plummets. Once we stop taking the drugs, our communities bounce back to something that's largely, but not entirely, like their original state. As in Sonnenburg's fibre experiment, each knock leaves the ecosystem slightly dented. As more knocks land, the dents deepen.

Ironically, this collateral damage can pave the way for more disease. Remember that a rich, thriving microbiome acts as a barrier to invasive pathogens. When our old friends vanish, that barrier disappears. In their absence, more dangerous species can exploit the uneaten nutrients and ecological vacancies that remain.[32] *Salmonella*, which causes food poisoning and typhoid fever, is one such opportunist. *Clostridium difficile*, which causes severe diarrhoea, is another. These weedy species bloom to fill the gaps left by a shrinking microbiome, dining heartily on the scraps that would normally be eaten by now-missing competitors. This is why *C. difficile* mostly affects people who have been taking antibiotics, and why most infections happen in hospitals, nursing homes, or other healthcare settings. Some call it a man-made disease, associated with the very institutions that are meant to keep us healthy. It is the unintended consequence of an indiscriminate approach to killing microbes, akin to blitzing a weed-infested garden with pesticides and hoping that flowers will grow in their stead; often, you just get more weeds.[33]

Even subtler doses of antibiotics can have unforeseen consequences. In 2012, Martin Blaser gave antibiotics to young mice, at doses too low to treat any disease. Still, the drugs changed the rodents' gut microbes, fostering communities that were better at harvesting energy from food. The mice became fatter. Next, Blaser's team fed mice with low doses of penicillin either at birth or at weaning, and found that the former group put on more weight after they stopped getting the drugs. Their microbiomes normalised but they still became heavier, and when the researchers transplanted these microbial communities into germ-free mice, the recipients also put on weight. This tells us a couple of important things. First, there's a critical window in early life during which antibiotics can have particularly potent effects. Second, those effects depend on changes in the microbiome, but endure even when it largely returns to normal. The second point is important; the first is arguably old news. Farmers have been inadvertently doing the same experiment since the 1950s, by fattening their livestock with low doses of antibiotics. No matter the drug or the species, the result is always the same: the animals grow faster and end up heavier. Everyone knew that these 'growth promoters' worked but no one really understood why. Blaser's work suggests one possible explanation: the drugs disrupt the microbiome, leading to weight gain.[34]

Blaser has repeatedly suggested that the overuse of antibiotics could be 'fuelling the dramatic increase in conditions such as obesity', not to mention other modern plagues. Are they? The effects in his experiments are relatively small: the antibiotic-treated mice gained weight, but just 10 per cent more; the equivalent of a 70-kilogram person putting on seven extra kilograms, or two body mass index (BMI) units. It also goes without saying that mice are not people, and studies in humans are far more equivocal about the link between antibiotics and obesity. One of Blaser's own showed that infants who get doses of antibiotics aren't any likelier to be overweight by the age of seven. And even the animal studies are inconsistent: in other mouse experiments, scientists have seen that high doses of some antibiotics, given early, can actually stunt growth or reduce body fat.

It is similarly plausible that early antibiotic exposure could increase the risk of allergies, asthma, and autoimmune diseases by altering the microbiome at a critical point – but, as with obesity, the risks are still hazy and imprecise. The benefits of antibiotics are much clearer. In the words of Nobel laureate Barry Marshall, 'I never killed anyone by giving them antibiotics but I know of plenty who died when they didn't get 'em.'[35] Before antibiotics, alarming numbers of people died from simple scratches, bites, bouts of pneumonia, or childbirth. After antibiotics, these potentially life-threatening events became controllable. Everyday life became safer. And medical procedures that would have carried a lethal risk of infection became feasible or common: plastic surgery; C-sections; surgery of any kind that involves bacteria-rich organs like the gut; treatment that suppresses the immune system, like cancer chemotherapy and organ transplants; anything involving catheters, stents, or implants, such as kidney dialysis, cardiac bypasses, or hip replacements. Much of modern medicine is built upon the foundations that antibiotics provide, and those foundations are now crumbling. We have used these drugs so indiscriminately that many bacteria have evolved to resist them, and some nigh-invincible strains can now shrug off every medicine we throw at them.[36] At the same time, we have utterly failed to develop new drugs to replace the ones that are becoming obsolete. We are heading into a terrifying post-antibiotic era.

The problem with antibiotics is less their use than their overuse, which both disrupts our microbiome and foments the rise of antibiotic-resistant bacteria. The solution is not to demonise these drugs but to deploy them judiciously, in situations when they are actually needed and in full knowledge of the risks and benefits. 'Up to this point, we've been viewing antibiotics as just a positive. A doctor might say: it probably won't help you but it won't hurt,' Blaser says. 'But once you think that it might hurt, you have to recalculate things.' For Rob Knight, those calculations became clear when his young daughter came down with a *Staphylococcus* infection. 'I thought: on the one hand, this infection, which could be life-threatening and is causing her a whole lot of pain right now, could evaporate,' he says. 'On the other hand, she could be one BMI fatter at eight. We try to keep her off antibiotics in general but when they work, they're amazing.'

The same kinds of decision apply to other microbial disruptors. Decent sanitation has been an unquestionable public health good, sparing us from many infectious diseases. But we have taken it too far. 'Cleanliness has moved up from being merely next to godliness into a religion in itself,' said Theodor Rosebury. 'We are becoming a nation of tubbed, scrubbed, deodorized neurotics.' He wrote that in 1969.[37] Things are worse now. If I search a certain major online retailer for 'antibacterial', I can find hand wipes, soaps, shampoos, toothbrushes, hairbrushes, detergents, crockery, bedding – even socks. Triclosan, an antibacterial chemical, comes infused into a wide range of consumer goods, including toothpastes, cosmetics, deodorants, kitchen utensils, toys, clothes, and building materials. We have taken cleanliness to mean a world without microbes, without realising the consequences of such a world. We have been tilting at microbes for too long, and created a world that's hostile to the ones we need.

Martin Blaser isn't just worried that some people are short of important microbes. He is deeply concerned that some of these species may be disappearing altogether. Take *Helicobacter pylori*, his favourite bacterium. Blaser was partly responsible for ruining its reputation in the 1990s. Scientists already knew that it caused stomach ulcers, but he and others confirmed that it increases the risk of stomach cancer, too. Only later did he realise the microbe's beneficial side: it reduces the risk of reflux (a condition where stomach acid gurgles back into the throat), oesophageal cancer, and perhaps asthma. Blaser now speaks about *H. pylori* with affection. It is one of the oldest of our old friends, having infected humans for at least 58,000 years.

It's now on the endangered list. Its reputation as a pathogen led to serious and resoundingly successful attempts to eradicate it. ('The only good *Helicobacter pylori* is a dead *Helicobacter pylori*,' said one opinion piece from *The Lancet*.) Once ubiquitous, *H. pylori* is now found in just 6 per cent of children in Western countries. Over the last half-century, 'this ancient, persistent, nearly universal and dominant inhabitant of the human stomach has been essentially disappearing,' writes Blaser. Its loss means that fewer people suffer from ulcers and stomach cancer – clearly, a good thing. But if Blaser is right, the same loss may have precipitated a rise in reflux and oesophageal cancer. Which matters more, the pros or the cons? Neither, it seems. In a large study of almost 10,000 people, Blaser showed that the presence or absence of *H. pylori* had absolutely no effect on a person's risk of dying at any given age. Does it matter, then, that *H. pylori* is vanishing? Perhaps not, but Blaser contends that its decline is a harbinger of other similar disappearances. *H. pylori,* being easy to detect, is the canary in a coal mine. It warns us that other microbes might be going missing right under our noses.[38]

B. infantis, the infant coloniser that we nourish through breast milk, might also be in jeopardy. David Mills's team recently noticed that *B. infantis* is present in 60 to 90 per cent of infants from developing countries like Bangladesh or Gambia, but just 30 to 40 per cent of infants in developed nations like Ireland, Sweden, Italy, and the USA.[39] Bottle-feeding can't explain this difference, since almost all the babies represented in the team's data set were breastfed. C-sections aren't responsible either, since most of the Bangladeshi infants – the ones *most* likely to carry *B. infantis* – were born through this route. In lieu of a solid explanation, Mills has a speculative one. He notes that *B. infantis* seems to disappear from the gut during adulthood, which means that mothers may not be able to pass it on to their children. This wasn't a problem for most of human history because women would help to raise and nurse each other's babies. 'There were always small children being nursed at all times, and they were passing *B. infantis* between themselves and their mothers,' says Mills. But as parenting became more isolated, those chains of transmission snapped. Perhaps this is why the microbe has started disappearing from Western populations, even among breastfed babies. Breast milk can't nourish it if it isn't around in the first place. Whether this is true or not, it certainly seems that *B. infantis* is edging towards the microbial endangered species list.

This work underscores an important principle: we will only learn if developed countries truly lack important microbes by studying a broad swath of humanity. Until recently, most microbiome research had focused on people from WEIRD countries – that is, Western, Educated, Industralised, Rich, and Democratic. These nations account for just an eighth of the world's population; focusing on them is like trying to understand how cities work by studying London or New York and ignoring Mumbai, Mexico City, São Paulo, and Cairo. Recognising this problem, microbiologists have now analysed the microbiomes of rural communities from Burkina Faso, Malawi, and Bangladesh. Others have worked with hunter-gatherers, including the Yanomami in Venezuela, the Matsés in Peru, the Hadza in Tanzania, the Baka in the Central African Republic, the Asaro and the Sausi in Papua New Guinea, and the Pygmies of Cameroon.[40] All of these groups still live traditional lifestyles. They find or catch all of their food. They are rarely, if ever, exposed to modern medicine. They are still modern people with modern microbes living in today's world, but they at least hint at what microbiomes look like without all the trappings of industrialised life.

All of these people have microbiomes that are far more diverse than those in the West. Their multitudes are more multitudinous. They also contain species and strains that are undetectable in Western samples. For example, both the Hadza and the Matsés have high levels of *Treponema*, a group that includes the bacterium responsible for syphilis. Their strains aren't related to those that cause disease, but to harmless relatives that digest carbohydrates. And though these strains are present in hunter-gatherers and apes, they are absent from industrialised populations. Perhaps they are part of an ancient package of microbes that our ancestors shared but that people from developed countries have lost contact with. Studies of fossilised faeces also suggest that people from pre-industrial times had a much richer set of gut microbes than today's city dwellers.

Have we become less healthy, as a result? There's some evidence that a diverse microbiome is better at resisting invaders like *C. difficile*, and that low diversity often accompanies diseases. In one study, a large European team, led by Oluf Pedersen, measured diversity by counting the number of bacterial genes in the guts of almost 300 people.[41] Compared to volunteers with high gene counts, those with low counts were more likely to be obese, and to show signs of inflammation and metabolic problems. Then again, their dwindled communities may be a consequence of poor health, rather than its cause. As yet, no one has shown that people with less diverse microbiomes are more prone to *acquiring* disease. And there are cases where people with diverse microbiomes are *more* likely to carry certain intestinal parasites.[42]

There are also signs that the human microbiome has been shrinking since well before the antibiotic era began, or even before the Industrial Revolution. While rural villagers have more diverse gut microbiomes than urban city dwellers, chimpanzees, bonobos, and gorillas have even more diverse communities; since we diverged from our fellow apes, the human microbiome has been slowly contracting.[43] Perhaps we have simply become better at clearing intestinal parasites. Also, our diets have changed. Gorillas, chimps, and bonobos eat a lot of plants. Rural villagers do too, but they cook their food, breaking it down with heat and taking some of the digestive responsibilities away from their microbes. Americans take this digestive independence even further by eating fewer plants and stripping fibre from the ones they do eat. Animals end up with the microbiome they need, and as our needs have shrunk, so has our pool of partners.

But those changes took place over millennia, giving hosts and microbes time to get used to new arrangements. The worry is that we are now changing our microbiomes at an accelerated pace, disrupting age-old contracts in a matter of generations. Both sides will eventually get used to the new status quo, but that could take many generations more. 'It's the medium term where we're seeing the problem,' says Sonnenburg. He means now.

Blaser shares this concern, writing that, 'The loss of microbial diversity on and within our bodies is exacting a terrible price'. He talks about a looming disaster, 'so bleak, like a blizzard roaring over a frozen landscape, that I call it "antibiotic winter"'.[44] He exaggerates;

it's clear that we *are* changing our microbiomes but there are still only slight hints of the terrible extinctions that Blaser warns about. Still, if forestalling them means stepping beyond the current evidence and raising some hackles, he's okay with that. He has cast himself as a microbiological Cassandra, sounding dramatic prophecies of imminent doom. And, like Cassandra, he invites sceptics.

In 2014, Jonathan Eisen awarded Blaser an Overselling the Microbiome Award for telling *Time* magazine that 'antibiotics are extinguishing our microbiome and changing human development'.[45] The award is an online plaque meant to (dis)honour any scientist or journalist who exaggerates the state of microbiome research and presents speculation as fact. Past winners, and there have been at least 38, have included the *Daily Mail* and the *Huffington Post*. 'I personally think antibiotics may be contributing to messing up the microbiome in many people and that this in turn might be contributing to the increase in a variety of human ailments,' Eisen wrote. 'But "extinguishing"? Not even close.'

The award can seem like a churlish wrist-slap, especially since Eisen himself is cheerful, good-natured, and an enthusiastic ambassador for microbes. But despite his zeal, Eisen practises a modest restraint and recognises that there is still a phenomenal amount to learn about our microbial companions. And he is concerned that the pendulum of scientific attitudes is swinging from germophobia, where all microbes must be vanquished, towards microbomania, where microbes are heralded as the explanation for – and the solution to – all our ills.

His agitation is well founded. There is a long-standing urge in biology to search for unifying causes behind complex diseases. The Ancient Greeks believed that many ailments were caused by an imbalance of four bodily fluids or 'humours' – blood, phlegm, black bile, and yellow bile – and this framework persisted well into the nineteenth century. The concept that diseases were caused by 'bad air' or miasma lasted just as long, until it was eventually dethroned by germ theory. More recently, in the 1960s, many cancer scientists were convinced that all tumours were caused by viruses, after just one carcinogenic virus was discovered in chickens.[46] Scientists will talk about Occam's razor – the principle that favours simple, elegant explanations over convoluted ones. I think the truth is that scientists, like everyone else, find simple explanations psychologically soothing. They reassure us that our messy, confusing world can be understood, and perhaps even manipulated. They promise to let us eff the ineffable, and control the uncontrollable. But history teaches us that this promise is often illusory. The cancer virus believers kicked off a lengthy quest that soaked up over a decade and half a billion dollars, and yielded nothing. We later discovered that several viruses *can* cause cancer, but they explain just a small fraction of all cases. The unifying cause – the one thing that rules them all – turned out to be just a small piece of a broader puzzle.

These lessons in humility are worth remembering when we think about the medical implications of the microbiome, or the absurdly long list of conditions that have been linked to it.[47] A non-exhaustive directory would include: Crohn's disease, ulcerative colitis, irritable bowel syndrome, colon cancer, obesity, type 1 diabetes, type 2 diabetes, coeliac disease, allergies and atopy, kwashiorkor, atherosclerosis, heart disease, autism, asthma, atopic dermatitis, periodontitis, gingivitis, acne, liver cirrhosis, non-alcoholic fatty liver disease, alcoholism, Alzheimer's disease, Parkinson's disease, multiple sclerosis, depression, anxiety, colic, chronic fatigue syndrome, graft-versus-host disease, rheumatoid arthritis, psoriasis, and stroke. A contributor to a satirical website called *The Allium* once wrote, 'In fact, nothing else is important to our health, except the microbiome – it can defeat cancer, cure hunger, poverty, restore amputated limbs, everything.'[48]

Satire aside, even the sincerely proposed links are mostly correlations. Researchers often compare people with the condition to healthy volunteers, find microbial differences, and stop. Those differences hint at a relationship but they don't reveal its nature or its direction. However, the studies I've described on obesity, kwashiorkor, IBD, and allergies go one step further. In trying to work out *how* microbial changes lead to health problems, and by showing that transplanted microbes can reproduce those problems in germ-free mice, they strongly hint at a causal effect. Still, they provide more questions than answers. Did the microbes set symptoms in motion or just make a bad situation worse? Was one species responsible, or a group of them? Is it the presence of certain microbes that matters, or the absence of others, or both? And even if experiments show that microbes *can* cause diseases in mice and other animals, we still don't know if they *actually do so* in people. Beyond the controlled settings of laboratories and the atypical bodies of lab rodents, are microbial changes really affecting our everyday health? How far can they account for the rise of twenty-first-century diseases? How do they compare to other potential causes of 'modern plagues', like pollution or smoking? When you move away from the one-microbe-one-disease model and into the messy, multi-faceted world of dysbiosis, the lines of cause and effect become much harder to untangle.

Speaking of which, what counts as dysbiosis? How can you tell if an ecosystem is in disarray? A bloom of *C. difficile* that causes unstoppable diarrhoea is a clear problem, but most other communities are not so easily classified. Is a gut without *B. infantis* in a state of dysbiosis? If your microbiome has fewer species than a hunter-gatherer's, is it dysbiotic? The term is great at conveying the ecological nature of disease but it has also become microbiology's version of art or pornography: hard to define, but you know it when you see it. And many scientists seem unhelpfully quick to label *any* change in the microbiome as a dysbiosis.[49]

This practice makes little sense because the microbiome is highly contextual.[50] The same microbes can have very different relationships with their hosts in different situations. *H. pylori* can be both hero and villain. Beneficial microbes can trigger debilitating immune responses if they bypass the mucus wall and penetrate the lining of the gut. Seemingly 'unhealthy' communities can be normal, even necessary. For example, gut microbiomes go through a huge upheaval by the third trimester of pregnancy and end up looking like those belonging to people with metabolic syndrome – a disorder that involves obesity, high blood sugar and a higher risk of diabetes and heart disease.[51] This isn't a problem: packing fat and building up blood sugar makes sense when you are nourishing a growing foetus. But if you looked at these communities in isolation you might conclude that their owners were on the verge of chronic disease, when they were merely on the verge of motherhood.

Even when the microbiome changes, it can do so for inexplicable reasons. Over a single day, vaginal communities can change dramatically and rapidly, flitting in and out of states that are supposedly conducive to disease, but with neither clear causes nor ill effects. If you tried to determine a woman's health by analysing her vaginal microbes, the results would be hard to interpret and might be outdated by the time they arrived. The same is true for other body parts, too.[52]

The microbiome is not a constant entity. It is a teeming collection of thousands of species, all constantly competing with one another, negotiating with their host, evolving, changing. It wavers and pulses over a 24-hour cycle, so that some species are more common in the day while others rise at night. Your genome is almost certainly the same as it was last year, but your microbiome has shifted since your last meal or sunrise.

It would be easier if there was a single 'healthy' microbiome that we could aim for, or if there were clear ways of classifying particular communities as healthy or unhealthy. But there aren't. Ecosystems are complex, varied, ever-changing and context-dependent – qualities that are the enemies of easy categorisation.

To make matters worse, some of the early microbiome discoveries are almost certainly wrong. Remember how obese people and mice have more Firmicutes and fewer Bacteroidetes than their lean counterparts? This result, the F/B ratio, is one of the most famous in the field – and it's a mirage. In 2014, two attempts to re-analyse past studies found that the F/B ratio is *not* consistently connected to obesity in humans.[53] You can tell the difference between obese and lean microbiomes within any single study, but there are no consistent differences *across* studies. This doesn't refute a connection between the microbiome and obesity. You can still fatten germ-free mice by loading them with microbes from an obese mouse (or person). *Something* about these communities affects body weight; it's just not the F/B ratio, or at least not consistently so. It is humbling that, despite a decade of work, scientists are barely any closer to identifying microbes that are clearly linked to this condition, which has received more attention from microbiome researchers than any other. 'I think that everybody is coming to the realisation that, unfortunately, a really compelling simple biomarker, like the percentage of a certain microbe, is not going to be enough to explain something as complicated as obesity,' says Katherine Pollard, who led one of the re-analyses.

These conflicting results naturally arise in the early days of a field because of tight budgets and imprecise technology. Researchers run small, exploratory studies comparing handfuls of people or animals in hundreds or thousands of ways. 'The problem is that they end up being like the Tarot,' says Rob Knight. 'You can tell a good story with any arbitrary combination.' Imagine that I pulled ten people off the street who are wearing blue shirts and ten who are wearing green shirts. If I ask them enough questions, I guarantee you that I can find at least a couple of striking differences between the two groups. The blue-shirts might prefer coffee while the green-shirts prefer tea. The green-shirts have bigger feet than the blue-shirts. I might propose that blue shirts produce coffee cravings and shrink one's feet. But if I accosted two groups of a million people each, I'd have a much harder time finding random differences between them, and I'd be more confident that the differences I *did* see were meaningful. Then again, it takes time and effort to accost a million people. Human geneticists faced the same problem. In the early twenty-first century, when technology hadn't quite caught up with ambition, they identified many genetic variants that were linked to diseases, physical traits, and behaviours. But once sequencing technology became cheap and powerful enough to analyse *millions* of samples, rather than dozens or hundreds, many of these early results turned out to be false positives. The human microbiome field is going through the same teething problems.

It doesn't help that the microbiome is so variable that the communities in lab mice can differ if they belong to different strains, come from different vendors, were born to different mothers, or were reared in different cages. These variations could account for phantom patterns or inconsistencies between studies. There are also problems with contamination.[54] Microbes are everywhere. They get into everything, including the chemical reagents that scientists use in their experiments.

But these problems are now being ironed out. Microbiome researchers are getting increasingly savvy about experimental quirks that bias their results, and they're setting standards that will shore up the quality of future studies. Sick of the never-ending stream of correlations, they are calling for experiments that will show causality, and tell us *how* changes in the microbiome lead to disease. They are looking at the microbiome in even greater detail, moving towards techniques that can identify the strains within a community, rather than just the species. Rather than just sequencing DNA, they are studying RNA, proteins, and metabolites; DNA reveals which microbes are present and what they are capable of, but the other molecules tell you what they are actually doing. Researchers are using machine-learning programs to identify complex communities of microbes that might be involved in diseases, rather than focusing on just one or two species in isolation.[55] They are making use of the falling cost of sequencing to run bigger studies.

They are also setting up *longer* studies. Rather than capturing a single screenshot of the microbiome, they are trying to watch the entire movie. How do these communities change with time? How many knocks can they absorb before they topple? What makes them resilient or unstable? And does their degree of resilience predict a person's risk of disease?[56] One team is recruiting a group of 100 volunteers who will collect weekly stool and urine samples for nine months, while eating specific diets or taking antibiotics at fixed times. Others are leading similar projects with pregnant women (to see if microbes contribute to pre-term births) and people at risk of developing type 2 diabetes (to see if microbes affect their progression to full-blown disease). And Jeff Gordon's group has been charting the normal progression of microbes in a healthy developing baby, and how it stalls in kids with kwashiorkor. Using stool samples collected from Bangladeshi children over their first two years, the team has created a score that measures the maturity of their gut communities and will hopefully predict if symptomless infants are at risk of developing kwashiorkor.[57]

The ultimate goal of all of these projects is to spot the signs of disease as early as possible, before a body turns into the equivalent of an algal reef: a degraded ecosystem that is very hard to repair.

'Professor Planer!' says Jeff Gordon. 'How are you?'

He means Joe Planer, one of his students, who is standing in front of a standard laboratory bench complete with pipettes, test tubes, and Petri dishes, all of which have been sealed in a transparent, plastic tent. It looks like one of the isolators from the germ-free facility but its purpose is to exclude oxygen rather than microbes. It allows the team to culture the many gut bacteria that are extremely intolerant of the gas. 'If you write the word oxygen on a piece of paper and show it to these bugs, they'll die,' jokes Gordon.

Starting off with a stool sample from a Malawian child with kwashiorkor, Planer used the anaerobic chamber to culture as many of the microbes within it as possible. He then picked off single strains from these collections, and grew each one in its own compartment. He effectively turned the chaotic ecosystem within a child's gut into an orderly library, dividing the teeming masses into neat rows and columns. 'We know the identity of the bacteria in each well,' he says. 'We'll now tell the robot which bacteria to take and combine in a pool.' He points to a machine inside the plastic, a mess of black cubes and steel rods. Planer can program it to suck up the bacteria from specific wells and mix them into a cocktail. Grab all the *Enterobacteriaceae*, he might say, or all the *Clostridia*. He can then transplant these fractions back into germ-free mice to see if they alone can confer the symptoms of kwashiorkor. Is the whole community important? Will the culturable species do? A single family? A single strain? The approach is both reductionist and holistic. They're breaking down the microbiome, but then recombining it. 'We're trying to work out which actors are responsible,' says Gordon.

A few months after I saw Planer working with the robot, the team had narrowed down the kwashiorkor community to just 11 microbes that replicate many of the disease's symptoms in mice.[58] This cabal included some familiar faces like B-theta and *Bacteroides fragilis,* none of which were harmful on their own. They only caused a problem when acting together – and even then, only when the mice were starved of nutrients. The team also created culture collections from healthy twins who didn't develop kwashiorkor, and identified two bacteria that counteract the damage inflicted by the deadly 11. The first is *Akkermansia,* which seemingly multitasks as a guardian against both malnutrition *and* obesity. The second is *Clostridium scindens,* one of those *Clostridia* that tamp down inflammation by stimulating regulatory T cells.

Opposite the tented bench, there is a blender that can take foods representative of different diets and pulverise them into rodent-friendly chow. On a piece of sticky tape affixed to the blender someone has written 'Chowbacca'. Gordon's lab can now explore the behaviour of *Akkermansia* and *Clostridium scindens,* either in test tubes or in the germ-free mice and work out which nutrients they need. This allows the team to compare the effects of the same microbes on a Malawian diet, or an American one, or the special microbe-nourishing sugars from breast milk (and Gordon is working with Bruce German and David Mills to do just that). Which foods nourish which microbes? And which genes do the microbes switch on? The team can take any one microbe and can create a library of thousands of mutants, each of which contains a broken copy of a single gene. They can put these mutants in a mouse to see which genes are important for surviving in the gut, liaising with other microbes, and both causing or protecting against kwashiorkor.

What Gordon has built is a causality pipeline – a set of tools and techniques that, he hopes, will more conclusively tell us how our microbes affect our health and take us from guesswork and speculation to actual answers. Kwashiorkor is just the start. The same techniques could work for any disease with a microbial influence.

We aren't just talking about human diseases, either. Many zoo animals get sick for unknown reasons.[59] Cheetahs come down with gastritis caused by their equivalent of *H. pylori*. Marmosets – small, adorable monkeys – suffer from the accurately named marmoset wasting syndrome. Are these also diseases of dysbiosis? Could these animals be suffering from microbiome problems caused by unusual diets, overly sanitised artificial environments, unfamiliar medical treatments, or quirks of captive breeding programmes? If animals lose native microbes, how would they fare if they were ever released back into the wild? Would they have the right digestive bacteria? Would their immune systems be properly calibrated to handle diseases without vets to fall back on? And since we know that microbes can affect behaviour (and that germ-free rodents are less anxious than most), would they have the necessary caution to survive in a predator-filled world?

It's the right time to be asking these countless questions. Our planet has entered the Anthropocene – a new geological epoch when humanity's influence is causing global climate change, the loss of wild spaces, and a drastic decline in the richness of life. Microbes are not exempt. Whether on coral reefs or in human guts, we are disrupting the relationships between microbes and their hosts, often pulling apart species that have been together for millions of years. Scientists like Gordon and Blaser are working hard to understand, and perhaps forestall, the end of these long partnerships. But others are more interested in how they began.

6. THE LONG WALTZ

On 15 October 2010, a retired engineer named Thomas Fritz set out, chainsaw in hand, to cut down a dead crab apple tree growing outside his home in Evansville, Indiana. The tree came down easily enough, but as Fritz dragged the debris away he tripped and a pencil-sized branch went straight through the fleshy web between his right thumb and index finger. Fritz was a volunteer firefighter with medical training; he knew how to dress a wound. But despite his efforts, the hand became infected. By the time he visited his doctor, two days later, a cyst had formed. Fritz took a course of antibiotics, to no avail. His hand only started healing five weeks later, after a surgeon extracted a few pieces of bark that were stubbornly lodged in his flesh.

The misadventure would have ended there had Fritz's doctor not collected some fluid from the wound. The extract made its way to a facility at the University of Utah, where mysterious microbial samples go to be identified. The lab's automated instruments identified the bacteria in Fritz's wound as *E. coli*, but medical director Mark Fisher didn't buy it. The DNA wasn't a good match. When he checked the sequences more closely, he realised that they were almost identical to a bacterium called *Sodalis*, which had been discovered as recently as 1999. And as luck would have it, its discoverer also worked at the university: Colin Dale, a British biologist.

Dale was sceptical. Fisher assured him that the microbe was growing in an agar dish in the lab. No, Dale countered, that must be a mistake. As far as anyone knew, *Sodalis* lived only in the bodies of insects. Dale had first found it in a blood-sucking tsetse fly, and then in weevils, stinkbugs, aphids, and lice. It nestled inside the cells of these animals, and had lost too many genes to live anywhere else. It couldn't possibly be growing in a dish, let alone in an infected hand or on a dead tree branch. And yet, the DNA wasn't lying. Many of the genes in the bacterium from Fritz's hand were identical to those in *Sodalis*. Dale called the new strain HS, for 'human *Sodalis*'. 'I suspect that HS is widespread but we don't go checking dead trees,' he says.

Think about all the coincidences in this story. The wild microbe just happened to sit on the right branch, impale the right person, and end up in the right lab, down the road from the person who discovered its domesticated insect cousin. It seemed like an absurd confluence of improbability. And then it happened again. This time, the victim was a kid who was climbing a tree. Like Fritz, he fell and impaled himself on a branch. Unlike Fritz, he didn't become infected. His first symptom appeared a decade later, when a mysterious cyst formed at the site of the old wound. Doctors removed it and sent a sample to the University of Utah. And then there were *two* strains of HS.[1]

Forget Fritz and forget the kid: they're fine, and perhaps more cautious now about arboreal safety. Let's talk about HS. Scholars of symbiosis get a little glint in their eyes when they discuss it, because it gives us a rare look at one of the most fundamental but uncertain aspects of partnerships between animals and bacteria: their beginnings. Usually, by the time we learn about these relationships the partners have been waltzing together for millions of years. But what did they look like when they first took hold? What made them do so? How did they carry on dancing together, and how did they change in the process? These are vexing questions. The first steps of the long waltz are almost always lost in deep time, and have left few footprints for us to follow.

HS is an exception. It shows what *Sodalis* might have looked like before it became an indentured part of an insect's body, back when it was a free-living microbe hanging out in the environment and capable of infecting an animal host if it got the right chance. It's a missing link. A symbiont-in-waiting. Scientists had long predicted that such ancestral microbes exist, but few thought they would actually find one. Dale found two. He has since given HS the formal name of *Sodalis praecaptivus* – '*Sodalis* before captivity'.[2]

So, picture HS, sitting on plants and who knows what else, going about its life. If it gets into an errant gardener or a falling child, it starts growing. More likely, it gets into an insect that lives on the plant. In fact, based on its genes, Dale speculates that it's a pathogen that causes diseases in trees, and spreads between them on the mouthparts of insects. Already, it depends on these animals to reach new hosts. It might then evolve to provide them with benefits, like nutrients or protection from parasites. Eventually, it might move from the gut or salivary glands of its hosts into its very cells. Then, rather than passing from one insect to another via a tree, it starts moving from mother to offspring. It becomes a permanent part of its host's body. In these comfy environs, in the way that insect symbionts do, it loses genes that it no longer needs and becomes *Sodalis*. These events probably played out several times over, producing the different versions of *Sodalis* that exist in the various insect groups.[3]

It's likely that many symbioses started this way, with random environmental microbes – some parasitic and others more benign – that somehow sneaked into animal hosts. Such incursions are common and inevitable. The ubiquity of bacteria means that almost everything we do brings us into contact with new species.

You don't need to impale yourself on a branch. Sex works: when aphids mate, they can pass along microbes that help them fend off parasites or withstand higher temperatures. Eating something will also do the trick. Woodlice can pick up microbes from their peers by cannibalising them. Mice can pick up bacteria from their neighbours by eating their droppings. Two bugs can pass microbes through their backwash if they both sip from the same plant. The average human swallows around a million microbes in every gram of food they eat. Since microbes are everywhere, virtually every source of food, whether a patch of water, the stem of a plant, or the flesh of another animal, is a potential source of new symbionts.[4]

Parasites offer another possible route into the body. Many wasps lay their eggs in the bodies of other insects via sharp tubes that they plunge into victim after victim. In doing so, the wasps act as living, flying, dirty needles, spreading potentially beneficial microbes from one host to another just as a mosquito's snout might spread malaria or dengue fever. We know that events like this happen because scientists have actually witnessed them in the field, and reproduced them in the lab.[5] Contaminated food and water, unprotected sex, dirty needles: these are all routes that we associate with *disease*. But any road that a pathogen can travel down is one that beneficial symbionts can also use to reach new hosts.

Of course, the journey isn't everything. Once a bacterium arrives in a new destination, it needs to make itself at home, and there's no guarantee it will succeed. It has to deal with the immune system, rival microbes, and other threats. Maybe only one of every hundred horizontal jumps leads to a stable partnership. Maybe it's more like one in a million. We have no way whatsoever of knowing. But in a single field, there could be a million aphids drinking from the same plants, and a million wasps buzzing around and stabbing the aphids with contaminated daggers. In such numbers, even improbable events become commonplace, and even the implausible becomes plausible, like impaling yourself on a tree branch and picking up a symbiont.

The newly arrived microbes might stick around if they are capable parasites, but some guarantee their residency by providing a benefit. They don't even need any special adaptations. The world is full of microbes that are preadapted to symbiosis by dint of what they naturally do. If a plant-eater ingested microbes that could break down complex fibres in plants and in doing so release otherwise inaccessible chemical by-products that their cells can burn for energy, the microbes would fit in immediately. By getting on with their usual activities, in a purely selfish way, they incidentally benefit their hosts. These 'by-product mutualisms' are the perfect first handshake.[6] Both partners get something out of the relationship, without either having to invest in it. The host can then evolve traits that solidify the partnership, from cells that house the tiny partners to molecular anchor-points for them to latch themselves to. And the most important of these traits – the one that does more than any other to seal a symbiosis – is inheritance.

In a European meadow, a honeybee buzzes between flowers under a hot summer sun. Suddenly, another black-and-yellow insect dives in, snatches the bee out of the air, and paralyses it with a sting. The attacker is a beewolf – a large, powerful, and aptly named wasp. She drags her victim back to an underground burrow, and buries it alongside one of her own eggs and several more bees – all immobilised, but still alive. When the young grub hatches, it will devour this living larder, which its mother so carefully stocked.

Bees are not the only gifts that beewolf mothers provide for their young. Martin Kaltenpoth was studying the behaviour of beewolves when he noticed that one of his specimens was leaking white fluid from its antennae. He had seen this substance before. After a beewolf digs her burrow, and before she adds an egg, she presses her antennae against the soil and squeezes a white paste out of them, like toothpaste from a tube. She then shakes her head from side to side to daub this secretion against the burrow's ceiling. The paste is an exit sign: it tells the young beewolf where to start digging when it is ready to leave the burrow. But when Kaltenpoth examined the paste under a microscope, he was stunned to see that it also swarmed with bacteria. A wasp that secretes microbes from its antennae? No one had heard of such a thing. Stranger still, the bacteria were all identical. Every beewolf had the same strain of *Streptomyces* in its antennae.

That was a huge clue. *Streptomyces* are microbes that excel at killing other microbes; this one group is the source of two-thirds of our own antibiotics. And a young beewolf certainly needs antibiotics. Once it finishes eating its stockpile of bees, it encases itself in a silken cocoon and stays that way throughout the winter. For nine long months, it is trapped in a warm, humid chamber that's perfect for nurturing pathogenic fungi and bacteria. Kaltenpoth reasoned that its mother's antibiotic paste might stop the youngster from contracting a lethal infection. Indeed, when he watched the grubs carefully, he saw that they would incorporate the bacteria from the paste into the fibres of their cocoons, tucking themselves into self-woven quilts of antibiotic-producing microbes. When Kaltenpoth deprived the young wasps of the white paste, almost all died from fungal infestations within a month.[7] If he gave them access to the paste, they usually survived. And come the spring, when new adult wasps emerge from their cocoons, they take up into their antennae the same *Streptomyces* that guarded them over the winter. Off they fly to dig their own burrows, capture their own bees, and pass their life-saving microbes to their own young.

These acts of transmission, where animals hand microbes to their offspring in a generational relay, are among the most critical in the world of symbiosis, because they braid together the fates of hosts and symbionts.[8] They ensure that the long waltz is actually long, that it continues through time, that new generations of animals and microbes will take hold just as their parents did. And they create an evolutionary pressure for the dancers to become even more closely entwined. The microbes face an immense evolutionary pressure to develop abilities that help their hosts, since that gives them an even bigger pool of partners to dance with. And the animals are driven to evolve ever more efficient ways of faithfully passing their microbial heirlooms to their offspring.

The most reliable route, and the one that creates the most intimate symbioses, involves adding microbes directly to egg cells. Mitochondria, those former bacteria that power our cells, are already in the egg cells of animals, so they pass from mother to child with no extra effort. Other microbes have to be imported – a strategy used by deep-sea clams, marine flatworms, and countless species of insects. These animals are accompanied by microbes from their very first moments as a single fertilised egg. They are never alone.

Even if the egg route isn't an option, there are other ways of ensuring that your offspring are colonised by the right microbes. Many insects use a similar strategy to beewolves: they provide a nearby glut of microbes for their hatchlings to use. The stinkbug family excels at this, and few people know stinkbugs better than Takema Fukatsu, an infectiously enthusiastic entomologist hell-bent on studying every insect that exists.[9] He has shown that one stinkbug packages its microbes into hardy, weatherproof capsules, which it lays alongside its eggs and which the hatchling bugs then eat. Another species packages the eggs themselves inside a jelly that's also laden with microbes. And one Japanese species, a handsome red-and-black insect that's easy on the eye and hard on crops, has the most extreme strategy of all. While most insects abandon their young to fate, this one zealously guards her clutch of eggs. She sits on them like a hen, and even collects fruit to feed the nymphs once they hatch. She can somehow sense when that's about to happen, and she pre-empts the fateful moment by secreting copious amounts of bacteria-laden mucus from her backside. The white liquid coats the eggs, which end up looking like a ball of jellybeans frosted with the world's most revolting icing. When the hatchlings emerge, they swallow the mucus and become colonised by the freshest possible gut microbes. Put aside your disgust for a second and think about how significant that moment is: in that first mouthful, each young bug transforms from an individual into a colony of multitudes; from a sterile body into a thriving ecosystem.

The blood-sucking tsetse fly, which spreads sleeping sickness between humans, also provisions its young with microbes, but does so *inside its own body*. It's an insect that's trying very hard to be a mammal. Rather than laying eggs, it gives birth to live young. And rather than hedging its bets with a horde of offspring, it devotes its energies to a single grub, which it raises inside a uterus and feeds with a milk-like fluid. The milk is full of both nutrients and microbes (including *Sodalis*), so when the grotesquely huge youngster wriggles out of its poor mother – believe me, human birth has *nothing* on tsetse birth – it already has all the bacterial partners it needs.[10]

Other animals wait until their young are hatched or born before feeding them with microbes. When a baby koala is six months old, it weans off its mother's milk and moves on to eucalyptus leaves. But first, it nuzzles mum's backside. She, in response, releases a fluid called pap, which the joey swallows. Pap is full of bacteria that will allow the koala joey to digest tough eucalyptus leaves, and contains up to 40 times more of these microbes than regular faeces. Without this initial meal, all the joey's later ones would be hard to stomach.[11]

Humans, you will be delighted to know, do not have pap. Our egg cells have no bacteria in them either (discounting mitochondria), and our mothers don't cover us in mucus. Instead, we unite with our first microbes at the moment of birth. In 1900, the French paediatrician Henry Tissier asserted that the womb is a sterile chamber that keeps babies and bacteria apart. This isolation ends when we pass through the birth canal and encounter vaginal bacteria. These are our first colonisers – the pioneers of the empty ecosystems inside us. Much like a Japanese stinkbug, we emerge into the world slathered in mum's microbes. In recent years, a few studies have challenged this concept by reporting traces of microbial DNA in supposedly sterile tissues like amniotic fluid, umbilical cord blood, and the placenta – but these results are highly controversial.[12] It's not clear how these microbes get there, whether their presence matters, or if they actually exist – the DNA could have come from dead cells, or from bacteria that contaminated the experiments. Tissier's sterile womb hypothesis might be wrong, but it certainly hasn't toppled yet.

Even if animals don't inherit microbes vertically from their parents, there are still ways for them to 'catch' the right symbionts through horizontal routes. Many animals regularly seed their surroundings with expelled microbes, which their offspring can pick up.[13] Others go for a more direct approach. Termites, in the words of Greg Hurst, 'go in for anal-licking, or proctodeal trophollaxis to give it its posh name'. Like koalas, they need microbes to digest their food – in this case, wood – and they get theirs by sucking fluid from their relatives. But, unlike koalas, termites lose the lining of their guts, and all the microbes within, every time they moult their outer shells. So they regularly need to lick their sisters' backsides to replenish their supply. We might find these habits unsavoury, but we are unusual in our distaste.

Many familiar animals, including cows, elephants, pandas, gorillas, rats, rabbits, dogs, iguanas, burying beetles, cockroaches, and flies, regularly eat each other's faeces – a practice known as coprophagy.

For skin microbes, simple contact can suffice. Animals as diverse as salamanders, bluebirds, and humans tend to harbour similar communities of bacteria if they live in close quarters. People who share the same house end up with more similar skin microbes than friends who live apart. Likewise, baboons who come from the same troop (and thus groom each other) have more similar gut microbes than outsiders, even if the two groups live in the same place and eat the same diet. And the most wonderful example of this convergence comes from a study of American roller derby players. The players share skin bacteria with their teammates, and different teams have their own distinctive communities. But during a match, as the two teams clash and jostle on the track, their skin microbes temporarily converge. Contact breeds conformity. Sometimes, the long waltz involves hip-checks.[14]

Many of these routes depend on some sort of social contact. They only work if parents stick with their offspring, or if different generations mingle in large groups. Japanese stinkbugs care for their young, so they can be there to infuse them with the right bacteria. Termites live in dense colonies, where new workers can lick the right microbes from their sisters. There's a reason for this pattern, says Michael Lombardo. He argues that some animals came to live in large groups because they could more easily pick up beneficial symbionts from their neighbours. That's not the *only* factor behind the evolution of sociality, or even the main one; sociable animals can also hunt as a team, find safety in numbers, or navigate effectively. Lombardo simply thinks that microbial transmission is another plausible benefit, and one that is traditionally ignored. When people think about contagious microbes, they tend to think of pathogens first. Herds, flocks, and colonies make it easier for diseases to spread. But they also create opportunities for *beneficial* symbionts to find new hosts.[15]

The seemingly infinite range of transmission routes through which animals pick up microbes from one another all serve the same imperative: the need to move microbes from one generation of hosts to the next. Whether stinkbug or koala, beewolf or baboon, animals have ways of ensuring that they continue the long waltz with more or less the same partners. Sometimes, this involves strict vertical inheritance from parent to offspring, which ties hosts to the same microbes for countless generations. At the other end are looser horizontal transfers from peers or shared environments; this ensures some continuity but allows animals to more freely swap their symbionts or pick up new ones. But even at this looser end of the spectrum, animals are still selective. They have a world of partners to choose from, but they won't dance with just anyone.

Your local pond is home to a fascinating and oddly charismatic creature that you have probably never seen before. Finding it is simple: scoop up some duckweed or other floating plants, put them in a jar with some water . . . and wait. If you carefully check the plants, you might notice a small green or brown blob, just a few millimetres wide, stuck to their stems or the undersides of their leaves. Give it time and a little light, and the blob will slowly stretch into a long stalk crowned with tentacles. Fully extended, it looks like a thin, gelatinous arm with long, splayed fingers.

This is a hydra: a relative of sea anemones, corals, and jellyfish. It is named after the terrifying, swamp-dwelling, multi-headed serpent that belaboured Hercules in Greek mythology. The name is comically absurd, given the creature's tiny size – but also oddly apt. Hydra the monster terrorised villagers with poisonous breath and blood, while hydra the animal kills water fleas and shrimp with stinging cells that fire venomous harpoons. The monster could grow two heads for every one that was cut off; the real hydra is also an expert at regeneration. Cut off a limb? No problem. Turn it inside-out? It will cope.

The hydra is incredibly attractive to biologists who want to understand how animals develop and grow. It is easy to collect, nurture, and breed. It is also mostly transparent, so a light microscope will reveal its inner workings. By the time the developmental biologist Thomas

Bosch came across it in 2000, scientists had been studying hydra for centuries. Leeuwenhoek himself had sketched the animal in one of his notebooks. Others had sussed out how it grows from a single cell into an adult, and how it re-grows its severed body parts. Bosch himself became ensnared by this animal for his whole career. 'I always forbid my students from using the term primitive,' he says. 'The hydra have led a beautiful, successful way of life for 500 million years.'

But even Bosch thought it strange that hydra have survived for such a long time, especially given their simple architecture. The human body is so complex that most of it is never exposed to the outside world; the only points of contact are the layers of cells that line your gut, lungs, and skin. These layers are called epithelia and, among their many functions, they block microbes from penetrating deeper into the body. But there's no 'deeper into the body' for a hydra. It consists of just two layers of cells with a jelly-like filling, and so its outsides and insides are both in constant contact with water. It has no barrier to separate its tissues from its environment – no skin or shell, cuticles or coverings. A hydra is about as exposed as it is possible for an animal to be. 'It's just a slimy epithelium sitting in a hostile environment,' says Bosch. So, why isn't a creature like this constantly plagued by infections? How does it stay healthy?

To answer this question, Bosch first had to figure out what microbes live in or around hydra. His student Sebastian Fraune did this by pulping their bodies, pulling out any bacterial DNA, and sequencing everything. He analysed two closely related species and, to his surprise, found that they harboured distinct communities of microbes. It was as if he was looking at the wildlife of different continents.

That was surprising because these hydra had come from lab stocks that had been raised in plastic containers for over thirty years. For decades, they had been immersed in the same carefully composed water, fed on the same food, and kept at the same temperature. Human prisoners kept in such stultifyingly standardised conditions would struggle to remember their identity. But each hydra – an animal without a brain – was still somehow assembling the appropriate microbial community for its species. It seemed implausible and, at first, Bosch didn't believe the result. But Fraune repeated the study and got the same results. He sequenced more species of hydra and found that each had its own distinct microbiome, which matched those of wild individuals that he collected from local lakes.[16]

'That was a real turning point for me,' says Bosch. 'I had always been thinking through the traditional lens of microbiology, in which tissues must defend against bad guys.' Instead, his experiments clearly showed that the various species of hydra were actively sculpting their own microbiome.

This is a trend that pervades the animal kingdom: we don't just dance with any old bacterium that happens to show up. New microbes constantly intrude into our lives, but each species chooses specific partners from the hodgepodge of candidates. The bacteria in the human gut, for example, almost all belong to four major groups, out of the hundreds that exist in the wild. Even hydras, simple and exposed though they are, have ways of allowing some bacterial species to colonise their surface while excluding others. Our bodies, whether big or small, complex or simple, create conditions in which only some microbes can thrive. Over time, and because of the continuity of inheritance, this selectivity becomes starker, as hosts and symbionts adapt to each other. We are picky.[17]

As a result, each species ends up with its own distinctive community. You can tell a human microbiome from a mouse or zebrafish microbiome, or even from a chimp or a gorilla microbiome. Even whales and dolphins that share the same oceans, and constantly scrub their skins by swimming and breaching, maintain their own species-specific skin communities. The beewolves that we met earlier are so selective about the bacteria in their antennae that if they end up with the wrong strains they won't produce the white mucus that passes those microbes to the next generation. Somehow, if they sense that they've got the wrong partners, they'll cut the chain of inheritance and end the long waltz.[18]

Microbes have their own preferred partners, too, and many have adapted to colonise specific hosts. Some strains of the bee symbiont

Snodgrassella are adapted to honeybees and others to bumblebees, and neither can colonise their non-native hosts. Similarly, the gut microbe *Lactobacillus reuteri* comes in strains that have adapted to humans, mice, rats, pigs, and chickens. If you shove them all into a mouse, the rodent strains will outgrow the others. These kinds of microbe-swapping experiments can be very instructive. John Rawls carried out the most influential ones when he exchanged the microbiomes of two stalwarts of laboratory science: mice and zebrafish. Rawls bred sterile, germ-free versions of both animals, and then infused them with microbiomes from conventionally raised individuals of the opposite species. Would a zebrafish just accept a mouse's gut microbes, and vice versa? The answer was yes. But Rawls found that the animals didn't just stick with the hands they were dealt. Instead, they reshaped their new communities to more closely match their native ones. The mice partly mousified the fish microbiomes, and vice versa.[19]

That's not to say that every individual in a particular species has an identical microbiome. There's a lot of variation. Think of it this way: an animal's genes are like set designers in a theatre – they create the stage upon which specific microbes can perform.[20] Our environment – friends and footsteps, dirt and diet – then affects the actors that take the stage. And random chance lords over the whole production, which is why even genetically identical mice that live in the same cage end up with slightly different microbiomes. The composition of our microbiome is a bit like height, intelligence, temperament, or risk of cancer: a complex trait that is controlled by the collective action of hundreds of genes, and by even more environmental factors. The big difference is that our genes don't directly create the microbiome as they do our height or the size of our brains. They set conditions which, in turn, select for certain species over others.

In his classic book *The Extended Phenotype*, Richard Dawkins introduces the idea that an animal's genes (its genotype) do more than sculpt its body (its phenotype). They also indirectly shape the animal's environment. Beaver genes build beaver bodies, but since those bodies go on to make dams, the genes are also redirecting the flow of rivers. A bird's genes create a bird, but they also make a nest. My genes made my eyes, hands, and brain, and in doing so they also made this book. All of these things – dams, nests, and books – are what Dawkins calls extended phenotypes. They are products of a creature's genes that extend beyond its body. In a way, that's what our microbiomes are. They too are shaped by animal genes, which create environments that encourage specific microbes to grow. Although they lie *inside* their owners, they are just as much an extended phenotype as a beaver dam.

But even this comparison doesn't entirely work because microbes – unlike the dam or this book – are themselves alive. They have their own genes, some of which are important or essential to their hosts. They aren't just extensions of a host's genome, any more than the host is an extension of the microbes' genomes! So, argue some scientists, maybe it makes no sense to conceptually separate them. If animals are picky about their microbes, and microbes are picky about their hosts, and both are locked in partnerships that endure through generations, maybe it makes more sense to think of them as unified entities. Maybe we should think of them as one.

We have already seen that some bacteria become so integrated into their hosts that it's hard to see where one species ends and the other begins. Many insect symbionts are like this, including the many lineages of *Hodgkinia* in the cicadas. Mitochondria certainly count: as we've seen, these cellular batteries were once free-living bacteria that became permanently enclosed within a larger cell. This process, known as endosymbiosis, was first proposed in the early twentieth century, but it only became accepted several decades later, largely thanks to the outspoken American biologist Lynn Margulis. She turned endosymbiosis into a coherent theory, which she expounded in a genre-hopping paper that contained an impressive mix of evidence from cell biology, microbiology, genetics, geology, paelaeontology, and ecology. It was a bravura piece of scholarship. It was also rejected around 15 times before seeing print in 1967.[21]

Margulis was dismissed and ridiculed by her peers, but she gave as good as she got. Rebellious, and contemptuous of dogma, she was the consummate scientific iconoclast. 'I don't consider my ideas controversial,' she once said. 'I consider them right.' She was certainly right about mitochondria and chloroplasts, but thanks to other oversold claims, she is often viewed with both the utmost respect and cautious scepticism. One biologist told me that he heard her mention his name in a talk. Great, he thought, Lynn Margulis knows my name! Then, she added, '. . . is completely wrong'. Phew, he thought, if Lynn Margulis thinks I'm wrong, I must be on to something.

Endosymbiosis influenced Margulis's view of the world throughout her career. She was drawn to the connections *between* living things, and she realised that every creature lives in communities with many others. In 1991, she coined a word to describe this unity: holobiont, from the Greek for 'whole unit of life'.[22] It refers to a collection of organisms that spend significant parts of their lives together. The beewolf holobiont is the wasp plus all the bacteria in its antennae. The Ed Yong holobiont is me plus my bacteria, fungi, viruses, and more.

When Israeli couple Eugene Rosenberg and Ilana Zilber-Rosenberg heard the term, they were enchanted by it. They had been studying corals, and had come to view these animals as collective entities whose fate depended on the algae in their cells and the other microbes that live around them. It made sense to think of them as unified communities. You could only understand the health of a reef, they realised, by accounting for the whole coral holobiont.

Rosenberg pushed the holobiont concept into the world of genes. Evolutionary biologists had come to treat animals and other organisms as vehicles for their genes. The genes that create the best vehicles – say, the fastest cheetahs, or the hardiest corals, or the most resplendent birds of paradise – are more likely to be passed to the next generation. Over time, these genes become more common. Their animal vehicles do too, but the genes are what natural selection really acts upon. They are, in the parlance, the 'units of selection'. But whose genes are we talking about? An animal doesn't just depend on its own genes but also on those of its microbes, which are often many times more numerous. Likewise, the microbes depend on the genes of their hosts to build the bodies that will carry them into future generations. To Rosenberg, it made no sense to think of these collections of DNA separately. He believed that they work as a single entity – a hologenome, which 'should be considered as the unit of natural selection in evolution'.[23]

To understand what that means, remember that evolution by natural selection depends on just three things: individuals must *vary*; those variations must be *heritable*; and those variations must have the potential to affect their *fitness* – that is, their ability to survive and reproduce. Variation, inheritance, fitness: if all three boxes are ticked, the engine of evolution whirrs into action, pumping out generations that are successively better adapted to their environment. An animal's genes certainly meet this trinity of criteria. But Rosenberg noted that an animal's *microbes* do, too. Different individuals can carry different communities, species, or strains of microbes – so, there's variation. As we have seen, there are many ways in which animals can pass microbes down to their offspring – so, there's inheritance. And as we shall see, the microbes bestow important abilities that influence the success of their host – so, they can affect fitness. Tick, tick, tick, and the engine starts up. Over time, the holobionts that can best meet life's challenges will pass their hologenomes – the total of their genes plus those of their microbes – to the next generation. The animals and their microbes evolve as one. It's a more holistic take on evolution, one that redefines what it means to be an individual and emphasises the indivisibility of microbes from animal life.

Any attempt to rewrite the fundamentals of evolutionary theory is bound to raise some hackles, and the hologenome idea is no exception: few concepts in this book are more likely to make mild-mannered symbiosis researchers snipe and jeer at each other. I find it ironic that this theory, which is the epitome of cooperation and togetherness, can deeply divide people who spend their entire time thinking about cooperation and togetherness.

Many like it as a bold statement. It elevates the neglected microbes to the same level as their hosts, draws a huge conceptual circle around them all, and adds flashing arrows pointing to the circle for good measure. It says microbes are important, and don't you forget it. 'Each animal is an ecosystem with legs,' says John Rawls. 'We can use holobiont or something else, but some term needs to capture the concept and I've not heard anything better.'

Forest Rohwer is more measured. After Margulis, he reintroduced the word 'holobiont' into popular use, but uses it simply to describe organisms that live together. 'It's just regular symbiosis,' he says. 'It mixes and matches depending on outside pressures, and it gives you properties that can be positive or negative.' And he isn't very keen on the hologenome idea. It has a somewhat hokey vibe, he feels, in which hosts and microbes harmoniously skip into a brighter future together. Evolution doesn't work like that. As we know, even the most harmonious of symbioses are tinged with antagonism. Rohwer feels that Rosenberg, by positioning the hologenome as the fundamental unit of selection, is glossing over those conflicts. Rosenberg seems to be saying that evolution acts to maximise the success of the whole – and that's not what happens. It acts upon the parts as well, and those parts are often at odds. Nancy Moran, an evolutionary biologist who studies aphids and their symbionts, agrees. 'I, more than anyone, would argue that symbionts are super-important, and much more so than people have believed,' she says, 'but the hologenome concept is being used to mask a lot of very fuzzy thinking.'

The nature of the hologenome is not clear, either. A symbiont like *Sodalis,* which lives in the cells of tsetse flies and is inherited vertically, is such an inextricable part of its hosts that its genes can easily count as part of the tsetse hologenome. The beewolves have their own *Streptomyces* strains and hydra have their carefully chosen multitudes; there, too, the concept fits reasonably well. But not all animals are similarly picky. Among cowbirds, cardinals, and perhaps other songbirds, individuals have utterly dissimilar gut microbes; there can be more variation within a single species than exists between *all mammals*.[24] There, the effect of the animals' genes, though it exists, seems to be overshadowed by the influence of the environment. If an animal's microbial partners can be so inconstant, does it really make sense to speak of a hologenome as a united entity? And what about all the species that transiently show up in our bodies? When Thomas Fritz impaled his hand, did the genes of strain HS count towards his hologenome? Does my hologenome include the microbes in the sandwich I just ate?

Seth Bordenstein from Vanderbilt University, who has taken up the mantle of chief hologenome evangeliser, says that none of these objections are fatal. He emphasises that the hologenome framework doesn't mean that every microbe in an animal's body is important to it. Some might be random residents, some temporary passers-by. But there should always be a small fraction that matters. 'It could be that 95 per cent of the microbes are neutral, and just a few key ones live stably with you throughout your lifetime, affecting your fitness in some way,' he says.[25] The former would be ignored by natural selection and the latter would be favoured. Some microbes might have negative effects – say, a passing cholera bacterium – and natural selection would purge these from the hologenome, just as it would a detrimental mutation from a genome. In this way, the theory accommodates conflict, too. The hologenome idea isn't necessarily about togetherness and co-operation, as its critics (and some of its proponents) suggest. It merely says that microbes and their genes are part of the picture. They affect their hosts in ways that matter to natural selection, and in ways we can't ignore when thinking about animal evolution. 'It's not a perfect framework, but I think it's the best we have right now for thinking about how the microbiota and the individual can come together,' says Bordenstein. His detractors would argue that the concept of symbiosis has already been doing that for centuries.[26]

If there's one thing that everyone agrees on, it's that the time for metaphors is over, and the time for mathematics is at hand. The gene-centric view of evolution has been so successful in part because evolutionary biologists can use equations to model the rise and fall of genes, and the benefits and costs of mutations. They can frame their abstract ideas with the precision of numbers. The hologenome supporters cannot, to the detriment of their argument. 'We're in the early phases and people think it's a touchy-feely topic without a lot of rigour,' says Bordenstein. That's a fair charge, he concedes, and one he hopes that others will remedy.

Rosenberg is undeterred. He thinks that old-school evolutionary biologists are too inured by decades of host-focused thought to appreciate the importance of microbes. ('I have been accused of being bacteriocentric even by my friends,' he says.) And having recently retired, he is happy to let others take up arms in the intellectual battle. 'I have closed my lab and opened my mind,' he says. But before that happened, he had one last contribution to make.

A few years ago, the Rosenbergs stumbled across an old paper from 1989, in which a biologist named Diane Dodd showed that a fly's diet could affect its sex life. She reared one strain of fruit flies on starch and another identical strain on maltose, a type of sugar. After 25 generations, the 'starch flies' preferred to mate with other starch flies, while the maltose flies were biased towards their own kind. It was a weird result. By changing the flies' diet, Dodd had somehow altered their sexual preferences.

The Rosenbergs immediately said it had to be bacteria. An animal's diet affects its microbiome; the microbes affect its smell; and its smell affects its attractiveness. It all made sense, and it fitted nicely with the hologenome concept. If they were right, the flies were evolving not just by changing their own genes, but by changing their microbes – just as the resistant Mediterranean corals had presumably done. They repeated Dodd's experiment and got the same result: after just two generations, the flies were more attracted to individuals that were reared on the same diet. And if the insects swallowed a dose of antibiotics and lost their microbes, they also lost their sexual biases.[27]

The experiment was quirky but profound. If two groups of the same insect ignore each other and only mate within their social circles, they should eventually split into distinct species. These splits occur all the time in nature, and the forces that cause them can take many forms. They could be physical obstacles like mountains or rivers. They could be differences in timing, in the hours or seasons in which animals are active. They could be incompatible genes that prevent two animals from interbreeding. Anything that stops animals from mating, or that kills or weakens the offspring of those couplings, can create 'reproductive isolation' – a gulf that drives two species apart. And as Rosenberg had shown, bacteria can cause reproductive isolation, too. By acting as a living barrier that stops two populations from meeting up, microbes could potentially drive the origin of new species.

This is not a new concept. In 1927, the American Ivan Wallin described symbiosis as an 'engine of novelty'. He argued that symbiotic bacteria transformed existing species into new ones, which was the fundamental means through which new species arose. Lynn Margulis echoed his views in 2002, claiming that the creation of new symbioses between distinct organisms – which she called *symbiogenesis* – has been *the* main force behind the origin of new species. To her, the kinds of relationships you've seen so far in this book were not just pillars of evolution, but its very foundations. She failed to make her case, though. She listed plenty of examples of symbiotic microbes that led to important evolutionary adaptations, but, crucially, presented almost no evidence that they actually gave rise to new species, much less that they are the principal force behind those origins.[28]

Some evidence is now coming to light. In 2001, Seth Bordenstein and his mentor Jack Werren were studying two closely related species of parasitic wasp: *Nasonia giraulti* and *Nasonia longicornis*. They have existed as separate species for just 400,000 years and to the untrained eye, they look identical – both tiny, with black bodies and orange legs. But they cannot breed. The two wasps carry different strains of *Wolbachia*; when they mate, the clash between these rival strains kills most of the hybrids. When Bordenstein took *Wolbachia* out of the equation with antibiotics, the hybrids survived. He showed that, in these wasps, reproductive isolation is *curable* – clear evidence that microbes are keeping these newly minted species apart. He found even more compelling results in 2013 in experiments on two more distantly related wasps, which also fail to produce viable hybrids. This time, he showed that hybrids end up with very different gut microbes to either of their parents, and he reasoned that this mixed-up microbiome kills them because it's incompatible with their own genomes. It's death by distorted hologenome.[29]

Bordenstein billed the study as clear evidence that symbiosis can drive the origin of new species, as Wallin and Margulis had argued. But critics say that mismatched microbiomes have nothing to do with it, and something simpler is happening.[30] They argue that the hybrids have faulty immune systems that leave them vulnerable to *any bacteria*. You could give them any microbiome, and they'd still die. No matter who is right, it is still clear that hybrids have problems with their microbes, which enforces a rift between the two wasp species. That is interesting in itself. 'We've come upon these two stories in *Nasonia*, and I don't think that's just serendipitous,' Bordenstein says. 'It's because we asked the question of whether microbes cause reproductive isolation. How many other people haven't asked that question? How many other stories have we missed? I just don't think that just by dumb luck we found the only two examples.'

For now, speciation by symbiosis remains a plausible and exciting idea, which still needs to prove itself. The handful of cases that have been identified are indisputably fascinating in their own right. If you find a gold nugget, you don't need to tell people that you have taken Fort Knox – you still have gold. Likewise, you don't need to redefine evolutionary theory in order to appreciate that the fates of microbes can become deeply entangled with those of animals.

There's no denying that microbes help to build the bodies of their hosts, or that they are involved in the most personal aspects of our lives from immunity to smell to behaviour, or that their presence can make the difference between health and disease. To me, that is extraordinary enough. Whether you use terms like hologenome, symbiosis or something else, it's clear that microbes, from inauspicious beginnings as parasites or environmental layabouts, can find themselves in the bodies of animals and create powerful and sometimes essential bonds that cascade through the generations. Now, it's time to look at the consequences of these intimate partnerships, not for the growth or health of individual animals, but for the fates of entire species and groups. It's time to see exactly how successful animals can become when they tap into the power of microbial partners.

7. MUTUALLY ASSURED SUCCESS

I'm standing in a room the size of a small garden shed. There's enough room to swing a cat, just, but you'd get claw-marks on the walls. The door is heavy and imposing. The interior is white and spotless. The air is controlled by an extremely loud fan that rhythmically whirrs into life – think Darth Vader with a megaphone. There are plants everywhere. Pea shoots, fava beans, and alfalfa seedlings all sprout from small pots arranged in trays and arrayed on shelves. It looks like a weird greenhouse – weirder still because everything is covered. Some pots are capped by transparent plastic cups. Others sit in plastic cubes, accessible only through arm-wide portholes with fine muslin knotted over them. One particularly large box contains an unruly spray of shoots.

'We've just started breeding them so I don't even know if they're in here yet,' says Nancy Moran, the biologist, who owns this room in the University of Texas at Austin, and everything in it.

I stare at the shoots. Whatever it is that Moran can't see, I can't either.

'Oh, there they are,' she offers. 'They're on that stem.'

After a long pause, and just before I haplessly ask which stem exactly, I spot them. Tiny black wedges, no longer than a centimetre, like miniature doorstops stuck to the shoots. These are glassy-winged sharpshooters. The name evokes both glamorous fashion and gun-toting cowboys; the reality is neither. These are minuscule insects that stab plants with piercing mouthparts and suck the fluid from their vessels. After filtering out the meagre nutrients, they get rid of the remaining water by squirting it out of their backsides in a fine stream of droplets – hence their name. The sharpshooter drains the fluids from dozens of different plants, making it a potent agricultural menace – hence the muslin and the imposing door.

This room is full of such menaces. Another plant is currently being devoured by a type of leafhopper. Several shelves' worth of fava bean shoots are being eaten by pea aphids. Being green insects on green stems, they are also inconspicuous but I eventually spot them: little green lozenges with spindly legs, backwards-pointing antennae, and two spines protruding from their abdomens. Each aphid has its own private fiefdom – a single upstanding shoot, all to itself. Aphids, like sharpshooters, are serious pests. They can make plants wilt and die through the sheer weight of their infestations, and that's not even counting the viruses they carry. They are the bane of agriculture, unwelcome in any place where humans deliberately grow plants – except this room. Here, they are the point. Here, the plants exist to feed them. This is one of the few gardens in the world where the owner is deliberately trying to cultivate aphids and other insect pests.

These unassuming insects all belong to the Hemiptera – a diverse order that includes bed bugs, assassin bugs, scale insects, and leafhoppers, all of which are characterised by stabbing, sucking mouthparts. When most people say 'bug', they mean any small, crawling thing. If entomologists say 'bug', they mean a hemipteran. Most members of the group spend their entire lives drinking plant sap, and they are the only animals that do so exclusively. Butterflies or hummingbirds might take the occasional sip of sap, but only hemipterans specialise in the stuff. They owe this lifestyle to symbiotic bacteria. If all of these bacteria suddenly died, so would all the insects in the room I'm standing in. 'These groups basically exist because of their symbionts,' says Moran. And not just exist, but *thrive*: around 82,000 species of hemipterans have been described, and thousands more await discovery.

We've seen how individual animals rely on microbes for ordinary and essential aspects of their lives, like building organs or calibrating immune systems. We've also briefly seen that some microbes can bestow on their hosts more unusual abilities, from the illuminated camouflage of the bobtail squid to the regenerative skills of the *Paracatenula* flatworm. Now, we'll see how other microbe-given superpowers have turned some groups of animals into evolutionary winners, which can digest indigestible foods, withstand inhospitable places, survive fatal meals, and otherwise succeed where other species fail. The hemipterans are the perfect place to start.

A German zoologist named Paul Buchner started studying their symbionts in 1910, as part of a grand tour of the insect world.[1] Slicing and dicing his way through innumerable species, he came to realise that symbiosis between animals and microbes was not a rare phenomenon, as others believed at the time. It was the rule rather than the exception: 'a widespread, though always supplementary, device, enhancing the vital possibilities of the host animals in a multiplicity of ways'. His decades of work went into a magnum opus called *Endosymbiosis of Animals with Plant Microorganisms*,[2] which was finally translated into English and published just before Buchner's eightieth birthday. When Moran pulls a copy from her office shelves, she flips through its yellowing pages reverently. 'It's the Bible of the field,' she says.

Moran has been fascinated by bugs for decades. She was once *that* kid who collected insects and kept them in jars. She's now one of the leading figures in the field of symbiosis, and aphids have been the bedrock of her career. In 1991, she helped to sequence the genes of symbionts from eleven species of aphid – a huge task at a time when sequencing technology was still in its infancy and she and her colleagues were 'sending floppy disks back and forth'. They found that all the aphid symbionts belonged to the same unnamed species. The tradition in the field is to honour bigwig microbiologists by affixing their names to newly discovered microbes, like a kind of autograph. Simeon Burt Wolbach will forever be immortalised in *Wolbachia*.

Louis Pasteur lives on as *Pasteurella*. You've probably never heard of the obscure American veterinarian Daniel Elmer Salmon, but you'll probably know of his namesake *Salmonella*. Which name to graft onto the aphid's symbiont? There was never really any choice. She called it *Buchnera*.[3]

It is an ancient companion for aphids. The family tree of different *Buchnera* strains perfectly mirrors that of their aphid hosts. You could draw one and you'd immediately have the other.[4] This means that *Buchnera* colonised aphids just the once (or, at least, that only one such infection was ever successful). That pioneering event happened between 200 and 250 million years ago, back when the dinosaurs were starting out and before mammals and flowers existed. What has *Buchnera* been doing for all of that time? Buchner had guessed the answer: he supposed that insect symbionts were mostly there for nutritional reasons, helping their hosts to digest their food. That's certainly the case for many of the insects he studied, but for *Buchnera* the truth is slightly different. It doesn't break down the aphid's meals. It supplements them.

Aphids feed on phloem sap – a sweet fluid that flows through plants. It is a superb food source in many ways: high in sugar, low in toxins, largely untapped by other animals. But it's also woefully deficient in several nutrients, including ten essential amino acids that animals need to survive. A shortfall of any one of these would be devastating. A deficit of all ten is intolerable – unless something else can compensate. There is now overwhelming evidence that *Buchnera* is that something.[5] Scientists have treated aphids with antibiotics that killed *Buchnera* and found that the insects needed artificial supplements of amino acids to survive. They used radioactive chemicals to trace the flow of nutrients from microbe to host, and proved that amino acids flow in that direction. And they showed that *Buchnera*'s genome, despite being extremely small and degenerate, retains many of the genes for making essential amino acids.

Many, but not all. Building amino acids is a complicated affair, which involves shepherding starter ingredients through a series of

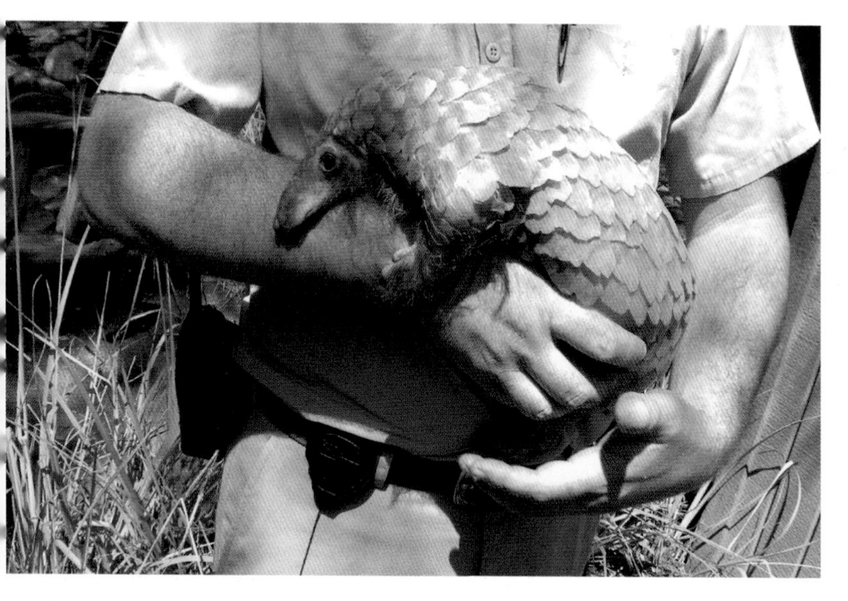

At San Diego Zoo, Baba the pangolin is about to get his skin bacteria swabbed. Like each and every one of us, Baba is a teeming mass of microbes.

The adorable Hawaiian bobtail squid houses a single species of luminous bacteria, which hide it from predators with their silhouette-cancelling glow.

Antony van Leeuwenhoek's microscopes looked like glorified door hinges, but they were the best of his day and they allowed him to become the first human in history to see bacteria.

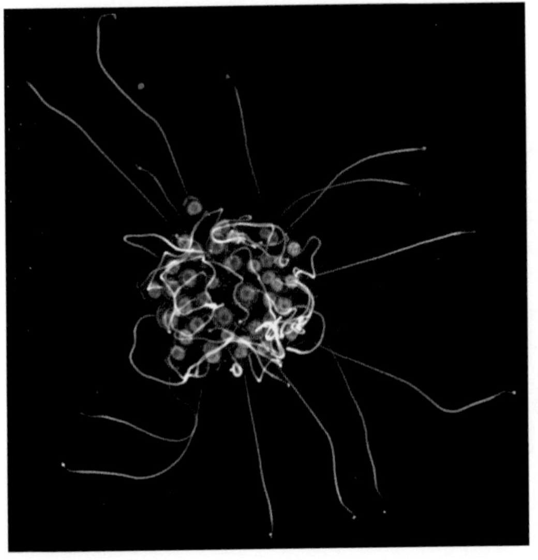

These single-celled choanoflagellates form rosette-shaped colonies when they sense a particular bacterial molecule; the first animals on earth may have done something similar.

Like many amphibians, this mountain yellow-legged frog is threatened by an apocalyptic fungus. But the bacteria on its skin might be its salvation.

Inside the body of this thirteen-year periodical cicada, a bacterium called *Hodgkinia* has split into two species – each just half of a former whole.

As adults, *Hydroides elegans* worms secrete white tubes that coat ship hulls in a centimetre-thick carpet. But without bacteria, the worms would never reach adulthood at all.

This germ-free mouse that I'm holding was raised in a sterile bubble, and is one of the only animals on the planet that has never been exposed to bacteria.

When it's not eating delicious peanuts, the desert woodrat can stomach the toxic leaves of the creosote bush because microbes in its gut neutralise the poisons.

The fearsome beewolf protects its larvae by painting their burrows in antibiotic-producing microbes.

The loss of sharks and other large predators can harm coral reefs by changing the communities of microbes that grow upon them.

Giant tube worms thrive in hellish hydrothermal vents, 2,400 metres below the ocean surface. They have no mouths or guts, because the bacteria in their bodies produce all the food they need.

Some people carry gut microbes that are uniquely efficient at digesting this *Porphyra* seaweed, because those microbes have stolen seaweed-busting genes from marine bacteria.

Me exchanging bacteria with the venerable Captain Beau Diggley.

Unusually among animals, the citrus mealybug is a living matryoshka doll. It has bacteria living inside its cells, and those bacteria have more bacteria living inside them.

What looks like a beautiful autumnal forest is actually a scene of terrible devastation. By forming partnerships with microbes, mountain pine beetles have destroyed millions of acres of evergreen trees across North America.

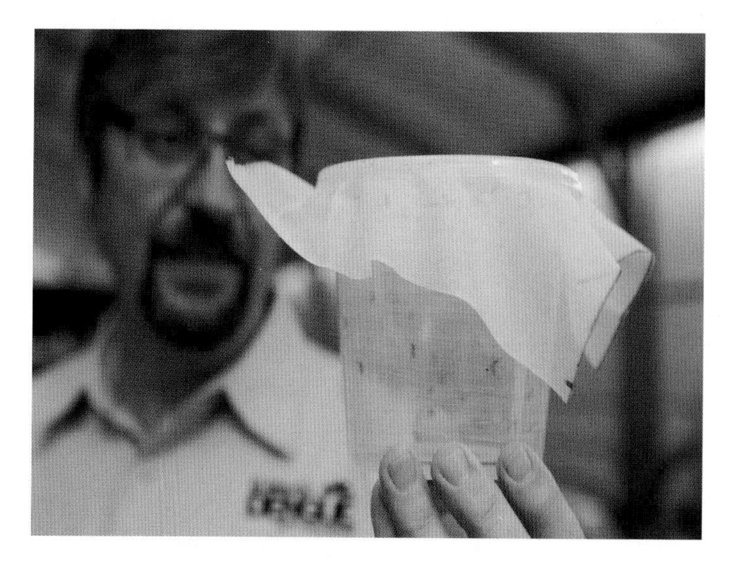

These mosquitos normally spread the virus behind dengue fever. But Scott O'Neill has transformed the mosquitos into dengue-fighters by loading them with *Wolbachia*, a bacterium that blocks the virus and spreads rapidly through insect populations.

A body-snatching braconid wasp lays an egg inside a doomed caterpillar. The wasp uses domesticated viruses to suppress its victim's immune system.

chemical reactions, each catalysed by a different enzyme. Imagine a production line in a car factory, where a conveyor belt snakes past a series of machines. One fixes the seats; another adds the chassis; the next attaches the wheels. At the end of the pathway, a car emerges. The biochemical pathways that create amino acids work in a similar way, but neither aphids nor *Buchnera* can build all the necessary enzyme machines on their own. Instead, they cooperate to set up the production lines, which wind in and out of two factories, one nested within the other. Only together can they subsist on phloem sap.[6]

The link between sap-sucking and supplementary symbionts is made clearer by the hemipterans that have abandoned both. A few have taken to eating plant cells whole, and since their diet has no shortage of amino acids, they have discarded their symbionts. There is no room in these relationships for nostalgia or sentimentality; the brutal contract of natural selection ensures that if one partner is unnecessary, it gets dumped. This diktat applies to genes too, and explains why hemipterans landed themselves in a nutritionally precarious predicament in the first place. They are animals, and all animals evolved from single-celled predators that ate other things. Their food gave them many of the nutrients they needed, so they lost the genes for making these nutrients for themselves. We – that is, aphids, pangolins, humans, and the rest – are saddled with their legacy. None of us can make those ten essential amino acids, and we eat to fill the gap. And if we want to eat a specialised and impoverished diet like phloem sap, we need help.

Enter bacteria. They have repeatedly allowed hemipterans to overcome a limitation that restrains the entire animal kingdom, and feast on a food that little else could exploit.[7] As plants colonised the land, so did these plant-sucking bugs. Today, they include some 5,000 species of aphids, 1,600 species of whiteflies, 3,000 plant lice, 8,000 scale insects, 2,500 cicadas, 3,000 spittlebugs, 13,000 planthoppers, and more than 20,000 leafhoppers – and those are just the ones we know about. Thanks to their symbionts, hemipterans have become exemplars of success.

Hemipterans are far from the only animals with nutritional symbionts. Some 10 to 20 per cent of insects depend on such microbes, which provide vitamins, amino acids for making proteins, and sterols for making hormones.[8] All of these living supplements allow their owners to subsist on deficient diets, from sap to blood. Carpenter ants – a diverse group with around 1,000 species – carry a symbiont called *Blochmannia* that allows them to live on a largely vegetarian diet and dominate the canopies of tropical forests.[9] Mini-vampires like lice and bed bugs (along with non-insects like ticks and leeches) rely on bacteria for the B-vitamins that are missing in their bloody meals.

Time and again, bacteria and other microbes have allowed animals to transcend their basic animalness and wheedle their way into ecological nooks and crannies that would be otherwise inaccessible; to settle into lifestyles that would be otherwise intolerable; to eat what they could not otherwise stomach; to succeed against their fundamental nature. And the most extreme examples of this mutually assured success can be found in the deep oceans, where some microbes supplement their hosts to such a degree that the animals can eat the most impoverished diet of all – nothing.

In February 1977, a few months before the *Millennium Falcon* blasted outwards into space, an equally adventurous vehicle called *Alvin* travelled downwards into the oceans. It was a submersible, big enough to house three scientists, small enough to stop them from stretching their arms, and sturdy enough to dive to incredible depths. It entered the water 250 miles north of the Galapagos Islands, where two tectonic plates drift away from each other like estranged lovers. Their separation created a rift in the Earth's crust, which was a likely spot for finding the first ever hydrothermal vents – sites where it was believed volcanically superheated water would belch forth through the ocean floor.

The *Alvin* team descended. The blue of the surface gave way to black, the all-consuming black of the abyssal ocean. Blacker than black. Black punctuated only by the occasional flashing of bioluminescent creatures and, eventually, the submersible's lights. At a depth of 2,400 metres, about a mile and a half straight down, the team found the vents they had predicted, but also something they had not – life, in extreme abundance. Huge communities of clams and shellfish clung to rocky chimneys. Ghostly white shrimps and crabs clambered over them. Fish swam past. And strangest of all, the rocks were encrusted with hard white tubes that ended in the crimson plumes of giant worms. They looked like tubes of lipstick that had been pushed out too far, or something even more sexually suggestive. They were actually giant worms.

In this supposedly lifeless underworld, untouched by the sun, buffeted by water that can reach 400 degrees Celsius, and compressed by the enormous pressure of the ocean above, the *Alvin* team had discovered a hidden ecosystem as rich as any rainforest. It was, as Robert Kunzig wrote in *Mapping the Deep*, 'like being born and raised in Labrador, in complete ignorance of the outside world, and then one day parachuting into Times Square'. The team were so unprepared to find life that there wasn't a single biologist among them – they were all geologists. When they collected specimens and brought them back to the surface, the only preservative they had was vodka.[10]

One of the giant tube worms made its way to Meredith Jones at the Smithsonian Museum of Natural History, who named it *Riftia pachyptila*. He found it so intriguing that he visited the Galapagos Rift himself in 1979 to collect even more, from a site so ridden with the red-plumed things that it was called the Rose Garden. In an old black-and-white photo, Jones, white-haired and moustached, holds one of his *Riftia* specimens. He looks tender, almost affectionate; the worm looks like a badly packed string of sausages. It's also huge – bigger than any deep-sea worm that had been discovered, and probably as long as Jones was tall. And bizarrely, it had no mouth, no gut, and no anus.

How does a worm survive if it cannot eat? The most obvious hypothesis was that it absorbed nutrients through its skin like a tapeworm, but that idea was quickly quashed – it couldn't possibly do so fast enough. Then, Jones noticed a big clue. The worm's trophosome, a mysterious organ that made up half its body weight, was packed with crystals of pure sulphur. Jones mentioned this in one of his lectures at Harvard University. And in the audience, an important thought stirred in the head of a young zoologist named Colleen Cavanaugh. On hearing him describe the trophosome, she had a bona fide Eureka moment. By her own account, she leapt up and proclaimed that the worms had bacteria in their bodies, which were using sulphur to make energy. Jones reportedly told her to sit down. Then, he gave her a worm to study.

Cavanaugh's epiphany was both right and revolutionary.[11] Under the microscope, she found that *Riftia*'s trophosome was full of bacteria, about a billion of them for every gram of tissue. Another scientist had shown that the trophosome was rich in enzymes that can process sulphide compounds, like the hydrogen sulphides that are common in the undersea vent environment. Cavanaugh put two and two together and suggested that these enzymes came from the bacteria, which used them to make food in a way that was utterly different to anything known at the time.

On land, life is powered by sunlight. Plants, algae, and some bacteria can harness the sun's energy to make their own food, by refashioning carbon dioxide and water into sugars. This process, in which carbon is shunted from inorganic matter into edible substances, is called *fixing carbon*, and using the sun's energy to do so is called *photosynthesis*. This is the basis of all the food webs that we're familiar with. Every tree and flower, vole and hawk, ultimately depends on solar power. But in the deepest oceans, sunlight isn't an option. You could filter the meagre snow of organic matter raining down from above but to really thrive, you need a different source of energy. For *Riftia*'s bacteria, that's sulphur, or rather the sulphides that spew out of the vents. The bacteria oxidise these chemicals and use the liberated energy to fix carbon. This is *chemo*synthesis: making your own food using *chemical* energy instead of light or solar energy. And rather than producing oxygen as a waste product, as photosynthetic plants do, these *chemo*synthesising bacteria churn out pure sulphur. Hence the yellow crystals in *Riftia's* trophosome.

Chemosynthesis explains why the worms are gutless and mouthless: their symbionts provide them with all the food they need. Unlike aphids or sharpshooters, which rely on bacteria for amino acids, these worms rely on their symbionts for *everything*.

Scientists soon found similar symbioses throughout the deep oceans. It turns out that a huge variety of animals play host to chemosynthetic bacteria, which use either sulphides or methane to fix carbon.[12] The regenerating flatworm *Paracatenula* is one of them. There are clams, worms, and armoured snails with chemosynthetic symbionts in their cells and shrimps with colonies growing on their gills and mouths. There are nematode worms that are so smothered in these microbes, they look like they're wearing fur coats. There are yeti crabs that grow gardens of bacteria on their bristly claws, which they wave around in a comical dance.

Many of these creatures live at hot hydrothermal vents. Others gather around cold seeps, which release much the same chemicals but at lower temperatures and a more leisurely pace. Some tube worms, related to *Riftia*, colonise the wood of wrecked ships and sunken trees, subsisting on the sulphides within the rotting timbers. Dead whales, raining upon the ocean floor like manna from heaven, also create sulphide-rich environments that support temporary but teeming communities of chemosynthetic creatures. Some of these, like *Osedax mucofloris*, the gutless bone-eating snot-flower worm, specialise in whale falls.

For these animals, life in the deep oceans is the tail end of an evolutionary return trip that spans billions of years. Life on Earth originated at deep-sea vents, and first took the form of chemosynthetic microbes. (Fittingly, one of the sites at the Galapagos Rift is called the Garden of Eden.) These ancestral microbes eventually evolved into endless forms most beautiful and most wonderful, spreading out of the depths and into the shallows. Some gave rise to more complex forms of life, like animals. And some of these, by partnering with chemosynthetic bacteria, managed to descend back into the abyss, to a world that would otherwise be too low in nutrients to support them. All the animals that live at hydrothermal vents today, *Riftia* included, evolved from shallow-water species that became hosts for deep-sea microbes. By internalising those bacteria, the animals gained a ticket to the Hadean depths from which all life emerged.

Chemosynthesis may have originated in the deep, but it is not restricted to it. Cavanaugh has found chemosynthetic bacteria in clams living in shallow, sulphide-rich mud off the coast of New England. Others discovered similar partnerships in mangrove swamps, marshes, sewage-contaminated mud, and even the sediments around coral reefs – ecosystems that are almost synonymous with shallow water. Nicole Dubilier, a former member of Cavanaugh's team, studies chemosynthesis in a place that's as far removed from the belching hydrothermal vents as you can imagine: the postcard-perfect Tuscan island of Elba.

The sun drenches Elba in light, and its energy isn't wasted. In the bays offshore, huge beds of seagrass grow. But even here, where photosynthesis is apparently king, chemosynthesis abounds. If Dubilier dives down to the seagrass and stirs up some sediment, bright white threads will wriggle out. These are worms called *Olavius algarvensis*, close relatives of the earthworm. They are a few centimetres long, half a millimetre wide, and bereft of mouths and guts. 'I think they're beautiful,' says Dubilier. 'They're white because the symbiotic bacteria beneath their skin are full of sulphur globules. You can easily pick them out.' These bacteria are chemosynthetic, as are those in many local nematodes, clams, and flatworms. In this Mediterranean mud, the diversity of sulphide-powered organisms matches that of the deep. 'In Italy!' says Dubilier. 'We had to go to deep-sea, exotic vents to realise that chemosynthetic symbioses are in our own backyards. With every field trip, we discover new species and new symbioses.'

Elba may seem idyllic, but it poses challenges for chemosynthetic life. Remember that *Riftia*'s bacteria liberate energy by oxidising sulphides. Elba's sediments are very low in sulphides, so chemosynthesis, as we know it, shouldn't work there. So how do the *Olavius* worms make a living? Dubilier found the answer in 2001, when she realised that they have *two* different symbionts: one big, one small, and both mingling beneath their skin.[13] The small bacterium grabs sulphates, which are plentiful in Elba sediments, and converts them into sulphides. The big bacterium then oxidises the sulphides to power chemosynthesis, much like *Riftia*'s microbes. In the process, it produces sulphates that its smaller neighbour can reuse. The two microbes feed each other in a cycle of sulphur, which then feeds the worm – a symbiosis à trois. By adding the small sulphate-grabbing bacteria to their existing partnerships, *Olavius* worms managed to colonise mud that would be too impoverished for their usual chemosynthetic partners.

Dubilier has since discovered that this alliance is even *more* complicated. *Olavius* actually has *five* symbionts – two that process sulphates, two that deal with sulphides, and a fifth corkscrew-shaped member of unknown function. 'It'll probably take us another thirty years to fully understand it,' Dubilier says, laughing. She's lucky, though. Since she studies shallow-water symbioses, she doesn't depend on stifling submersibles to gather her subjects. She can just dive in sunny Elba, or in sites like the Caribbean and the Great Barrier Reef. It's hard stuff, this science business, but someone's got to do it.

For Ruth Ley, collecting microbes was harder. The problem wasn't the stool samples she was after – in the microbiome world, you quickly get used to handling poo. The problem wasn't the zoo animals that she was collecting from, either – there were always cages, walls, and keepers with sticks standing between her and claws or teeth. No, the problem was the red tape.

Ley is a microbial ecologist who wanted to compare the gut bacteria of different mammals to see how their diets and evolutionary histories shaped their microbiomes. She needed a wide menagerie and a lot of poo, and she found both at the nearby Saint Louis Zoo. In gaps between other experiments, Ley would hop over with gloves, bags, and a bucket of dry ice. A friendly keeper drove her around and distracted the animals while she sneaked in and bagged their dung. 'I just kept on going back until someone realised that we were running around picking up poo, and decided that it had to be all bureaucratic,' she says. Out went the friendly keeper and the informal adventures; in came an official liaison, a poo-collection form, and a fusty attention to protocol. For example: one winter's day, Ley noticed that the hippos had relieved themselves on the floor of their enclosure. 'There was this huge pile of dung!' she says. 'But they kept on saying that hippos weren't on the form. Then, the guy who was shovelling the stuff said: This will all be in the alley out back in ten minutes. You can get some then.' She did.

She also got dung from bears (black, polar, and spectacled), elephants (African and Asian), rhinos (Indian and black), lemurs (black, mongoose, and ring-tailed), and pandas (giant and red). Over four years of visits, she collected the droppings of 106 individuals from 60 species. Each sample she dried in an oven, pulped in a blender, and pulverised with a mortar and pestle. The smell was memorable. The reward was DNA, which allowed her to catalogue the microbes that lived in the guts of its maker.

Ley found that each mammal had its own distinctive set of gut microbes, but these communities clustered into certain groups depending on their owner's ancestry and, in particular, their diet.[14] The plant-eating herbivores typically had the highest diversity of bacteria. The meat-eating carnivores had the lowest. The omnivores, with their broad diets, were in the middle. There were exceptions: the gut microbes of red and giant pandas are more like those of their carnivorous relatives – bears, cats, and dogs – than the herbivores they surely are.[15] Still, the general pattern held, and it has both a simple explanation and a profound implication.

First, the explanation. Plants are by far the most abundant source of food on land, but it takes more enzymatic power to digest them. Compared to animal flesh, plant tissues contain more complex carbohydrates, such as cellulose, hemicellulose, lignin, and resistant starches. Vertebrates don't have the molecular chops for breaking these apart. Bacteria do. The common gut bacterium B-theta has over 250 carb-busting enzymes on its own; we have fewer than 100, despite owning a genome that's 500 times bigger. By sundering plant carbohydrates with their broad toolkits, B-theta and other microbes release substances that directly nourish our own cells. Collectively, they provide 10 per cent of our energy intake, and a whopping 70 per cent of a cow's or sheep's. To eat a diet of plants, animals need microbes in both great diversity and great abundance.[16]

Now, the implication. The very first mammals were carnivores – small, scurrying, scourges of insects. Shifting from meat to plants was an evolutionary breakthrough for our group. The sheer abundance and variety of plants allowed herbivores to diversify much faster than their carnivore kin, and spread into niches that had been vacated by the demise of the large dinosaurs. Today, the majority of living mammal species eat plants, and most orders have at least some herbivorous members. Even the Carnivora – the order that includes cats, dogs, bears, and hyenas – count the bamboo-eating pandas among their number. So, mammalian success was founded on vegetarianism, and that vegetarianism was founded on microbes. Time and again, different groups of mammals swallowed plant-breaking microbes from their environments, and used their enzymes to mount assaults on leaves, shoots, stems, and twigs.

It's not enough to have the right microbes. They need room and time to work. Plant-eating mammals gave them both. They enlarged parts of their guts into fermentation chambers, partly to house their digestive comrades and partly to slow the passage of food so they could do their thing. In elephants, horses, rhinos, rabbits, gorillas, pigs, and some rodents, these chambers sit at the end of the gut. These 'hindgut fermenters' can use their own enzymes to extract as much nutrition from their food as possible before giving their microbes a shot. Other mammals, like cows, deer, sheep, kangaroos, giraffes, hippos, and camels, are foregut fermenters – they house their microbes either ahead of their stomachs or in the first of several chambers. They sacrifice some nutrients to their bacteria, but they then digest these partners-in-digestion. 'That's why you put the bag up the front: you eat the bacteria too,' says Ley. 'It's very smart. You can get away with eating straw and still have all the nutrition you want.' Some foregut fermenters, like cattle, give their microbes even more time to work by ruminating – a distasteful but effective cycle of regurgitating, re-chewing and re-swallowing one's stomach contents.

The position of the fermentation chamber also influences the kinds of microbes that mammals have mustered. Ley found that microbiomes of foregut fermenters are more similar to each other than to those of hindgut fermenters, and vice versa. These similarities transcend the boundaries of ancestry. The kangaroo is a hopping Australian marsupial and the okapi is a stripe-trousered giraffe-ish creature from Africa – but both are foregut fermenters and have broadly similar microbiomes. The pattern holds for hindgut fermenters as well.[17]

In other words, microbes shaped the evolution of the mammalian gut, and the shape of the mammalian gut influenced the evolution of microbes.[18]

This theme became even clearer with Ley's next study. Together with Rob Knight, she compared the sequences of her zoo microbes with those from other animals, and from diverse habitats like soils, seawater, hot springs, and lakes. They found that, in terms of microbial diversity, the vertebrate gut is unlike anything else. It's even more different from the lakes, springs, and the rest than these environments are from each other. There is, as the team described it, a 'gut/non-gut dichotomy'.[19] 'It was a huge surprise,' says Knight. 'The first time someone ran that analysis, I thought they had just done it wrong.' The reasons for the dichotomy are unclear, but Knight ventures that the gut is a unique habitat for microbes – dark, starved of oxygen, awash with fluids, patrolled by immune cells, and *extremely* rich in nutrients. Not all bacteria can survive here, but those that do find a welter of ecological opportunities, and they run with it. One representative gets in and goes nuts, diversifying into a throng of related strains and species. The result is a family tree with a long, deep trunk and a wide but shallow spray of branches, more like a palm than an oak.

You see the same thing on islands. A pioneering animal lands, having been blown over by a freak storm, having been carried over on a floating log, having been transported by boat. It flies, scuttles, or slithers off and its descendants slowly start colonising the island's various habitats, differentiating into new species as they go. So arose the honeycreepers of Hawaii, the finches of the Galapagos, the snails of French Polynesia, the anoles of the Caribbean . . . and, perhaps, the microbes of our guts.

The team showed that the gut microbiomes of plant-eating vertebrates were particularly distinct from anything else – from environmental communities, from carnivores, from other body parts, or from invertebrates. Guts may be special, but a vertebrate gut is especially special, and a vertebrate gut that's full of plants is special squared. A bolus of shoots and leaves, with its wide array of digestible carbohydrates, is like an island that offers a multitude of food sources. It provides myriad ways for colonists to eke out a living and encourages their diversification into new forms.[20] Repeatedly, microbe-powered digestion has made for successful vegetarians – and not just among mammals.

Among insects, the champion plant-eaters are termites. In 1889, Joseph Leidy, an extraordinary American naturalist, cut open the guts of termites to find out what they were eating. As he watched the dissected insects under a microscope, he was shocked to see small specks fleeing from the corpses like 'a multitude of persons from the door of a crowded meeting-house'. He billed them as 'parasites' but we now know that these tiny evacuees are protists: eukaryotic microbes that are more complex than bacteria but still consist of a single cell. The protists can make up half the weight of their termite host, and they are abundant for a reason: they have enzymes that digest the tough cellulose in the wood that termites eat.[21]

Protists are mostly found in the guts of the earliest termite groups: the disparagingly named 'lower termites'. The high-falutin' 'higher termites' evolved later; they rely more on bacteria, which they house in a series of stomachs that are almost cow-like in their organisation.[22]

The even more grandiosely named macrotermites are the newest arrivals on the scene, and they have the most sophisticated strategy for destroying wood: agriculture. Inside their cavernous nests, they farm a fungus, which they feed with bits of wooden shrapnel. The fungus splits cellulose into smaller components, creating a compost that the termites then eat. Inside their guts, bacteria digest the fragments even further. The termites themselves contribute very little to this assault; their main role is to harbour the bacteria and cultivate the fungus. Without either partner, they starve. A macrotermite queen takes things even further. She is enormous. Her torso is the length of a fingernail but her abdomen is a palm-sized, pulsating, egg-laying sac, so grossly distended that she cannot move. She also has a distinct lack of gut microbes. Instead, she relies on her worker daughters (and *their* microbes) to feed her. Her entire colony – thousands of workers, billions of microbes, and giant nests laced with wood-breaking fungus – functions as her gut.[23]

You can see how successful this strategy is by travelling to Africa. The macrotermites there build enormous mounds. Some can tower for up to 9 metres, scraping the skies with Gothic ensembles of spires and buttresses. The oldest one on record – now abandoned – is 2,200 years old. The mounds provide homes for many other animals, while the termites themselves provide food for others. They also drive the flow of nutrients and water through their environment by consuming fallen and decaying plants. They are ecosystem engineers. In the savannah, they secretly run things; or rather, their microbes do. If these plant-breaking gut bacteria didn't exist, the African landscape would be radically changed. Not only would the termites disappear, but so would the immense herds of grazers and browsers – the antelopes, buffaloes, zebras, giraffes, and elephants that are synonymous with African wildlife.

I once went to Kenya in the middle of the great wildebeest migration – the annual marathon where millions of these cow-like antelopes travel long distances in search of greener pastures. At one point, we stopped our jeep for over half an hour to let an impossibly long line of them cross in front of us. Without microbes to extract as much nourishment as possible from tough indigestible mouthfuls, these herbivores wouldn't exist. We wouldn't, either. It's hard to imagine that without domesticated ruminants, humanity would ever have gone far beyond hunting, gathering, and basic farming, much less invent international flight and safari tours. Instead of slack-jawed tourists watching a herd of fermentation chambers running past on thundering hooves, there would have been a clear horizon, and silence.

For thirty weeks, Katherine Amato kept to the same routine. She'd wake up before dawn, drive over to Palenque National Park in Mexico, and listen. As dawn trickled through the trees, the branches would resound with deep, guttural, and extremely loud hoots. These calls came from the throats of Mexican howler monkeys: large, black, tree-dwelling monkeys with prehensile tails and powerful voices. All day, Amato tracked them by following their howls, and kept pace with them on the ground while they clambered through the treetops. She was interested in their gut microbiomes, so she needed to collect their dung. Conveniently, howler monkeys all defecate at the same time: 'When one starts to go, you know that it's all coming,' says Amato.

Why bother? Because howler monkeys eat very different foods throughout the year. For roughly half the year, they mostly eat figs and other fruit: high in calories and a cinch to digest. When the fruit runs out, they mostly subsist on leaves and flowers: low in calories and tougher to break down. Some scientists had suggested that the monkeys cope with this dietary shortfall through sloth, but Amato didn't see that; her howlers were equally active at all times of the year. Their gut microbes, however, change. In particular, they churned more short-chain fatty acids (SCFAs) during the fruitless months. Since these substances nourish the monkeys' cells, the microbes were effectively providing their hosts with more energy at a time when they were getting fewer calories from their food. They provided the monkeys with nutritional stability, despite the vagaries of the seasons.[24]

To talk about animals as if each species eats one thing constantly, as I have been, is a simplification. In reality, our diets vary from season to season and even day to day. A howler monkey might dine well on figs one month, and munch its way through dissatisfying leaves the next. A squirrel might eat nuts aplenty one season and nothing at all the next. I might wolf down a croissant today, and prod at a salad tomorrow. And with each new meal or mouthful, we select for microbes that are best at digesting whatever we've just eaten. They react with incredible speed. One study asked ten volunteers to stick to two strict diets for five days each: one rich in fruit, vegetables, and grains, and the other heavy in meat, eggs, and cheese. As their diets changed, so did the recruits' microbiomes – and quickly. Within a single day, they could flip between a carbohydrate-busting, plant-eating mode, and a protein-busting, meat-eating one.[25] In fact, these two kinds of community looked a lot like the gut microbes of herbivorous and carnivorous mammals, respectively. They were recapitulating millions of years of evolution in less than a week.

In this way, our gut microbes make us more flexible eaters. That might not matter so much for people in developed countries, or zoo animals, both of whom are regularly and plentifully fed. But it could make all the difference to our hunter-gatherer ancestors, or to wild animals like Amato's howler monkeys. They must cope with seasonal menus. They encounter feasts and famine. They'll be forced to try unfamiliar foods. A rapidly adapting microbiome helps cope with all of these challenges. It provides flexibility and stability in a changing and uncertain world.

This flexibility may be a boon for animals but it's a curse for us. The western corn rootworm is a North American beetle that's a serious pest. The adults lay eggs in cornfields and, the following year, their larvae feast on the roots of the plants. This life cycle creates a vulnerability: if farmers plant corn and soybean in alternate years, the adults lay eggs among corn but the larvae hatch among soybean – and die. This practice, known as crop rotation, has been very effective at thwarting the rootworm. But some 'rotation-resistant' strains have developed a microbial countermeasure. Their gut bacteria have become better at digesting soybeans. This allows the adults to break their ancient dependence on corn and lay eggs in soybean fields, too. Now, their larvae hatch into fields of gold. Thanks to their rapidly adapting microbiomes, these pests can continue to pester us.[26]

As a general rule, organisms don't line up to be consumed. They defend themselves. Animals have the option to fight or flee, but plants, being more passive, rely more on chemical defences. They fill their tissues with substances that deter plant-eaters – poisons that harm, sterilise, cause weight loss, initiate tumours, trigger abortions, lead to neurological disorders, and just plain kill.

The creosote bush is one of the most common plants in the deserts of the American Southwest. It succeeds because it is extremely resistant – to drought, to ageing, and to animals. It slathers its leaves in a resin that contains hundreds of chemicals, which collectively account for up to a quarter of its dry mass. This cocktail gives the plant a distinctively pungent smell that becomes especially apparent when raindrops hit the leaves; it is said that creosote smells of rain, but maybe it is more that rain smells of creosote. Either way, it is fine to catch a whiff of the resin. Swallowing it is another matter. The resin is highly toxic to the liver and the kidney. If a lab rat eats too much, it dies. But if a desert woodrat eats the leaves, nothing happens. It eats more. And more. In the Mojave Desert, this rodent is so content to nibble away at creosote leaves that during the winter and spring it eats little else. Every day, it swallows quantities of resin that would kill another rodent many times over. How does it cope?

Animals have many ways of getting around plant poisons, but each solution has a cost. They could eat the least toxic parts, but fussiness restricts opportunities. They could swallow neutralising substances like clay, but antidotes take time and effort to find. They could make their own detoxifying enzymes, but that takes energy. Bacteria offer an alternative. They are masters of biochemistry, and can degrade everything from heavy metals to crude oil. A few plant poisons? Not a problem. As far back as the 1970s, scientists had suggested that microbes in an animal's digestive tract ought to be able to pre-detoxify any toxins in its diet before the gut can absorb them.[27] By relying on such microbes to disarm their food, animals could save themselves the bother of investing in their own countermeasures. Ecologist Kevin Kohl suspected that bacteria might explain the woodrat's fortitude, and several millennia of climate change had given him an obvious way of testing his hunch.

Around 17,000 years ago, the climate of the southern United States started getting warmer and the creosote bush, which had originated in South America, moved in. It made itself at home in the warm Mojave, where it came within the reach of the woodrat. But it never managed to push northwards into the colder Great Basin Desert. The woodrats there had never tasted creosote before; they mainly fed on juniper. If Kohl's hunch was right, the experienced Mojave woodrats should be full of detoxifying gut bacteria, which would be absent in the naïve Great Basin rodents. Kohl captured individuals from both deserts and found exactly that. When confronted by an influx of creosote toxins, the gut bacteria in the naïve rodents shrank away, while those in the experienced animals switched on toxin-degrading genes, and flourished. To confirm that the experienced woodrats actually rely on their microbes, Kohl dosed their food with antibiotics. When he fed the rodents with normal laboratory chow, they did fine. When he gave them resin-laced meals, they suffered. As their gut microbes died, they became *less* tolerant of creosote resin than even their naïve Great Basin cousins, and they lost so much weight that Kohl had to pull them out of the experiment early. In just a couple of weeks, he had reversed 17,000 years of evolutionary experience and turned the veteran creosote-eating rodents into complete amateurs.[28]

He also did the opposite. He took the faecal pellets of the experienced rodents, pulped them in a blender, and fed them to the naïve ones to give them an infusion of detoxifying microbes. Suddenly, these individuals could happily eat creosote. Their new-found powers were obvious in their urine: creosote toxins darken and discolour the urine of a woodrat, but these formerly naïve rodents were destroying so much of the poison that their wee was golden and consommé-clear. Naïve no more, they had gained millennia of experience in just a few meals.

Something similar probably happened when creosote first appeared in the Mojave. A woodrat stumbles across the new shrub and decides to take a nibble. Its mouthful doesn't quite agree with it; then again, food is scarce in winter and beggars can't be choosers. Another nibble, then. With each mouthful, it takes in microbes that live on the surface of the creosote; perhaps these have already evolved ways of breaking down the resin cocktail. Having eaten these microbes, the rat is itself better equipped. Later, it scurries away and defecates, leaving a small microbe-filled pellet behind – which another woodrat finds and eats. The ability spreads. Eventually, the rats unlocked the ability to eat what would soon become the most common plant in the Mojave. Maybe this readiness to pick up new microbes from one another explains why these rodents are so versatile and successful.[29]

There are many such cases of microbes allowing their hosts to dine on potentially lethal meals.[30] Lichens – those icons of symbiosis – are loaded with a poison called usnic acid. But reindeer, which feed heavily on lichen, are so good at breaking down usnic acid that there's barely a trace of the substance in their waste; gut microbes, presumably, are responsible. Many plant-eating mammals, from koalas to woodrats, carry microbes that degrade tannins – bitter-tasting compounds that give texture to red wine but inflict damage upon livers and kidneys. The coffee berry borer beetle has gut microbes that can destroy caffeine – a substance that gives a delightful jolt to coffee-drinkers but poisons any pest which tries to subsist on the beans. Any pest, that is, except the coffee berry borer. With its caffeine-busting bacteria, it has become the only animal that can feed solely on coffee beans, and one of the biggest threats to the global coffee industry.

These tricks are all part and parcel of life as a herbivore: defusing as well as digesting, surviving not just *on* the food you eat but *in spite of* it. By combining microbial abilities with whatever strategies they themselves have, plant-eaters get to exploit the abundant greenery around them. Meanwhile, the plants take a hit but generally don't seem to suffer too badly. Creosote bushes get pummelled by woodrats but they're still masters of the Mojave. Lichens get nibbled by reindeer but still smother the tundra. Eucalyptus trees lose leaves to koalas but you can't walk through Australia without bumping into one. Even coffee – *thankfully* – will be fine. But sometimes, microbial detoxification goes too far. Sometimes, the plants lose big.

If you fly over the western forests of North America, there's a good chance that you'll spot large expanses of red or bare-branched trees. It might look like a picturesque autumn scene, but it's actually a tableau of disaster. These trees are *pines*. Their needles aren't meant to redden. They're evergreens – or at least they would be, if they weren't dying in droves. Their killer? The mountain pine beetle, a charcoal-black insect no bigger than a grain of rice. It infiltrates pine trees and carves long galleries beneath the bark, laying eggs as it goes. When the eggs hatch, the larvae tunnel inwards and feed on phloem sap. One beetle does very little, but thousands can infest a single tree. Peel off a piece of bark and you'll see their handiwork – a labyrinth of tunnels extending down the trunk. The beetles drain so much of the tree's nutrients that it starts to die. So does the tree next to it. So do all of their neighbours. Acres upon acres of trees redden and perish.[31]

The beetles have even smaller accomplices: two species of fungus that accompany them wherever they go and that behave like dietary supplements, much like *Buchnera* to aphids. While the beetles are restricted to nutrient-poor layers just under the bark, the fungi can grow deeper into the tree, tapping into otherwise inaccessible stores of nitrogen and other essential substances. It then pumps these back up to the surface within reach of the larvae. 'These beetles are living on junk food, so the fungi provide them with nutrients,' says Diana Six, an entomologist who has been studying the beetles for years. When the beetle larva eventually pupates, the fungi produce spores – hardy reproductive capsules. After the adult beetle emerges, it packs the spores into suitcase-like structures in its mouth and carries them off to the next unfortunate pine.

Beetle outbreaks come and go but the current one, fuelled by a warming climate, is ten times bigger than any other. Since 1999, the beetles and their attendant fungi have killed more than half the mature pines in British Columbia and affect 3.8 million acres in the United States. They have even hopped over the cold Canadian Rockies, which long fenced them into the west coast, and are now spreading east. A continuous belt of lush vulnerable forests lies in front of them.

The trees, however, don't go gently into the good night. When attacked, they mass-produce compounds called terpenes that, at high concentrations, can kill both the beetles and the fungi. The beetles supposedly thwart this defence through brute force: they swarm in such overwhelming numbers that the trees can't manufacture enough terpenes to hold them all back. But this explanation made no sense to entomologist Ken Raffa. If it was true, the trees should produce a burst of terpenes that would quickly become exhausted as the beetle onslaught proceeded. That's not what happens; the trees actually maintain their chemical defences at high levels for at least a month. If anything, the beetle larvae have to deal with even more toxins than their parents. So how do they do it?

Raffa's team found that, besides fungi, the beetles also associate with bacteria like *Pseudomonas* and *Rahnella,* which turn up throughout their range and on all of their host trees. They get everywhere. They're on the insects' exoskeletons, on the walls of their galleries, in their mouths and guts. They're a select bunch, far narrower in membership than the diverse gut communities of termites, and probably unsuited to any feats of digestion. They do, however, possess a large suite of genes for degrading terpenes, and they destroy these chemicals effectively in lab conditions. Different species tackle different compounds and, together, they defuse the lot.[32]

It's tempting to declare the matter solved: the bacteria disarm the tree's defences, and the beetles carry them from one trunk to another.

But, as we've already seen, the world of symbiosis is a complicated one, and simple narratives, though satisfying, are often wrong. For a start, the same bacteria also exist on healthy, uninfected conifers, so they might be part of the *tree*'s microbiome. When the beetles attack and terpene production ramps up, these bacteria go to town on the sudden chemical feast. They dine well, but they inadvertently harm their host tree and help the invading beetles. The beetles also have a limited set of terpene-breaking enzymes, so how much do the bacteria contribute? Do they take on the bulk of the detoxifying work, or do they share their duties with the insects, just as aphids and *Buchnera* cooperate to make amino acids? And, crucially, do they actually improve the beetles' odds of survival?

For now, this much is clear: a massive alliance of animals, fungi, and bacteria descends upon a forest, and the trees, despite their best chemical defences, start to die. Their demise is testament to the power of symbiosis – a force that allows the most innocuous of organisms to topple the mightiest ones. You need to squint to see the beetles and to bring out a microscope to see their microbes, but the consequences of their mutually assured success are visible from the sky.

Thanks to the powers bestowed by their microbes, hemipterans have evolved to suck the sap from the world's plants, while termites and grazing mammals have come to chew their stems and leaves. Tube worms have colonised the deepest oceans, woodrats can spread through the American deserts, and mountain pine beetles have wreaked continental ruin in evergreen forests.[33]

In contrast to these ostentatious examples, the two-spotted spider mite wreaks a subtler brand of havoc. Like the beetle, this tiny red arachnid, barely bigger than a full stop, also kills plants by descending upon them in untold numbers. It is a global pest, succeeding thanks to its knack for resisting pesticides and its catholic tastes: it feeds on more than 1,100 plant species, from tomatoes to strawberries, maize to soy. Such a wide-ranging palette implies a certain skill at detoxification: each plant wields its own cocktail of defensive chemicals, and the spider mite needs ways of disarming all of them. Fortunately, it packs an arsenal of detoxification genes, which are variously activated depending on the plant it decides to drink from.

Microbes, it seems, will not be the heroes of this particular story. Unlike the desert woodrat or the mountain pine beetle, the spider mite doesn't rely on gut bacteria to render its meals more palatable. It has everything it needs in its own genome. But even in their absence, bacteria matter.

Many of the plants that the spider mite targets can release hydrogen cyanide when their tissues are breached. This substance is extraordinarily inimical to life. Exterminators have poisoned rats with it. Whalers added it to their harpoons. The Nazis used it in concentration camps. But the spider mite is impervious. One of its genes can make an enzyme that converts hydrogen cyanide into a harmless chemical. The same gene is present in the caterpillars of various butterflies and moths; they, too, shrug at cyanide. Neither the spider mite nor the caterpillars invented the cyanide-busting gene for themselves, nor did they inherit it from the common ancestor.

The gene came from bacteria.[34]

8. ALLEGRO IN E MAJOR

When you were born, you inherited half your genes from your mother and half from your father. That's your lot. Those inherited bits of DNA will remain with you for all of your life, with no further additions or omissions. You can't have any of my genes, and I can't acquire any of yours. But imagine a different world where friends and colleagues can swap genes at will. If your boss has a gene that makes her resistant to various viruses, you can borrow it. If your child has a gene that puts him at risk of disease, you can swap it out for your healthier version. If distant relatives have a gene that allows them to better digest certain foods, it's yours. In this world, genes aren't just heirlooms to be passed on vertically from one generation to the next, but commodities to be traded horizontally, from one individual to another.

This is exactly the world that bacteria live in. They can exchange DNA as easily as we might exchange phone numbers, money, or ideas. Sometimes, they sidle up to one another, create a physical link, and shuttle bits of DNA across: their equivalent of sex. They can also scrounge up discarded bits of DNA in their environment, left by their dead and decaying neighbours. They can even rely on viruses to move genes from one cell to another. DNA flows so freely between them that the genome of a typical bacterium is marbled with genes that arrived from its peers. Even closely related strains might have substantial genetic differences.[1]

Bacteria have been carrying out these horizontal gene transfers, or HGT for short, for billions of years, but it wasn't until the 1920s that scientists first realised what was happening.[2] They noticed that harmless strains of *Pneumococcus* could suddenly start causing disease after mingling with the dead and pulped remains of infectious strains. *Something* in the extracts had changed them. In 1943, a 'quiet revolutionary' named Oswald Avery showed that this transformative material was DNA, which the non-infectious strains had absorbed and integrated into their own genomes.[3] Four years later, a young geneticist named Joshua Lederberg (who would later popularise the word 'microbiome') showed that bacteria can trade DNA more directly. He worked with two strains of *E. coli*, each of which was unable to make different nutrients. Unless they received supplements, these bacteria would die. But when Lederberg mixed the two strains together, he found that some of their daughters could survive without any help. It became clear that the two parental strains had horizontally exchanged genes that compensated for each other's shortcomings. The daughters then vertically inherited a complete working set, and thrived.[4]

Sixty years on, we know that HGT is one of the most profound aspects of bacterial life. It allows bacteria to evolve at blistering speeds. When they face new challenges, they don't have to wait for the right mutations to slowly amass within their existing DNA. They can just borrow adaptations wholesale, by picking up genes from bystanders that have already adapted to the challenges at hand. These genes often include dining sets for breaking down untapped sources of energy, shields that protect against antibiotics, or arsenals for infecting new hosts. If an innovative bacterium evolves one of these genetic tools, its neighbours can quickly obtain the same traits. This process can instantly change microbes from harmless gut residents into disease-causing monsters, from peaceful Jekylls into sinister Hydes. They can also transform vulnerable pathogens that are easy to kill into nightmarish 'superbugs' that shrug off even our most potent medicines. The spread of these antibiotic-resistant bacteria is undoubtedly one of the greatest public health threats of the twenty-first century, and it is testament to the unbridled power of HGT.

Animals aren't so fast. We adapt to new challenges in the usual slow and steady way. Individuals with mutations that leave them best suited to life's challenges are more likely to survive and pass on their genetic gifts to the next generation. Over time, useful mutations become more common, while harmful ones fade away. This is classic natural selection – a slow and steady process that affects *populations*, not *individuals*. Hornets, hawks, and humans might gradually accumulate beneficial mutations, but *that* individual hornet, or *this* specific hawk, or *those* particular humans can't pick up beneficial genes for themselves. Except sometimes, they can. They could swap their symbionts, instantly acquiring a new package of microbial genes. They can bring new bacteria into contact with those in their bodies, so that foreign genes migrate into their microbiome, imbuing their native microbes with new abilities. And on rare but dramatic occasions, they can integrate microbial genes into their own genomes, as the two-spotted spider mite from the previous chapter did when it picked up a cyanide-detoxifying gene.[5]

Excitable journalists sometimes like to claim that HGT challenges Darwin's view of evolution, by allowing organisms to escape the tyranny of vertical inheritance. ('Darwin was wrong' proclaimed an infamous *New Scientist* cover – wrongly.) This is not true. HGT adds new variation into an animal's genome but once these jumping genes arrive in their new homes, they are still subject to good ol' natural selection. Detrimental ones die along with their new hosts, while beneficial ones are passed on to the next generation. This is as classically Darwinian as it gets – vanilla in its flavour, and exceptional only in its *speed*.

We've seen that microbes help animals to take up exciting new evolutionary opportunities. Now we'll see that they sometimes help us to take up those opportunities *very quickly*. By partnering with microbes, we can quicken the slow, deliberate adagio of our evolutionary music to the brisk, lively allegro of theirs.

Along the coasts of Japan, a reddish-brown seaweed clings to tide-swept rocks. This is *Porphyra*, better known as *nori*, and it has filled Japanese stomachs for over 1,300 years. At first, people ground it into an edible paste. Later, they printed it into flattened sheets, which they wrapped around morsels of sushi. This practice continues today and nori's popularity has spread all over the world. Still, it has a special tie to Japan. The country's long legacy of nori consumption has left its people especially well equipped to digest the sea vegetable.

Like other marine algae, nori contains unique carbohydrates that aren't found in land plants. We don't have any enzymes that can break down these substances, and neither do most of the bacteria in our guts. But the sea is full of better-equipped microbes. One of these, a bacterium called *Zobellia galactanivorans*, was discovered just a decade ago, but has been eating seaweed for much longer. Picture *Zobellia*, centuries ago, living in coastal Japanese waters, sitting on a piece of seaweed and digesting it. Suddenly, its world is uprooted. A fisherman collects the seaweed and uses it to make nori paste. His family wolfs down these morsels and, in doing so, they swallow *Zobellia*. The bacterium finds itself in a new environment. Cool salt water has been substituted by gastric juices. Its usual coterie of marine microbes has been replaced by weird and unfamiliar species. And as it mingles with these exotic strangers, it does what bacteria typically do when they meet up: it shares its genes.

We know that this happened because Jan-Hendrick Hehemann discovered one of *Zobellia*'s genes in a human gut bacterium called *Bacteroides plebeius*.[6] The discovery was a total shock: what on earth was a marine gene doing in the gut of a landlubbing human? The answer involves HGT. *Zobellia* isn't adapted to life in the gut, so when it rode in on morsels of nori, it didn't stick around. But during its brief tenure, it could easily have donated some of its genes to *B. plebeius*, including those that build seaweed-digesting enzymes called porphyranases. Suddenly, that gut microbe gained the ability to break down the unique carbohydrates found in nori, and could feast on this exclusive source of energy that its peers couldn't use. It seems to have made a habit of this. Hehemann found that it is full of genes whose closest counterparts exist in marine microbes rather than in other gut-based species. By repeatedly borrowing genes from sea microbes, it has become adept at digesting sea vegetables.[7]

B. plebeius isn't alone in thieving marine enzymes. The Japanese have been eating nori for so long that their gut microbes are peppered with digestive genes from oceanic species. It's unlikely that such transfers are still going on, though: modern chefs roast and cook nori, incinerating any hitchhiking microbes. The diners of centuries past only managed to import such microbes into their guts by eating the stuff raw. They then passed their gut microbes, now loaded up with seaweed-busting porphyranase genes, to their children. Hehemann saw signs of the same inheritance going on today. One of the people he studied was an unweaned baby girl, who had never eaten a mouthful of sushi in her life. And yet, her gut bacteria had a porphyranase gene, just as her mother's did. Her microbes came pre-adapted for devouring nori.

Hehemann published his discovery in 2010 and it remains one of the most striking microbiome stories around. Just by eating seaweed, the Japanese diners of centuries past booked a group of digestive genes on an incredible voyage from sea to land. The genes moved horizontally from marine microbes to gut ones, and then vertically from one gut to another. Their travels may have gone even further. At first, Hehemann could only find the genes for porphyranases in Japanese microbiomes and not North American ones. That has now changed: some Americans clearly have the genes, even those who aren't of Asian ancestry.[8] How did that happen? Did *B. plebeius* jump from Japanese guts into American ones? Did the genes come from other marine microbes stowing away aboard different foods? The Welsh and Irish have long used *Porphyra* seaweed to make a dish called laver; could they have acquired porphyranases that they then carried across the Atlantic? For now, no one knows. But the pattern 'suggests that once these genes hit the initial host, wherever that happens, they can disperse between individuals', says Hehemann.

This is a glorious example of the adaptive speed that HGT confers. Humans don't need to evolve a gene that can break down the carbohydrates in seaweed; if we swallow enough microbes that can digest these substances there's every chance that our own bacteria will 'learn' the trick through HGT.

When Eric Alm from MIT read about Hehemann's discovery, he wondered if he could find similar examples. He searched the genomes of over 2,200 species of bacteria for long stretches of DNA that were virtually identical even though the surrounding genes were very different. These islands of similarity, floating in oceans of difference, were unlikely to have moved from mother microbe to daughter; they must have been horizontally transferred, and recently at that. Alm's team found over 10,000 of these swapped sequences – a testament to how common HGT is.[9] They also showed that such swaps are exceptionally common in the human body. Pairs of bacteria from the human microbiome were 25 times more likely to have swapped genes with each other than pairs from other environments.

This makes perfect sense: HGT depends on proximity, and our bodies engineer proximity on a huge scale by gathering microbes into dense crowds. It is said that cities are hubs of innovation because they concentrate people in the same place, allowing ideas and information to flow more freely. In the same way, animal bodies are hubs of genetic innovation, because they allow DNA to flow more freely between huddled masses of microbes. Close your eyes, and picture skeins of genes threading their way around your body, passed from one microbe to another. We are bustling marketplaces, where bacterial traders exchange their genetic wares.

With so many microbes living in our bodies, surely, sometimes, their genes should make their way into animal hosts.[10] For the longest time, the consensus was that they don't, and that animal genomes were impenetrable sanctuaries, isolated from the genetic promiscuity of microbes. Yet in February 2001, this view took a small hit, when the first full draft of the human genome was published. Of the thousands of genes identified, 223 were shared with bacteria but not with other complex organisms like flies, worms, or yeast. These genes, as the scientists behind the Human Genome Project wrote, 'appear likely to have resulted from horizontal transfer from bacteria'. But just four months later, this bold claim started to evaporate. Another group of researchers showed that these special genes were probably present in some very early organisms and then lost in later lineages, creating an illusion of HGT where none had actually happened.[11] This riposte had a chilling impact. It cast a pall of disbelief over the very possibility of HGT between bacteria and animals.

It took a few years for the scepticism to start cracking. In 2005, a microbiologist named Julie Dunning-Hotopp found genes from that ubiquitous bacterium *Wolbachia* within the genome of the Hawaiian fly *Drosophila ananassae.*[12] At first, she thought that these genes were coming from living *Wolbachia* cells stowing away in the insects' bodies. But even when she treated the flies with antibiotics, the bacterial genes remained. After months of frustration, she finally realised that the genes had seamlessly integrated into the fly's DNA. She then found similar patterns in the genomes of seven other animals, including wasps, a mosquito, a nematode worm and other flies. It looked as if *Wolbachia* had liberally sprayed the tree of life with its DNA. Many of the fragments were small, with one exception: to her astonishment, Dunning-Hotopp found that *D. ananassae* was harbouring *Wolbachia*'s complete genome. At some point in the recent past, *Wolbachia* had shunted *all* of its genetic material into this particular host. Everything that it was, the sum of its genetic identity, hopped over into the fly. This is one of the most dramatic examples of HGT ever found, and perhaps the ultimate expression of the hologenome concept: the genes of an animal and a microbe, fused into a single entity.

Dunning-Hotopp published her results with a clear statement: genes move from bacteria to animals. More than that: they move between the most common symbiotic bacterium into some of the most abundant animals. Somewhere between 20 and 50 per cent of insect species have evidence of HGT from *Wolbachia* – and that's a *lot* of insects! 'The view that [these] transfers are uncommon and unimportant needs to be re-evaluated,' she wrote.[13]

Well, certainly, they're not uncommon.[14] But are they important? The mere presence of a guitar in someone's bedroom doesn't make them Slash. Likewise, the mere presence of a gene in a genome means nothing; it might just be lying around, unused. Many of the *Wolbachia* fragments found in flies are probably like this: genetic flotsam, drifting through genomes to little effect. A small proportion of those *Wolbachia* genes *are* switched on, but even that isn't evidence that they are functional; there's always a certain amount of noisy activity in a cell, where genes are spontaneously turned on without being put to actual use. There's really only one way of proving that the introduced genes do something useful, which is to find what that something is. In a few cases, such proof exists.

The root-knot nematodes are microscopic worms that parasitise plants, and they're so effective that they regularly ruin some 5 per cent of the world's crops. They kill through vampirism: they stick their mouthparts into the cells of plant roots and suck out the innards. This is more difficult than it seems. Plant cells are surrounded by tough walls of cellulose and other sturdy chemicals, and the nematodes must first deploy enzymes that soften and break these barriers before they can slurp the nutritious soup inside. They build these enzymes using instructions encoded within their own genome, and a single species can wield more than 60 plant-infiltrating genes. That's odd. Such genes are the province of fungi and bacteria; animals shouldn't have them, let alone in such extreme numbers . . . and yet the nematodes obviously do.

The nematodes' plant-penetrating genes are clearly bacterial in origin.[15] They are unlike any genes in other nematodes; instead, their closest counterparts exist in microbes that grow on plant roots. And unlike most horizontally transferred genes, whose roles in their new homes are non-existent or uncertain, the nematodes' acquisitions have a clear purpose. The nematodes switch them on in their throat glands to make a demolition squad of enzymes, which they then spew into roots. That's the foundation of their entire lifestyle. Without these genes, these little vampires would be ineffective parasites.

No one knows how the root-knot nematodes originally picked up their bacterial genes in the first place, but we can make an educated guess. These nematodes are closely related to another type that lives near plant roots and eats bacteria. If these other nematodes consumed microbes that could infect or infiltrate plants, they could have slowly acquired genes that allowed them to do the same. Over time, these soil-dwelling, bacteria-munching worms became a blight on plants and a nuisance to farmers.

The coffee berry borer beetle is another pest that owes its devastating powers to HGT.[16] This black speck of an insect, as we saw in the last chapter, uses gut microbes to detoxify the caffeine in coffee plants. But it has also incorporated a bacterial gene into its own genome, which allows its larvae to digest the lush banquets of carbohydrates within coffee beans. No other insect – not even very close relatives – has the same gene or anything like it; only bacteria do. By jumping into an ancient coffee borer, the gene allowed this unassuming beetle to spread across coffee-growing regions around the world, and become a royal pain in the espresso.

Farmers, then, have reasons to loathe HGT – but also reasons to celebrate it. For one group of wasps, the braconids, transferred genes have enabled a bizarre form of pest control. The females of these wasps lay their eggs in still-living caterpillars, which their young then devour alive. To give the grubs a hand, the females also inject the caterpillars with viruses, which suppress their immune systems. These are called bracoviruses, and they aren't just allies of the wasps: they are *part of* the wasps. Their genes have become completely integrated into the braconid genome, and are under its control. When a female wasp builds her viruses, she loads them with the genes they need to attack a caterpillar, while withholding those they need to reproduce or spread to different hosts.[17] The bracoviruses are domesticated viruses! They're entirely dependent on the wasps for their reproduction. Some might say they're not true viruses at all; they're almost like secretions of the wasp's body rather than entities in their own right. They must have descended from an ancient virus, whose genes wheedled their way into the DNA of an ancestral braconid and stayed there. This merger gave rise to over 20,000 species of braconid wasps, all of which have bracoviruses in their genomes – an immense dynasty of parasites that uses symbiotic viruses as biological weapons.[18]

Other animals have used horizontally transferred genes to *defend* themselves from parasites. Bacteria, after all, are the ultimate source of antibiotics. They have been at war with each other for billions of years, and have invented an extensive arsenal of genetic weapons for beating their rivals. One family of genes, known as *tae*, make proteins that punch holes in the outer walls of bacteria, causing fatal leaks. These were developed by microbes for use against other microbes. But these genes have found their way into animals, too. Scorpions, mites, and ticks have them. So do sea anemones, oysters, water fleas, limpets, sea slugs, and even the lancelet – a very close relative of backboned animals like ourselves.[19]

The *tae* family exemplifies the kind of genes that spread very easily through HGT. They are self-sufficient, and don't need a supporting cast of other genes to do their job. They are also universally useful, because they make antibiotics. Every living thing has to contend with bacteria, so any gene that allows its owner to control bacteria more effectively will find gainful employment throughout the tree of life. If it can make the jump, it's got a good chance of establishing itself as a productive part of its new host. These jumps are all the more impressive because we humans, with all our intelligence and technology, positively struggle to create new antibiotics. So flummoxed are we that we haven't discovered any new types for decades. But simple animals like ticks and sea anemones can make their own, instantly achieving what we need many rounds of research and development to do – all through horizontal gene transfer.

These stories portray HGT as an additive force, which infuses both microbes and animals with wondrous new powers. But it can also be subtractive. The same process that bestows useful microbial abilities upon animal recipients can make the microbes themselves wither and decay, to the point where they disappear entirely and only their genetic legacies remain.

The creature that best exemplifies this phenomenon can be found in greenhouses and fields around the world, much to the chagrin of farmers and gardeners. It's the citrus mealybug: a small sap-sucking insect, which looks like a walking dandruff flake or a woodlouse that's been dusted in flour. Paul Buchner, that super-industrious scholar of symbionts, paid a visit to the mealybug clan on his tour of the insect world. To no one's surprise, he found bacteria inside their cells. But, more unusually, he also described 'roundish or longish mucilaginous globules in which the symbionts are thickly embedded'. These globules languished in obscurity for decades until 2001, when scientists learned that they weren't just houses for bacteria. They were bacteria themselves.

The citrus mealybug is a living matryoshka doll. It has bacteria living inside its cells, and those bacteria have more bacteria living inside them. Bugs within bugs within bugs.[20] The bigger one is now called *Tremblaya* after Ermenegildo Tremblay, an Italian entomologist who studied under Buchner. The smaller one is called *Moranella* after aphid-wrangler Nancy Moran. ('It *is* a kind of a pathetic little thing to be named after you,' she told me with a grin.)

John McCutcheon has worked out the origins of this weird hierarchy – and it's almost unbelievable in its twists and turns. It begins with *Tremblaya*, the first of the two bacteria to colonise mealybugs. It became a permanent resident and, like many insect symbionts, it lost genes that were important for a free-living existence. In the cosy confines of its new host, it could afford to get by with a more streamlined genome. When *Moranella* joined this two-way symbiosis, *Tremblaya* could afford to lose even *more* genes, in the surety that the new arrival would pick up the slack. As long as a gene exists in one of the partners, the others can afford to lose it. These types of gene transfer are different to those that turned nematodes into plant parasites, or those that sprinkled antibiotic genes into tick genomes. Here, the recipients don't gain beneficial abilities. Here, HGT is more about evacuating bacterial genes from a capsizing ship. It preserves genes that would otherwise be lost to the inevitable decay that afflicts symbiont genomes.

For example, all three partners cooperate to make nutrients. To create the amino acid phenylalanine, they need nine enzymes.

Tremblaya can build 1, 2, 5, 6, 7, and 8; *Moranella* can make 3, 4, and 5; and the mealybug alone makes the 9th. Neither the mealybug nor the two bacteria can make phenylalanine on their own; they depend on each other to fill the gaps in their repertoires. This reminds me of the Graeae of Greek mythology: the three sisters who share one eye and one tooth between them. Anything more would be redundant: their arrangement, though odd, still allows them to see and chew. So it is with the mealybug and its symbionts. They ended up with a single metabolic network, distributed between their three complementary genomes. In the arithmetic of symbiosis, one plus one plus one can equal one.[21]

This explains another truly bizarre thing about *Tremblaya*'s genome: it's missing a class of supposedly essential genes that are among the oldest in existence. They're present in the last common ancestor of all living things, and are found in everything from bacteria to blue whales. They are as synonymous with life, and as indispensable to it, as genes can get. There should be 20 of them. Some symbionts have lost a few. *Tremblaya* has *none*. And yet, it survives because its partners – the insect that hosts it, and the bacterium inside it – compensate for its vanished genes.

Where did all the missing genes go? As we have seen, bacterial genes are often relocated to the genomes of their hosts. And sure enough, when McCutcheon checked the citrus mealybug's genome, he found 22 bacterial genes nestled among the insect DNA. But to his astonishment, *none* of these came from either *Tremblaya* or *Moranella*. Not one. Instead, they had come from three *other* lineages of bacteria, all of which can colonise insect cells, and *none* of which still exist in the citrus mealybug.[22]

This insect contains bits of *five* bacteria – two shrunken, co-dependent ones that are nested within its cells, and at least three more that must once have shared its body but have since vanished.

The genes they left behind, these ghosts of symbionts past, aren't sitting idly among the mealybug's DNA. Some make amino acids. Others help to make a large molecule called peptidoglycan. That's odd. Animals don't use peptidoglycan – it's a bacterial molecule that builds the thick outer walls that keep a bacterium's insides inside.[23] *Moranella*, however, has lost its own peptidoglycan-making genes. In order to make its walls, it must rely on the bacterial genes that the mealybug borrowed from its departed symbionts.

McCutcheon wonders if the mealybug can deliberately destabilise *Moranella* by withholding the supply of peptidoglycan. Deprived of this substance, *Moranella* eventually bursts. And when it does, it releases the proteins that it can make but that *Tremblaya* cannot. Remember that *Tremblaya* is missing a class of supposedly essential genes; perhaps this is how it copes. 'That's wild speculation,' says McCutcheon. 'It's an idiotic guess, but it's also my best one.' He talks with a mixture of awe, confusion, and faint embarrassment, as if his discoveries are so outlandish that he barely believes them himself. And yet, there they are.

The data might tell tales with preposterous plots, but they don't lie. They tell us that the citrus mealybug is a mash-up of at least six different species, five of which are bacterial and three of which *aren't even there*. It uses genes borrowed from former symbionts to control, cement, and complement the relationship between its two current ones, one of which lives inside the other.[24]

Not all insect symbionts are so tightly bound to their hosts. Aphids, for example, contain several species of bacteria besides the ever-present *Buchnera*. These 'secondary symbionts' are less stalwart in their allegiances. They are common in some aphid populations but rare or absent in others. Some aphids have all three; others have none.

When Nancy Moran noticed these patterns, she realised that these microbes couldn't be providing essential nutrients. If they were, they would be constant. Instead, they must offer some service that the insects only occasionally need. In many ways, they were behaving like variations in the human genome that affect our risk of disease. For example, some people have a mutation that causes their red blood cells to change shape, from a rounded pastille into a thin sickle. The mutation comes with a cost: inheriting two copies leads to a debilitating condition called sickle-cell disease. But it has benefits, too: a single copy makes its carriers very good at resisting malaria because their warped cells are harder for malarial parasites to infect. In equatorial Africa, where malaria is common, as many as 40 per cent of people carry the sickle-cell mutation. In areas where malaria is rare, the sickle-cell trait is too. The frequency of the mutation depends on the severity of the threat it protects against. Maybe, Moran reasoned, the aphids' secondary symbionts are the same. Maybe they protect the aphids against a natural enemy. If the enemy is rare, their services are not needed and their numbers fall. If the enemy is common, so are they.

But which enemy? Aphids are not short of them. Spiders ensnare them, fungi infect them, ladybirds and lacewings devour them. But arguably their biggest threat comes from parasitoids – body-snatchers that implant their young inside other insects. This grisly lifestyle is surprisingly common. One in every ten insect species is a parasitoid, including braconid wasps with their domesticated viruses. The latter group includes a slender, black creature called *Aphidius ervi*. It targets aphids, and so effectively that farmers regularly release the wasps over their crops. You can order them by their hundreds over the Internet for around £20.

Aphids vary in their ability to cope with these wasps. Some are completely resistant, others always succumb. Scientists had assumed that this variation lay within the aphids' own genes but Moran wondered if the symbionts were involved. She recruited graduate student Kerry Oliver to test the idea.[25] It was a long shot. At the time, the concept of symbionts protecting against parasites was unheard of, and so outlandish that Moran didn't really believe that the experiments would work.

Using a microscope, a needle, and very steady hands, Oliver extracted the symbionts out of different aphids and injected them into one particular strain. Then, he unleashed *A. ervi* on them. Within a week, the aphids' cages were littered with mummified corpses and newly emerged wasps. But one group was surprisingly resilient. They had still been implanted with wasp eggs but they carried a symbiont that was somehow killing the wasp larvae. When Oliver dissected these aphids, he usually found a dead or dying wasp inside. In other words, the team's crazy idea was right: one of the aphid's microbes was acting as a wasp-killing bodyguard. They named it *Hamiltonella defensa*.[26]

In retrospect, the existence of defensive microbes isn't surprising. Protecting their hosts from harm is an obvious way of guaranteeing their own success, and besides, bacteria are very good at making antibiotics. But *Hamiltonella defensa* doesn't make antibiotics. When *Hamiltonella*'s genome was sequenced, the real reason behind the bacterium's protective powers became apparent: about half of its DNA actually belonged to a virus. It was a phage – one of those spindly-legged, mucus-loving viruses that we met before. They typically kill bacteria by reproducing inside them and bursting fatally outward. But they can also opt for a more passive lifestyle, where they integrate their DNA into a bacterium's genome and stay there for many generations. Dozens of these phages now hide within *Hamiltonella*.[27]

The viruses are *Hamiltonella*'s fists; they give the bacterial bodyguard its punch. Oliver showed that when *Hamiltonella* carries a particular phage strain, it renders aphids almost totally wasp-proof. If the virus disappeared, the same bacterium became useless, and almost all of its aphid hosts succumbed to their parasites. Without the phage, *Hamiltonella* might as well have been completely absent for all the good it did. The phages could be poisoning the wasps directly: they certainly mass-produce toxins that can attack animal cells, but don't seem to harm the aphids. Alternatively, they could split *Hamiltonella* apart, causing the bacteria's own toxins to spill onto the wasps. Or maybe the viral and bacterial chemicals are working together. Whatever the case, it is clear that an insect, a bacterium, and a virus have formed an evolutionary alliance against a parasitic wasp that threatens them all.

This alliance is a varied one. Aphids differ in their ability to fend off wasps because they harbour different *Hamiltonella* strains, and *Hamiltonella* confers different degrees of protection depending on its resident phages. Like the sickle-cell trait, these microscopic partners come at a cost. For some reason, at certain temperatures, aphids that carry these bodyguards live shorter lives and raise fewer young. If there are lots of wasps around, this price is worth paying, but otherwise it's too expensive to bear – and out the symbiont goes. Similarly, if aphids are herded by ants (which farm them for the sweet liquid they excrete), they are less likely to carry *Hamiltonella*, since the ants provide all the anti-wasp protection they need. This is why *Hamiltonella* isn't a permanent fixture of the aphids' bodies. It sticks around when it is needed. Likewise, the phage isn't a permanent fixture of *Hamiltonella*. In the wild, it frequently disappears for reasons that are still unclear. Theirs is a dynamic partnership that, through natural selection, tunes itself to the level of threat around.

But how does *Hamiltonella* get into aphids in the first place? If the aphids dispense with it when life is easy, how does it rejoin them when the going gets tough? Moran found one possible answer: sex. Males carry *Hamiltonella* and other defensive symbionts in their semen. When they have sex, they can pass these microbes to the females, who can then inoculate their offspring. So females, by mating with the right partners, can suddenly become immune to wasp attacks, which makes *Hamiltonella* that rarest of things: a desirable venereal infection.[28]

An aphid that catches *Hamiltonella* through sex is not shunting bacterial DNA into its genome. It is, however, picking up a massive suite of bacterial genes that are still inside their original packaging. This is similar to HGT, except the G here stands for *genome,* not *gene.* And as with HGT, these whole-microbe acquisitions allow animals to adapt to new challenges very quickly, perhaps instantly.

Rather than gradually accruing mutations in their own genomes over many generations, they can pick up microbes that already have the right adaptations.[29] Rather than slowly training their existing staff to carry out new jobs, they just hire fresh employees who do those jobs already. Chances are, the right recruits are already out there: bacteria are infinitely more versatile than we are. They are metabolic wizards that can digest everything from uranium to crude oil. They are expert pharmacologists that excel at making chemicals that kill each other. If you want to defend yourself from another creature or eat a new source of food, there's almost certainly a microbe that already has the right tools for the job. And if there isn't, there soon will be: these things reproduce rapidly and swap genes readily. In the great evolutionary race, they sprint, while we crawl. But we can get a little closer to their blinding pace by forming partnerships with them. Bacteria, in other words, allow us to do decent impressions of bacteria.

That's what happened when the desert woodrats swallowed microbes that allowed them to detoxify the poisons in creosote bushes. It's what happens when the Japanese bean bug engulfs soil microbes that destroy insecticides, rendering it instantly immune to the rain of toxins being sprayed by farmers. And it's what aphids do *all the time.* Besides *Hamiltonella,* they have at least *eight* different secondary symbionts. Some protect against deadly fungi. Others help their hosts to tolerate heatwaves. One allows aphids to eat specific plants, like clover. One paints the aphids, changing them from red to green. These abilities are important. Across the aphid family, the acquisition of new symbionts tends to coincide with invasions into new climates or shifts to new types of plant.[30]

These changes are all fundamentally Darwinian. This point is worth repeating: taking any fast or instant evolutionary shifts as a refutation of the slow, gradual changes we associate with Darwin's vision is a fatal mistake because these quick shifts are *still* powered by gradualism. The woodrats might have been able to resist creosote by picking up the right *bacteria,* but those strains had to evolve the ability to break the insecticide on their own. From their perspective, evolution proceeded through the usual stepwise way; from the host's perspective, everything happened in a flash. That is the power of symbiosis: it allows gradual mutations in microbes to produce instant mutations in hosts. We can let bacteria do the slow work for us, and then quickly change ourselves by associating with them. And if these alliances are beneficial enough, they can spread with blinding speed.

A fruit fly buzzes through a North American forest and catches the whiff of a tasty treat: a mushroom, poking up through the leaf litter. It lands, feeds, and starts to lay eggs. And all the while, it inadvertently seeds the mushroom with parasitic nematodes, known as *Howardula*. These reproduce inside the mushroom before seeking out the fly larvae growing next to them. When the flies mature and leave to find more mushrooms, they carry a payload of worms.

When John Jaenike first started studying *Howardula* in the 1980s, he saw that the worms exert a heavy toll on fruit flies. The insects would die earlier, the males would struggle to find mates, and the females were completely sterilised. They were little more than worm vehicles. But as the new millennium rolled over, things changed and Jaenike started catching parasitised females that carried a full cargo of eggs. Jaenike is a *Wolbachia* man, and since this uber-microbe infects the flies he was studying, he naturally wondered if it was defending its hosts from the parasites. He was half right: the flies *were* under symbiotic protection but – for once! – *Wolbachia* wasn't part of the story. Instead, their guardian was a corkscrew-shaped microbe called *Spiroplasma*.

The story of the flies, the worms, and *Spiroplasma* is extraordinary, not because of its themes or characters, but because Jaenike watched it being written. When he went to museums and analysed fly specimens that had been collected in the 1980s he couldn't find a trace of *Spiroplasma*. But in 2010, he saw the bacterium in anywhere from 50 to 80 per cent of the flies in eastern North America. It was already sweeping westwards. By 2013, it had crossed the Rockies. 'It should get to the Pacific in ten years,' says Jaenike.[31]

Despite its recent ascendance, *Spiroplasma* isn't actually a new ally. Jaenike estimated that it first jumped into flies a few thousand years ago, but stayed at extremely low levels. That's why he couldn't find it in his surveys of specimens from the 1980s. It only became common recently, when the parasitic *Howardula* nematode left Europe and touched down in North America. When the parasite arrived, it spread like wildfire, riding through the forests in the bodies of its sterilised hosts. The flies needed a countermeasure and *Spiroplasma* rose to the occasion. It restored its hosts' ability to reproduce, and allowed them to outcompete their sterile peers. Since the flies could pass these little saviours to their offspring, the proportion of infected insects grew with each generation. And Jaenike had caught this spread at exactly the right moment. 'It made me doubt my sanity,' he says. 'What are the chances?'

But his colleagues started stumbling across more supposedly rare sweeps: another bacterium called *Rickettsia* that hurtled through sweet potato whiteflies in the USA within just six years, making the insects fitter and more fertile.[32] We usually only see the *consequences* of these events. We see worms, clams and other animals that live in the darkest oceans, herds of grazing mammals that prune the savannahs, and immeasurable hordes of bugs that suck the fluids from plants, all thriving in their respective niches thanks to microbial power. But these alliances are clearly forged often enough for scientists occasionally to catch their origins, provided they look in the right place at the right time.[33]

The world around us is a gigantic reservoir of potential microbial partners. Every mouthful could bring in new microbes that digest a previously unbreakable part of our meals, or that detoxify the poisons in a previously inedible food, or that kill a parasite that previously suppressed our numbers. Each new partner might help its host to eat a little more, travel a little further, survive a little longer.

Most animals can't tap into these open-source adaptations deliberately. The flies didn't seek out *Spiroplasma* to solve their worm problem. Woodrats didn't go looking for the creosote-defusing microbes so they could widen their diet. They must rely on luck to endow them with the right partners. But we humans aren't so restricted. We are innovators, planners, and problem-solvers. And we have one huge advantage that all other animals lack: *we know that microbes exist!* We have devised instruments that can see them. We can deliberately grow them. We have tools that can decipher the rules that govern their existence, and the nature of their partnerships with us. And that gives us the power to manipulate those partnerships intentionally. We can replace faltering communities of microbes with new ones that will lead to better health. We can create new symbioses that fight disease. And we can break age-old alliances that threaten our lives.

9. MICROBES À LA CARTE

It starts with a bite. A mosquito lands on a man's arm, sinks its mouthparts into his flesh, and begins to suck. As blood rushes into the insect, tiny parasites head in the other direction. They are the larvae of filarial nematodes. These microscopic worms swim through the man's bloodstream and travel to the lymph nodes in his legs and genitals. Over the next year, they mature into adults, and mate with each other to produce thousands of new larvae *every day*. A doctor with an ultrasound scanner could see them wriggling around, but the infected man has no reason to visit one; despite the millions of parasites inside him, he still isn't showing any symptoms. Eventually that changes. As the worms die, they trigger inflammation. They also block the flow of lymph, which accumulates under the man's skin. His limbs and groin swell to gigantic proportions. His thigh grows to the width of his entire torso. His scrotum becomes head-sized. He can't work; he'll be lucky if he can stand. He will carry that disfigurement, and the accompanying social stigma, for the rest of his life. The man might be a farmer in Tanzania, or a fisherman in Indonesia, or a cattle-herder in India. It doesn't matter; he is now one of millions of people who suffer from lymphatic filariasis.

This disease, also known as elephantiasis because of the grotesque swellings it inflicts, occurs throughout the tropics. It is the work of three species of nematode – *Brugia malayi*, *Brugia timori*, and especially *Wuchereria bancrofti*. Another related species – *Onchocerca volvulus* – causes a related condition called onchocerciasis. That worm spreads through the bites of blackflies not mosquitoes, and eschews lymph glands in favour of deeper tissues. There, the females, which can grow up to 80 centimetres long, entomb themselves in honeycombs of sturdy, fibrous flesh. They release larvae that migrate to the skin, where they cause unbearable itching; or the eye, where they can destroy the retina and optic nerve. That's why onchocerciasis goes by the simpler name of 'river blindness'.

These two diseases, collectively known as filariasis, are among the most widespread in the world: more than 150 million people have one or the other, and a further 1.5 billion are at risk.[1] Until recently, there were no cures. There were drugs that kept the symptoms under control by killing the larval nematodes, but they are useless against the astonishingly durable adults. And since these species can live for *decades* – an extraordinarily long existence for a nematode – the people who carry them must resign themselves to regular treatment. 'These are among the most debilitating of all tropical diseases,' says Mark Taylor, a sharp-suited and silver-haired parasitologist.

When Taylor started studying the filarial diseases in 1989, it was their severity that most intrigued him. There are many parasitic nematodes that infect humans, but they typically cause benign symptoms. Why do the ones behind filarial diseases inflict such incapacitating inflammation? It turns out that they have help – from a familiar ally. In the 1970s, researchers looked at these worms under a microscope and noticed bacteria-like structures inside them.[2] Then, the microbes were promptly forgotten until, in the 1990s, they were identified as *Wolbachia* – the same bacterium that has shunted its genome into Hawaiian fruit flies, that kills male blue moon butterflies, and that exists in two-thirds of the world's insect species.

Compared to its insect counterparts, the nematode version of *Wolbachia* is a shrunken degenerate. Having abandoned a third of its genome, it is permanently chained to its hosts. The reverse is also true. For reasons that are still unclear, the nematodes cannot complete their life cycles without their symbionts. They couldn't trigger intense disease, either. When the worms die, they release their *Wolbachia* into the people they infect. These bacteria can't infect human cells but they do trigger immune responses, of a different kind to those caused by the worm. According to Taylor, it's the combination of the two responses – against the worm *and* its symbiont – that leads to the intense symptoms of filariasis. Unfortunately, this means that killing the worms *worsens* the disease, because they release all their *Wolbachia* in their death throes. 'You get exploding nodules, and scrotal inflammation,' says Taylor, grimly. 'You don't want that. You want to kill the worms slowly, and it's hard to think how you'd do that with an anti-nematode drug.'

There is another option. Why not ignore the worms altogether? Why not go after the *Wolbachia?*

In lab tests, Taylor and others showed that removing the bacteria with antibiotics has fatal results for the worms. The larvae fail to mature. The existing adults stop reproducing. And after some time, their cells start to self-destruct. In this partnership, divorce is clearly not an option; if the bonds of symbiosis break, both partners die. The process is slow, taking up to 18 months, but a slow death is still a death. And since these worms have no *Wolbachia* to release, they can be slaughtered with impunity.

In the 1990s, Taylor and his colleagues took these ideas into the field. They wanted to see if they could use an antibiotic called doxycycline to eliminate *Wolbachia* from people with filariasis. One group tested the drug in Ghanaian villagers with river blindness, while another tried it on Tanzanians with lymphatic filariasis. Both trials were successful. In Ghana, doxycycline sterilised the female worms, and in Tanzania, it wiped out the larvae.[3] And at both sites, it killed the adult nematodes in around three-quarters of the volunteers, without triggering any catastrophic immune responses. That was huge. 'For the first time, we were able to cure people of filariasis,' says Taylor. 'We can't do that with standard drugs.'[4]

But doxycycline isn't quite a wonder drug. Pregnant women can't take it, nor can children. It also acts slowly, so people need to take several courses over many weeks; in rural, remote communities, it can be hard to get the drug to people over that whole period, much less persuade them to complete their course. As weapons go, doxycycline isn't a bad one. But Taylor thought he could do better.

In 2007, he set up an international team called A·WOL—the Anti-*Wolbachia* Consortium. With $23 million of funding from the Bill and Melinda Gates Foundation, their mission is to find new drugs that kill filarial nematodes by targeting their *Wolbachia* symbionts.[5] They have already screened thousands of potential chemicals and found one promising lead – minocycline. It proved 50 per cent more potent than doxycycline in lab tests, and the team immediately ushered it into trials in Ghana and Cameroon. Minocycline has its own problems: it is still inaccessible to kids and pregnant women, and is several times more expensive than doxycycline. But A·WOL has since screened another 60,000 compounds and identified dozens more promising candidates.

In the meantime, Taylor has found that the partnership between filarial nematodes and *Wolbachia* may be more precarious than it seems. He realised that when *Wolbachia* numbers start rising, at the times when they are supposedly most needed, the worms see the bacteria as invaders and try to destroy them.[6] 'The nematode sees *Wolbachia* as a pathogen,' he says. It needs the bacteria, but if they grow uncontrollably, they might burst their hosts like a kind of symbiotic tumour. So, the nematode must keep them in check. Even in this alliance, where either partner would die without the other, there is still conflict. And, in Taylor's eyes, there is opportunity. He has been searching for drugs that kill *Wolbachia*, when it turns out that the nematodes have already evolved ways of doing exactly that. If A·WOL can find chemicals that stimulate their symbiont-control programmes, they could turn the simmering tensions between host and symbiont into outright war, cajoling the nematodes into launching the means of their own destruction. The idea is ambitious and the stakes are high. If Taylor can break this symbiosis, which has been around for 100 million years, he could improve 150 million lives.

* * *

We have already seen how pliable the microbiome can be. It can change with a touch, with a meal, with a parasitic incursion or a dose of medicine, or simply with the passage of time. It is a dynamic entity that waxes and wanes, forms and re-forms. This flexibility underlies many of the interactions between microbes and their hosts. It means that symbioses can change in positive ways, as new microbial partners offer fresh genes, abilities, and evolutionary opportunities to their hosts. It means that partnerships can change in negative ways, as dysbiotic communities or missing microbes lead to disease. And it means that partnerships can change in *deliberate* ways – ways that we choose. Theodor Rosebury recognised this back in 1962. Our indigenous microbes are 'no less subject than the rest of our environment to manipulation for human benefit', he wrote. We should accept them as a natural part of our lives but acceptance 'need not be passive or resigned'.[7]

Fifty years on, passivity and resignation are nowhere to be seen. Today's microbiologists find themselves racing to rewrite the relationships between microbes and their animal hosts – from nematodes to mosquitoes to ourselves. Taylor is going for annulment: by depriving nematodes of their symbionts, he plans to doom them both and save those they plague. Other would-be microbiome manipulators are trying to introduce microbes to hosts in a bid to restore disrupted ecosystems or even forge new symbioses. They are developing cocktails of beneficial microbes that we can take to correct and forestall illnesses, packages of nutrients that will feed those microbes, and even ways of transplanting entire communities from one individual to another. This is what medicine looks like when you understand that microbes are not the enemies of animals, but the foundations upon which our kingdom is built. Say goodbye to dated and dangerous war metaphors, in which we are soldiers hell-bent on eradicating germs at whatever cost. Say hello to a gentler and more nuanced gardening metaphor. Yes, we still have to pull out the weeds, but we also seed and feed the species that bind the soil, freshen the air, and please the eye.

This concept can be hard to grasp, and not just because the idea of beneficial microbes is new to many. It's also counter-intuitive because so much of healthcare relies on basic arithmetic. Got scurvy? You are missing vitamin C, which you can add to your body via fruit. Got flu? You have a virus, which you need to remove from your airways by taking a drug. Add what's missing. Subtract what's unwanted. These simple equations still drive much of modern medical thought. By contrast, the maths of the microbiome are more complicated, because they involve large, changing networks of connected, interacting parts. To control a microbiome is to sculpt an entire world – which is as hard as it sounds. Remember that communities have a natural resilience: if you hit them, they bounce back. They are also unpredictable; if you tweak them, the consequences ripple outwards in capricious ways. Add a supposedly beneficial microbe, and it might displace competitors that we also rely on. Lose a supposedly harmful microbe, and an even worse opportunist might rise to take its place. This is why attempts at world-shaping have so far led to a few magnificent successes, but also many puzzling setbacks. In an earlier chapter we saw that fixing a microbiome is not as simple as removing 'bad bacteria' with antibiotics. In this one, we'll see that it isn't as simple as adding 'good bacteria' either.

The twenty-first century is a terrible time for frog-lovers. All over the world, these amphibians are hopping out of existence so quickly that even the most optimistic conservationists are furrowing their brows. A full third of amphibian species are at risk of extinction. Some of the reasons behind this decline are applicable to all wildlife: habitat loss, pollution, climate change. But amphibians are also plagued by a nemesis that's all their own: a doomsday fungus called *Batrachochytrium dendrobatis*, or Bd for short. It is a frog-killer par excellence. It thickens its victims' skin, stops them from absorbing salts like sodium and potassium, and triggers the equivalent of a heart attack. Since its discovery in the late 1990s, Bd has spread to six continents. It has shown up everywhere that amphibians exist. And everywhere it shows up, amphibians stop existing. The fungus can destroy entire populations in weeks, and has already consigned dozens of species to history. The sharp-snouted day frog is probably gone. The gastric brooding frog is no more. The Costa Rican golden toad has ribbited its last. Hundreds of others have been exposed. For good reason, Bd has been called 'the worst infectious disease ever recorded among vertebrates'.[8] Frogs, toads, salamanders, newts, caecilians: no group of amphibian is exempt. If a new fungus emerged that killed every mammal – every dog, dolphin, elephant, bat, and human – we would rightly panic. And biologists who work with amphibians are indeed panicking.

Bd is a harbinger of things to come. In 2013, scientists described a related fungus, *B. salamandrivorans*, which attacks salamanders and newts in Europe and North America. Since at least 2006, yet another fungus has been sweeping through North America's bats, causing a fatal disease called white nose syndrome and littering caves with millions of corpses. For decades now, corals have been hit by epidemic after epidemic.[9] These emerging infectious diseases of wildlife are emerging ever more quickly, and humans are at least partly to blame. On planes, boats, and boots, we carry pathogens around the world with unprecedented speed, overwhelming new hosts before they can acclimatise and adapt. The rise of Bd is a perfect example. Yes, it is virulent. Yes, it represses the immune systems of amphibians. But it's still just a fungus, and amphibians have been dealing with fungi for some 370 million years. This isn't their first rodeo. They are fumbling this particular ride because they have already been weakened by changing climate, introduced predators, and environmental pollutants. Add a destructive and quickly spreading disease into the mix and the future suddenly looks exponentially bleak.

But amphibian specialist Reid Harris has hope. Harris has discovered a possible way of protecting these animals from their fungal foes. In the early 2000s, he found that the red-backed and four-toed salamanders – two small, sinuous species from the eastern USA – are covered in a rich cocktail of antifungal chemicals.[10] These substances aren't made by the animals themselves but by the bacteria on their skin. They might help to protect the salamanders' eggs from fungi that would otherwise thrive in the humid underground nests. And as Harris later found, they can also stop Bd from growing. Perhaps, he thought, this explained why some lucky amphibian species seem to resist the killer fungus: their skin microbiomes act as symbiotic shields. And perhaps, he hoped, those microbes could help to save vulnerable species from the looming Amphibiageddon.

On the other side of the US, Vance Vredenburg was entertaining the same hope. He had been studying the mountain yellow-legged frogs of California's Sierra Nevada, and was becoming despondent as Bd hit the area. 'It was unbelievable,' he says. 'The fungus would go from not being there at all to wiping out an entire basin.' One after another, in quick succession, dozens of sites emptied of frogs. But not everywhere. In an alpine lake at Mount Conness, yellow-legged frogs were infected with Bd but still resolutely hopping around. Bd kills by overwhelming its hosts with tens of thousands of spores, but these frogs were carrying just a few dozen each. The supposedly lethal fungus that was filling other lakes with upturned cadavers was proving to be, at worst, a mild nuisance at Conness. At this site, and a few others, something was resisting Bd's advance. And when Vredenburg heard about Harris's experiments, he suddenly knew what. By swabbing the skins of the Conness frogs, he confirmed that they carried the same antifungal bacteria that Harris saw in his salamanders. One bacterial species stood out, both for its protective powers and for its colour: blackish-purple, ominous but darkly beautiful. It was called *Janthinobacterium lividum*. Everyone just refers to it as J-liv.[11]

In lab tests, Vredenburg and Harris confirmed that J-liv can indeed protect naïve frogs from Bd – but how? Does it kill the fungus directly by making antibiotics? Does it stimulate the frogs' own immune system? Does it reshape the frogs' native microbiome? Does it simply take up space in the skin, physically preventing the fungus from taking hold? And if it is so useful, why is it only found on some frogs and not others? And why is it relatively rare even when it is present? 'It would be great to figure out every little detail but we don't have time,' says Vredenburg. 'If we take time, the frogs won't be around any more. We're really working in a crisis.' Forget the details. What mattered was that the bacterium worked, at least in the cosy confines of a lab. Would it also work in the wild?

At the time, Bd was creeping fast across the Sierra Nevada, covering around 700 metres a year. By charting its advances, Vredenburg predicted that it would next hit Dusy Basin, a site some 11,000 feet above sea level, where thousands of yellow-legged frogs remained oblivious to the encroaching doom. It was the perfect place to put J-liv's powers to the test. In 2010, Vredenburg and his team hiked to Dusy Basin and grabbed every frog they could find. They found J-liv on the skin of one individual, and grew it into rich, thriving cultures. They then baptised some of the other captured individuals in this bacterial broth. The rest, they left in containers that just had pond water. After a few hours, they released all the frogs to fate and fungus.

'The results were phenomenal,' says Vredenburg. As predicted, Bd arrived that summer. The fungus took its usual toll on the frogs that had just been soaked in pond water – dozens of spores became thousands of spores, and each frog became an ex-frog. But in the animals that were dunked in J-liv, the fatal accumulation of spores not only plateaued early, it often reversed. A year later, around 39 per cent of them were still alive, while their peers were all dead. The trial had worked. The team had successfully protected a wild population of vulnerable frogs with a microbe. And they had established J-liv as a probiotic: a term that is most commonly linked to yoghurts and supplements, but really applies to any microbe that can be applied to a host to improve its health.

But conservationists can't very well catch and inoculate every amphibian that's threatened with Bd – that would be *all of them*. Instead, Harris is thinking about seeding soils with probiotics so that any passing frog or salamander would automatically dose itself. Alternatively, threatened frogs that are already being bred in captivity could be dosed in the lab before being released as a group. 'There is a lot of potential,' says Vredenburg, 'but this is not a silver bullet.

Like any complex problem, we can't expect it to be a winner all the time.' Indeed, Matthew Becker, one of Harris's former students, found that the same approach failed completely with captive Panamanian golden frogs. This species is a ghost in a bumblebee's colours: a gorgeous black-and-yellow creature that has already been exterminated from the wild by Bd. Today, it exists only in zoos and aquariums and cannot be reintroduced to Panama as long as Bd persists. J-liv, despite its initial promise, won't help with that.[12]

Perhaps that was predictable. We have seen how even closely related animals can carry very different microbiomes. There's no reason to suppose that a bacterium which colonises one species will thrive on another, or that there would ever be a universal probiotic that shields every amphibian. J-liv might live in salamanders and frogs throughout the USA, but it's not native to Panama and has no evolutionary history with the golden frog. With the acuity of hindsight, shoving an American microbe onto a Panamanian frog seems overly optimistic, not to mention a bit imperialist. Undaunted, Becker travelled to Panama to find a better probiotic. He surveyed the skin microbiomes of the golden frog's closest relatives, and found several indigenous species that stopped Bd from growing – at least in Petri dishes. Unfortunately, none of these native microbes would colonise the golden frogs either, and none of them bested the fungus in real conditions. There was one sign of hope: against all expectations, five of Becker's golden frogs were *naturally* resistant to Bd. Their skin microbes differed from those of the frogs that had died, and Becker is now trying to identify the protective bacteria within these communities. Harris is doing similar work in Madagascar, an amphibian Shangri-La that Bd has only just invaded. He is trying to find local bugs that can block Bd and persist on skins when artificially added. Becker and Harris aren't trying to create any new symbioses or introduce bacteria from one part of the world to another. 'We're just augmenting locally occurring bacteria,' Harris says.

Even if they identify good candidates, they still need to work out how to get these bacteria to stick on the frogs. A simple bath may not be enough. Timing might be important, since the transformation from tadpole to adult sweeps a frog's skin clean of microbes, like a fire burning through a forest. It creates a barren world that must be recolonised. This is the time when the animals are most at risk from Bd but it might also be the perfect moment to add probiotics. Perhaps these foreign microbes might more easily integrate into tumultuous, reassembling communities than into fixed, stable ones. Other subtleties probably matter, too. What about the microbes that already live on the skins of different amphibians: would they block or complement the incipient probiotics? What about the host's immune system: would it allow the boosted microbial populations to persist on the skin, or correct them towards a different state? The details, it turns out, *do* matter.[13] They could mean the difference between success or failure, preservation or extinction. And they matter in human guts just as much as in frog skins.

The word 'probiotic' means 'for life'. It is the opposite of antibiotics in both etymology and intention. Antibiotics are designed to remove microbes from our bodies, while probiotics represent a deliberate attempt to add them. In the early twentieth century, the Russian Elie Metchnikoff was one of the first scientists to champion this idea; for several decades he regularly drank sour milk in an effort to ingest lactic-acid-making bacteria, which he held responsible for extending the lives of Bulgarian peasants. But following his death, microbiologists Christian Herter and Arthur Isaac Kendall showed that the microbes that Metchnikoff idolised don't persist in the gut. Swallow them all you like – they won't hang around. Yet despite puncturing Metchnikoff's idea, Kendall defended the spirit of it. 'The time is coming when human intestinal lactic-acid bacteria will be used extensively in the correction of certain types of intestinal microbic disease,' he wrote. 'Science will discover and point out the conditions essential for success.'[14]

Science has certainly tried.[15] In the 1930s, Japanese microbiologist Minoru Shirota led the quest by looking for hardy microbes that could reach the gut without first being destroyed by the stomach's acids. He eventually homed in on a strain of *Lactobacillus casei*, grew it in fermented milk, and, in 1935, created the first bottle of the dairy drink called Yakult. Today, the company sells around 12 billion bottles a year, worldwide. Overall, the probiotics industry is a multi-billion-dollar business. Its products feed our stomachs along with our desire for 'natural' healthcare (even though many probiotics include proprietary microbes that have been altered and domesticated through generations of industrial culturing). In some products, the microbes grow in live cultures; in others, they are freeze-dried and packaged into capsules or sachets. Some include just one strain; others a mix. They are promoted as ways of improving digestion, boosting the immune system, and treating all kinds of disorders, digestive or otherwise.

Even the most concentrated probiotics contain just a few hundred billion bacteria per sachet. That sounds like a lot but the gut already holds at least a hundredfold more. Gulping down a yoghurt is like ingesting scarcity. Rarity, too: the bacteria in these products are not important members of the adult gut. They largely belong to the same category that Metchnikoff canonised – makers of lactic acid, like *Lactobacillus* and *Bifidobacterium*, which were chosen more for practical reasons than scientific ones. They are easy to grow, they are already found in fermented foods, and they can survive the trip through both a commercial packaging plant and a consumer's stomach. 'But most of them never arose in the human gut, and they don't have the factors that allow them to dwell for a long time there,' says Jeff Gordon. His team confirmed this by monitoring the gut microbiomes of volunteers who ate twice-daily servings of Activia yoghurt for seven weeks. The bacteria in the yoghurt neither colonised the volunteers' guts, nor changed the composition of their microbiomes. It's the same problem that Herter and Kendall identified in the 1920s, and that Matthew Becker and others saw when working on probiotics for frogs. They're like a breeze that blows through two open windows.[16]

Some would argue that this doesn't matter – the breeze can still rattle objects along its path. Gordon's team saw some signs of this:

the yoghurt they studied could nudge the microbes in mouse guts into activating genes for digesting carbohydrates, albeit temporarily. Wendy Garrett later found that a strain of *Lactococcus lactis* can help mice without sticking around – or even staying alive. When it enters a mouse's guts it bursts apart, and in its death it releases enzymes that can reduce inflammation. It might be a poor coloniser, but it can still do some good.

It can. But does it? The very word 'probiotics' is an answer of sorts. The World Health Organization defines them as 'live microorganisms which, when administered in adequate amounts, confer a health benefit on the host'. They are, *by definition*, healthy. There's a long caravan of studies that, at first glance, seem to support this claim. But many of these were done using isolated cells or laboratory animals, and their relevance to people isn't clear. Of the studies that involved actual humans, many included small numbers of volunteers, producing results that are prone to bias and statistical flukes.

Sifting through such research for strong and reliable work is a chore. Fortunately, the Cochrane Collaboration – a respected non-profit organisation that methodically reviews medical studies – has done exactly that. According to their verdicts, probiotics can shorten bouts of infectious diarrhoea, and reduce the risk of diarrhoea brought about by antibiotic treatments. They can also save lives from necrotising enterocolitis – a horrible gut disease that affects premature infants. And there ends the list. Compared to the hype, it's modest. There is still no clear evidence that probiotics help people with allergies, asthma, eczema, obesity, diabetes, more common types of IBD, autism, or any other disorders in which the microbiome has been implicated. And it's still not clear if the documented benefits happen *because of changes in the microbiome.*[17]

Regulatory agencies have taken note of these problems. For their purposes, probiotics are typically classed as foods rather than medicines. This means that manufacturers don't face the intimidating gauntlet of regulatory hurdles that pharmaceutical companies must leap over when developing a drug. But it also prevents them from saying that their products prevent or treat a specific disease – because that is what medicine does. If they cross that line, they face reprisals: in 2010, the US Federal Trade Commission sued Dannon (Danone in the UK) for claiming that Activia can 'relieve temporary irregularity', or help prevent colds and flu. This is why the language around probiotics tends to be nebulous to the point of meaninglessness, with brands talking about 'balancing the digestive system' or 'boosting immune defence'.

Even these claims have faced opposition. In 2007, the European Union instructed food and supplement companies to produce the scientific evidence behind the avalanche of exaggerated statements on their packaging. If they want to say that their products make people healthier, fitter, slimmer, they should be able to prove it. They tried – and did poorly. The EU's scientific advisory group rejected more than 90 per cent of the thousands of claims laid at their feet, including *all* the ones that dealt with probiotics. And since the very word implies a health benefit, the EU banned it from food packaging and advertising as of December 2014. Probiotics advocates argued that this dismissal ignored good science and had a chilling effect on the field, while sceptics felt that the EU was rightly forcing the industry to raise its game, and produce solid evidence for unsubstantiated claims.[18]

Yet despite the excessive hype, the *concept* behind probiotics is still sound.[19] Given all the important roles that bacteria play in our bodies, it *should* be possible to improve our health by swallowing or applying the right microbes. It's just that the strains in current use may not be the right ones. They make up just a tiny fraction of the microbes that live with us, and their abilities represent a thin slice of what the microbiome is fully capable of. We met more suitable microbes in earlier chapters. There's the mucus-loving bacterium, *Akkermansia muciniphila,* whose presence correlates with a lower risk of obesity and malnutrition. There's *Bacteroides fragilis*, which stokes the anti-inflammatory side of the immune system. There's *Faecalibacterium prausnitzii,* another anti-inflammatory bug, which is conspicuously rare in the guts of people with IBD, and whose arrival can reverse the symptoms of that disease in mice. These microbes could be part of the probiotics of the future. Their abilities are relevant and impressive. They are well adapted to our bodies. Some are already abundant – in healthy adults, one in every twenty gut bacteria is *F. prausnitzii*. These are not D-listers of the human microbiome like *Lactobacillus;* they are the stars of the gut. They won't be shy about colonising it.[20]

Then again, effective colonisation carries greater risk along with greater reward. So far, probiotics have enjoyed a mostly clean safety record,[21] but that might well be because they're poor at maintaining a foothold in the body. What would happen if more common gut residents are used? We know from animal studies that a dose of microbes in early life can have long-lasting effects on an individual's physiology, immune system, and even behaviour. And as we have seen, no microbe is inherently good; many species, including long-standing parts of the human microbiome like *H. pylori*, can play both positive and negative roles. *Akkermansia* has been touted as a saviour in many studies, but it also seems to be more common in cases of colorectal cancer. These are not products to be used lightly, without a more thorough understanding of how they change the microbiome, and the long-term consequences of those changes. As with the frogs, the details matter.

Amid the mixed news about probiotics, there have also been success stories. The most compelling of these began in Australia during the 1950s. At the time, the country's national science agency started looking for tropical plants that could feed its growing cattle population. One candidate seemed particularly promising: a Central American shrub called *Leucaena*. It grew easily, tolerated a lot of grazing, and was packed with protein. Unfortunately, it was also packed with mimosine, a toxin whose by-products cause goitre, hair loss, stunted growth and, occasionally, death. Scientists tried to breed these poisons out of *Leucaena* to no avail. The perfect plant had a fatal flaw. Then, in 1976, a government scientist named Raymond Jones stumbled upon a solution. While attending a conference in Hawaii, he noticed a pen full of goats that were eating lots of *Leucaena* with no apparent problems.

He suspected that the goats carried mimosine-detoxifying microbes in the first chamber of their stomachs – the rumen.

After several long flights, some involving thermos flasks full of rank rumen fluids and others involving live goats, Jones finally proved his hypothesis. In the mid-1980s, he introduced rumen bacteria from tolerant goats into vulnerable Australian livestock, and found that the recipients could dine on *Leucaena* with impunity. With the foreign microbes in their stomachs, animals that would once have become deathly ill on *Leucaena* could eat so much of the nutritious shrub that they put on weight at a record-breaking pace. Jones had done what bean bugs do when they swallow insecticide-busting bacteria from the environment, or what desert woodrats do when they acquire creosote-defusing microbes from each other: he had equipped animals with new microbes that neutralised a chemical threat. His colleagues eventually identified the specific mimosine-degrading bacterium from the Hawaiian goats, and named it *Synergistes jonesii* in his honour. As of 1996, farmers have been able to buy the bug as a 'probiotic drench': an industrially concocted cocktail of microbe-rich rumen fluids that they spray onto their herds. By allowing farmers to feed *Leucaena* to their animals without concern, this probiotic has transformed northern Australia's farming industry.[22].

Why was Jones triumphant when other manipulators have met with such frustration? You could argue that he was trying to fix a simple problem. He wasn't trying to cure IBD or stop a killer fungus. He just needed to detoxify one chemical. Chances were good that a single microbe could do the job. But even then, success isn't guaranteed.

Take the case of oxalate. It's found in beetroot, asparagus, and rhubarb, among other foods. At high concentrations, it stops your body from absorbing calcium, which congeals into a hard lump. That's one way in which kidney stones form. We can't digest oxalate; only microbes can do that. One species – a gut bacterium called *Oxalobacter formigenes* – is so good at it that it uses oxalate as its one and only source of energy. At first glance, this situation looks identical to the *Leucaena* dilemma. There's one chemical (oxalate), which causes a defined problem (kidney stones), and can be destroyed by one microbe (*Oxalobacter*). The solution, surely, is to swallow an *Oxalobacter* probiotic if you are prone to kidney stones. Unfortunately, such probiotics exist and they aren't very effective.[23] Why?

There are two likely answers, and both provide valuable lessons. First, it is not enough to infuse an animal with bacteria and hope for the best. Microbes are living things. They need to eat. *Oxalobacter* eats nothing but oxalate, and people with kidney stones often go on an oxalate-free diet. They can ingest the bacterium, but it will instantly starve.[24] Conversely, farmers are told to feed their herds on *Leucaena* for at least a week before drenching them in *Synergistes*. That way, the inoculated bacteria have enough food to digest.

Substances that selectively nourish beneficial microbes are called prebiotics – a term that could include oxalate or *Leucaena* but normally describes plant carbohydrates like inulin, which are purified and packaged as supplements.[25] These substances can increase the numbers of important microbes like *F. prausnitzii* or *Akkermansia*, and perhaps lower appetite and reduce inflammation. Whether they need to be taken as supplements is another matter. We have already seen that what we eat can substantially change the microbes in our gut, and prebiotics like inulin are in plentiful supply in onions, garlic, artichokes, chicory, bananas and other foods.

HMOs, those microbe-feeding sugars in breast milk, count as prebiotics too, in that they nourish *B. infantis* and other specialist microbes. Paediatrician Mark Underwood thinks they could help to save the lives of some of the most vulnerable people alive: premature babies. Underwood heads a neonatal intensive care unit at the University of California, Davis, where his team cares for up to 48 premature babies at any one time. The youngest are born at just 23 weeks; the lightest weigh just over a pound. They're usually delivered through C-sections, put on courses of antibiotics, and stuck in a supremely sanitised environment. Bereft of the usual pioneering microbes, they grow up with a very strange microbiome: low on the usual Bifs and high in opportunistic pathogens that grow in their place. They are the epitome of dysbiosis, and their strange internal communities put them at risk of the often fatal gut condition, necrotising enterocolitis, or NEC. Many doctors have tried to prevent NEC by giving probiotics to premature babies, with some success. But Underwood, after talking to people like Bruce German and David Mills, thinks that he can do better by infusing the infants with a combination of *B. infantis* and breast milk. 'The food you feed these bugs is as important as the bugs themselves, in getting them to grow and colonise a fairly hostile environment,' he says. He has already run a small pilot study, in which he showed that *B. infantis* does indeed colonise premature babies more effectively when its favourite food is on the menu.[26] He is now running a larger clinical trial to see if the *B. infantis* probiotic, when combined with milk prebiotics, could help to prevent NEC.

The second lesson from the *Synergistes* and *Oxalobacter* stories is that teamwork matters. No bacterium exists in a vacuum. Different species often form complex networks that feed and support each other in co-dependent ways. Even when it looks as if a single microbe can fix a problem, it might need a supporting cast just to stay alive. Maybe that's why the *Synergistes* probiotic works so well – it includes a lot of other stomach microbes. And maybe that's why the *Oxalobacter* probiotic *doesn't* work – it has no playmates. The same applies to other microbes. You could envision a *F. prausnitzii* sachet that will cure IBD or an *Akkermansia* pill that will make you skinnier, but I wouldn't hold my breath.

So, perhaps a smarter approach to making probiotics is to create a *community* of microbes that work well together. In 2013, Japanese scientist Kenya Honda found 17 *Clostridia* strains that can reduce inflammation in the gut; based on his work, Boston company Vedanta BioSciences has now created a multi-microbe cocktail for treating IBD.[27] As this book goes to print, the company should start putting its new probiotic to the test in clinical trials. Will it work? Who knows? But it certainly makes more sense to adjust a microbiome with a network of cooperating microbes than with any solitary strain. After all, the most successful method for manipulating the microbiome does exactly that.

* * *

In 2008, Alexander Khoruts, a gastroenterologist at the University of Minnesota, met a 61-year-old woman whom I will call Rebecca. For the previous eight months, she had suffered from relentless bouts of diarrhoea that left her dependent on adult nappies, stuck in a wheelchair, and four stone lighter. The culprit was *Clostridium difficile* – the bacterium that's informally known as C-diff. It is infamous for its staying power, often succumbing to antibiotics before rebounding in a newly resistant form. So it was with Rebecca: her doctors tried her on drug after drug, none of which worked. 'She was in desperate shape,' Khoruts recalls. She had exhausted all her options.

All but one. Casting his brain back to his days at medical school, Khoruts remembered learning about a technique called a faecal microbiota transplant, or FMT. It is exactly what it sounds like: doctors take stool from a donor and install it in a patient's guts, microbes and all. And apparently, that could cure C-diff infections. The idea seemed revolting, weird, and implausible. But Rebecca had no qualms. She just wanted – needed – to get better. She agreed to the procedure. Her husband donated a stool sample, which Khoruts pulverised in a blender. He then delivered a cupful of the slurry into Rebecca via a colonoscopy.

Within a day, her diarrhoea had stopped. Within a month, the C-diff had vanished. This time, it did not rebound. She had been cured – thoroughly, quickly, and enduringly.

Rebecca's case, though anecdotal, is also archetypal. The same leitmotifs appear in hundreds of similar stories involving FMT: a patient with untreatable C-diff; a desperate doctor; and a miraculous recovery. In some cases, the physicians hear about the procedure from their patients.[28] That was the case for Elaine Petrof from Queen's University in Kingston, Ontario. In 2009, she was unsuccessfully treating a woman for C-diff when her family members started repeatedly showing up with a small bucket of poo. 'I thought they were nuts,' she recalls. 'But after watching this woman deteriorate and being helpless to do anything, I thought: What do we have to lose? We did it and, lo and behold, it worked. She went from death's door to walking out of the hospital, looking great and basically cured.'

Faecal transplants are certainly gross, both conceptually and practically; someone, after all, has to use that blender.[29] But 'patients don't care about the ick factor,' says Petrof. 'They're willing to try anything. They'll often cut me off and go: Where do I sign?' Indeed, we humans are unusual in our aversion to faeces. Many other animals practise coprophagy, and will gamely swallow each other's dung and droppings to acquire microbes. In this way, bumblebees and termites spread bacteria that act as a colony-wide immune system to defend against parasites and pathogens.[30] FMT offers similar benefits in a more palatable way – since it involves no palates. Instead, bacteria can be administered via colonoscopy, enema, or a tube threaded through the nose and into the stomach or intestine.

The procedure works along the same principles as a probiotic, but rather than adding just one strain of bacteria, or even 17, it adds *all of them*. It's an *ecosystem transplant* – an attempt to fix a faltering community by completely replacing it, like returfing a lawn that's overrun by dandelions. Khoruts showed this process at work by collecting stool samples from Rebecca before and after her transplant.[31] Beforehand, her gut was a mess. The C-diff infection had completely restructured her microbiome, creating a community that 'looked like something that doesn't exist in nature – a different galaxy', says Khoruts. Afterwards, her microbiome was indistinguishable from her husband's. His microbes had stormed into her dysbiotic gut and reset it. It was almost as if Khoruts had done an organ transplant, throwing out his patient's diseased and damaged gut microbiome and replacing it with the donor's shiny new one. This makes the microbiome the only organ that can be replaced without surgery.

Faecal transplants have been taking place on and off for at least 1,700 years. The earliest record comes from a handbook of emergency medicine written in fourth-century China.[32] Europeans took much longer to cotton on: in 1697, a German physician recommended the technique in a book with the unparalleled name of *Heilsame*

Dreck-Apotheke – Salutary Filth-Pharmacy. It was rediscovered by an American surgeon named Ben Eiseman in 1958, but just one year later was overtaken by the introduction of vancomycin, an antibiotic that worked well against C-diff. As Khoruts once wrote, FMT 'retreated to merely anecdotal use, sporadically reported and providing amusing reading for a number of decades'. But it was never entirely forgotten. In the last decade, intrepid doctors have started using it, reticent hospitals started offering it, and success stories have accumulated.

This momentum came to a head in 2013 when a Dutch team led by Josbert Keller finally tested FMT in a randomised clinical trial – medicine's gold standard for sorting genuine treatments from quackery.[33] Keller's team recruited patients with recurring C-diff infections and randomly assigned them to receive either vancomycin or an FMT. They originally planned to recruit 120 participants, but they only made it to 42. By that point, vancomycin had cured just 27 per cent of the people who received it, while FMT had cured 94 per cent. The stools were patently so much better that the hospital deemed it unethical to continue giving people the antibiotic. They cut the trial short; from then on, *everyone* received FMT.

In medicine, a cure rate of 94 per cent among very sick patients, with no major side effects, is unheard of. Better still, FMT is incredibly cost-effective: vancomycin is expensive, but stools are free. In the eyes of many sceptics, the trial was enough to transform the procedure from a kooky alternative treatment to an impressive mainstream one, and from a desperate last resort to a front-line option. There's a popular saying among doctors: there's no such thing as alternative medicine; if it works, it's just called medicine. The growing acceptance of FMT among mainstream medics epitomises this idea. Khoruts has now used it to cure hundreds of people with C-diff. So has Petrof. There have been thousands of similar reports from around the world.

These successes emboldened doctors to try FMT on patients with other conditions. If it worked so well against C-diff, might it not also treat IBD, resetting that agitated ecosystem to a calmer state? Not easily, it seems. For IBD, success rates are lower and more inconsistent, while side effects and recurrences are more common.[34] What about other conditions? Could stool from a lean person help a fat person to lose weight? Again, the jury's out. Some doctors have reportedly used FMT to treat obesity, irritable bowel syndrome, autoimmune diseases, mental health problems, and even autism – but these anecdotes don't reveal if the patients recovered because of the infusion, or because of natural remission, lifestyle changes, the placebo effect, or something else. The only way to sort anecdotal myths from medical realities is through clinical trials, and several dozen are now under way. For example, the same team behind the Dutch C-diff trial also randomly assigned 18 obese volunteers to get an infusion of either their own gut microbes or those of a lean donor. The group that received the lean microbes became more sensitive to insulin – a sign of good metabolic health – but they didn't lose any weight.[35] Even through FMT, it is not easy to reset a microbial ecosystem.

C-diff is the exception that proves the rule.[36] People get it after taking antibiotics, and they typically control it by taking even *more* antibiotics. This pharmacological carpet-bombing clears many of the native bacteria from their guts. When a donor's microbes arrive in this wasteland, they find few competitors, and certainly few that are as well adapted to the gut as they are. They can easily colonise. If you wanted to design a disease that could be easily treated by FMT, you'd create something like a C-diff infection. You wouldn't create something like IBD, where a donor's bacteria would have to colonise a hostile, inflamed environment that's already full of indigenous, well-adapted microbes. To give these transplanted communities an edge, Khoruts wonders if doctors need to condition guts with antibiotics, to wipe the slate clean. Alternatively, they could put recipients on a prebiotic diet that helps the new microbes settle in. Either way, 'you can't just infuse microbes into people and expect a transplant to happen', says Khoruts. 'I think a lot of people thought that FMT is some magic bullet that could address their particular problem, without realising the complexities.'

Even for C-diff, FMT isn't straightforward. Stools must be rigorously screened for pathogens like hepatitis or HIV. Some doctors will also reject donors who have any kind of microbiome-related condition, including allergies, autoimmune diseases, or obesity. This time-consuming process rules out so many people that donors can be hard to find, and some practices have taken to freezing stool samples from anyone who passes muster.[37] The non-profit organisation OpenBiome runs one such stool bank. If potential donors pass a battery of screening tests, their poo is then filtered, piped into capsules, frozen, and delivered to hospitals in need.[38] Khoruts runs a similar service in Minnesota. In 2011, when his ur-patient Rebecca returned with a new case of C-diff, Khoruts cured her using a frozen stool sample. In 2014, she returned again, and this time Khoruts carried out an FMT by giving her a capsule to swallow. 'She was a pioneering patient more than once,' he says.

The act of swallowing a capsule of frozen poo speaks to the bizarre nature of FMT. Here is something that looks like a normal pill but that consists of a largely uncharacterised product, which comes from the backsides of volunteers rather than the conveyor belts of factories, and is different every single time. Unnerved by this variability, the US Food and Drug Administration decided in May 2013 to regulate stools as a drug – a move that forced doctors to fill out an extensive application before performing an FMT. Patients and physicians complained, saying that the lengthy process would stop people from getting timely care.[39] Six weeks later, the FDA waived the process for C-diff cases, but left it in place for other conditions. Some researchers find these regulatory teething problems unnecessary and frustrating. Others think they provide a valuable breather. Interest in FMTs has risen exponentially in recent years, and there is mounting pressure to try the technique for all kinds of condition.

The problem is that no one understands its long-term risks.[40] Animal experiments have clearly shown that transplanted microbiomes can make recipients more likely to develop obesity, IBD, diabetes, psychiatric problems, heart disease, or even cancer, and we still can't accurately predict if any particular microbial community will confer these risks. Such concerns might not matter to a 70-year-old patient with C-diff, who wants to be cured *right now*. But what about young adults in their twenties – a demographic in which C-diff is increasingly common? What about children? Emma Allen-Vercoe tells me that she has heard from both doctors and parents who have tried FMT on autistic children. 'It scares the hell out of me,' she says. 'This is adult poo, and a paediatric population. What if you're basically setting someone up for something as awful as colorectal cancer later on in life? I just think this is dangerous.'

FMTs are so simple that anyone can do them at home – and many have. Inspirational and instructional videos have appeared online, as have large communities of DIY-transplanters.[41] To be sure, these resources have helped many people with genuine needs who were turned away by dismissive doctors. But the ease of such transplants has also allowed misinformed people to act on their misinformation.[42] And outside a lab, where it's impossible to screen donors for pathogens, several people have presented with severe infections after doing their own transplants. 'It's the Wild West,' says Allen-Vercoe. 'Anyone is using anyone's stool.' Mindful of these problems, a group of leaders in the microbiome field recently urged researchers to formalise the technique, collect systematic data on both donors and recipients, and create a system for reporting unexpected side effects.[43]

Petrof agrees. 'I think everyone recognises that stool is a stopgap,' she says. 'We should ultimately go to defined mixtures.' By that, she means creating a specific community of microbes that duplicates the benefits of a donor's stool. FMT but without the F. A stool substitute. A sham-poo. Together with Allen-Vercoe, Petrof found the healthiest donor she could – a 41-year-old woman who had never taken antibiotics. The team cultured the woman's gut bacteria and removed any that even showed hints of virulence, toxicity, or antibiotic resistance. That left a community of 33 strains that, in a fit of whimsy, Petrof called RePOOPulate. When she tested the mixture on two patients with C-diff, both recovered within days.[44]

That was just a small pilot study, but Petrof is convinced that RePOOPulate represents the future of FMT; some commercial companies are also developing their own blends of transplantable microbes. You could see these mixtures as either a pared-down FMT or a souped-up probiotic. They all consist of well-defined strains that can be cooked up, again and again, according to the same standardised recipe. And surely, argues Petrof, that's better than the poorly characterised and highly variable communities in actual stools.[45] Implanting so many unknowns into a patient's guts is a gamble. By contrast, RePOOPulate is an exercise in precision. Still, these synthetic communities face the same problem that probiotics do: no single set of bacteria will treat all ills, or even all people with one particular ill. 'We don't think it's good to have one ecosystem for all. You wouldn't put a V8 engine in a Mini because you'd probably kill someone,' says Allen-Vercoe. Ideally, there would eventually be a series of RePOOPulates, perhaps tailored to different diseases. These are not one-size-fits-all solutions. They will need to be personalised.

For hundreds of years, doctors have used digoxin to treat people whose hearts are failing. The drug – a modified version of a chemical from foxglove plants – makes the heart beat more strongly, slowly, and regularly. Or, at least, that's what it usually does. In one patient out of every ten, digoxin doesn't work. Its downfall is a gut bacterium called *Eggerthella lenta*, which converts the drug into an inactive and medically useless form. Only some strains of *E. lenta* do this. In 2013, Peter Turnbaugh showed that just two of the bacterium's genes distinguish the problematic drug-deactivating strains from the neutral ones.[46] He thinks that doctors might be able to use the presence of these genes to guide their treatments. If they are absent from a patient's microbiome, fine – give them digoxin. If they are there, the patient needs to eat a lot of protein, since that seems to stop the genes from decommissioning the drug.

And that's just one drug. The microbiome affects many others.[47] Ipilimumab, one of the hottest new cancer drugs around, works by stimulating the immune system to attack tumours – but only if gut microbes are around. Sulfasalazine, which is used to treat rheumatoid arthritis and IBD, only works when gut microbes convert it into its active state. Irinotecan is used to treat colon cancer but some bacteria change it into a more toxic form, which has severe side effects. Even paracetamol (acetaminophen), one of the most familiar drugs in the world, is more effective in some people than others because of the microbes they carry. Again and again, we see that variations in our microbiome can drastically alter the effectiveness of our medicines – even drugs that consist of a single, well-characterised, inanimate chemical. Imagine then what happens when we take a probiotic or a faecal transplant, which consists of complex, poorly understood, and constantly evolving organisms. These are *living* drugs. Their odds of working or failing will depend on a patient's existing microbiome, which will itself vary with age, geography, diet, gender, genes, and other factors we still don't fully understand. These contextual effects have reared their head in studies of flies, fish, and mice; it would be foolish to think that they wouldn't apply to people.[48]

What we need, then, are *personalised* infusions. We cannot expect the same probiotic strains, or the same donor stools, to treat a variety of diseases. A better approach would be to customise probiotics according to the ecological vacancies in an individual's body, the quirks of their immune system, or the diseases that they are genetically vulnerable to.[49]

Doctors will also need to treat both the patient *and* their microbes at the same time. If someone with IBD took an anti-inflammatory drug, her microbiome might just send her back to the same inflamed state. If she opted for probiotics or an FMT, the new bugs might not survive her inflamed intestines. If she ate a high-fibre prebiotic diet, and she lacked fibre-digesting microbes in the first place, her symptoms might just get worse. Piecemeal solutions won't work. You won't fix a bleached coral reef or a bare meadow just by adding the right animals or plants; you might also need to remove invasive species, or control the influx of nutrients. So it is with our bodies. The *entire ecosystem* – host, microbes, nutrients, everything – must be manipulated through a multi-pronged approach.

Here is what that might look like. If people have high cholesterol levels, doctors might prescribe drugs called statins, which block a human enzyme that's involved in creating cholesterol. But Stanley Hazen has shown that gut bacteria make good targets, too. Some of them can transform nutrients like choline and carnitine into a chemical called TMAO, which slows the breakdown of cholesterol.[50] As TMAO levels build, so do fatty deposits in our arteries, leading to atherosclerosis – a hardening of arterial walls – and other heart problems. Hazen's team have now found a chemical that can stop this process by preventing the bacteria from making TMAO – all without harming them. Perhaps this chemical, or something like it, will sit alongside statins in tomorrow's medicine cabinet: two complementary drugs, one that targets the human half of the symbiosis, and another that nudges the microbial half.

And that's just a sliver of the full potential of microbiome medicine. Imagine we're ten, twenty, maybe thirty years into the future. You see a doctor. You've been feeling anxious, so she prescribes a bacterium that's been shown to affect the nervous system and repress anxiety. Your cholesterol is a little high, so she adds another microbe that makes and secretes a cholesterol-lowering chemical. The levels of secondary bile acids in your gut are unusually low, leaving you vulnerable to a C-diff infection – best to include a strain that produces these acids. Your urine contains molecules that are signs of inflammation, and since you also have a genetic predisposition to IBD, she adds a bug that releases anti-inflammatory molecules. She chooses these species not just for what they can do, but because she predicts that they will interact well with your immune system and your existing microbiome. She also adds a supporting cast of other bacteria chosen to prop up the therapeutic core, and she suggests some dietary plans that will nourish them effectively. And off you go, with a bespoke probiotic pill – a treatment designed to improve not just any old microbial ecosystem but *yours* in particular. As microbiologist Patrice Cani told me, 'The future will be à la carte.'

And in this à la carte future, we won't have to stop at picking the right bacteria for the job. Some scientists are picking the right *genes* for the job, and combining them into artisanal *bacteria*. Rather than just recruiting species with the right abilities, they are tinkering with the microbes themselves to endow them with new skills.[51]

In 2014, Pamela Silver from Harvard Medical School outfitted *E. coli*, the most thoroughly characterised of microbes, with a genetic switch that can sense an antibiotic called tetracycline.[52] In the presence of the drug, the switch flips, and, under the right conditions, activates a gene that turns the bacteria blue. When Silver fed these engineered bacteria to lab mice, she could tell if the rodents had taken a dose of tetracycline by collecting their droppings, growing the microbes inside them, and checking their colour. She had effectively turned *E. coli* into a tiny journalist that could sense, remember, and report on the goings-on in the gut.

We need such reporters because the gut is still a black box. It's a 28-foot-long organ and the most common way of studying it is to analyse what comes out at the end. That's a bit like characterising a river by sticking a sieve at its mouth. Colonoscopies offer a more detailed view, but they are invasive. So instead of pushing a tube up one end, why not send bacteria like Silver's *E. coli* down the other? When they emerge, they could fill us in on whatever they encountered during their travels. Forget tetracycline: that was just a proof-of-principle. Silver wants to program microbes to sense toxins, drugs, pathogens, or telltale chemicals that reflect the early stages of disease.

Her ultimate vision is to engineer bacteria that can detect problems in the body – and fix them. Imagine a strain of *E. coli* that senses the signature molecules produced by *Salmonella*, and reacts by releasing antibiotics that specifically kill this microbe. Now, in addition to being a mere reporter, it's also a park ranger. It could prevent food poisoning by patrolling the gut, staying inert if it sees no threat, and leaping into action if *Salmonella* appears. You could give it to children in poor countries, who are at risk of diarrhoeal diseases. You could give it to soldiers who deploy overseas. You could pass it around communities that are in the midst of an epidemic.

Other scientists are building their own microbial minions. Matthew Wook Chang has programmed *E. coli* to find and destroy *Pseudomonas aeruginosa*, an opportunistic bacterium that infects people with weak immune systems. When the engineered bacteria sense their prey, they swim towards them and release two weapons: an enzyme that breaks *P. aeruginosa* communities apart, and an antibiotic that specifically assaults the vulnerable fragments. Jim Collins for MIT is also programming gut bacteria to destroy pathogens. His hunter-killer microbes target *Shigella*, which causes dysentery, and *Vibrio cholerae*, which causes cholera.[53]

Silver, Chang, and Collins are practitioners of synthetic biology, a young discipline that applies the mindset of an engineer to the world of flesh and cells. Their lingo is clinical and detached: they treat genes as 'parts' or 'bricks' that can be assembled into 'modules' or 'circuits'. But their ethos is vibrant and creative: science writer Adam Rutherford compares them to hip-hop DJs in the 1970s, who ushered in a new movement of music by remixing existing riffs and beats into thrilling new combinations.[54] In similar fashion, synthetic biologists are remixing genes to usher in a new generation of probiotics.

'Applying these principles to a bacterium gives you a lot more flexibility,' says fibre specialist Justin Sonnenburg. A naturally occurring bacterium might be great at fermenting fibre, or talking to the immune system, or making neurotransmitters, but it's unlikely to excel at everything. For every new desirable quality, scientists will have to screen for new bugs. Or, they could simply load the circuits they want into a single synthetic microbe. 'The hope is to have a parts list, and that this will become a plug and play system where the results will be predictable,' says Sonnenburg.

Synthetic biologists aren't limited to sending microbes after pathogens. They might also train their creations to eliminate cancer cells or to convert toxins into medicines. Some are trying to supercharge our microbiome's natural ability to make antibiotics that control other microbes, or immune molecules that quell chronic inflammation, or neurotransmitters that affect our moods, or signalling molecules that influence our appetite. If that sounds like meddling with nature, remember that we already do everything on that list in much cruder ways, by swallowing pills like aspirin or Prozac. When we do, our bodies are flooded with fixed doses of the drugs. By contrast, synthetic biologists could program a bacterium to make the same drugs at the exact site of a problem and in the appropriate dose. These microbes can practise medicine with millimetre precision and millilitre finesse.[55]

At least, they could in theory. 'It's easy to have the circuits work on the whiteboard in your office,' says Collins. 'But biology is very messy and noisy. Engineering it isn't as easy as it is sometimes presented. The challenge is to get the circuits to function the way you'd like in the stressful environment of a host.' For example, it takes energy to switch on a gene, so a synthetic bacterium that's packed with complex circuits may be unable to compete against natural counterparts with leaner, lither genomes.

One solution, which Sonnenburg favours, to make engineered bacteria more competitive, is to stuff the synthetic genetic circuits into a common gut resident like B-theta, instead of the more familiar *E. coli*. The latter is easier to manipulate but is also a poor intestinal coloniser. B-theta, however, is exquisitely attuned to the gut and lives there in high numbers.[56] What better candidate for the job of human ecosystem ranger? Jim Collins is more circumspect. Given how much we still don't understand about the microbiome, he is unsettled by the prospect of engineering microbes that can permanently establish themselves in our bodies. That's why he is also focusing on building kill-switches that will force the microbes to self-destruct if something goes wrong, or if they leave their hosts. (Containment is a big issue for these bacteria, since they could potentially enter the environment every time someone flushes a toilet.) Silver is also working hard on safety measures. By tweaking the genetic code of her synthetic microbes, she hopes to erect a biological firewall that will stop them from exchanging DNA horizontally with their wild counterparts, as bacteria are wont to do. She also wants to create synthetic *communities* of microbes – teams of, say, five species that all depend on each other, so that if any one of them dies, the others will follow.

Whether these features will satisfy regulatory agencies or consumers is unclear.[57] Genetically modified organisms are always controversial and if probiotics and faecal transplants have told us anything, it's that the world doesn't know how to deal with this wave of living drugs. Synthetic biology will only increase that tension. Still, it is worth noting that none of these programmed bacteria are truly 'synthetic'. They have extraordinary skills and they contain genes that have been wired up in new combinations, but at heart, they're still *E. coli*, B-theta, and other familiar faces that we have lived with for millions of years. They're the same old symbionts with a modern twist.

What is arguably even more impressive is creating an entirely new symbiosis – uniting animals and microbes that have never before encountered each other. One team of scientists has spent over two decades doing exactly that. And the products of their quest can already be found buzzing through the skies of eastern Australia.

On 4 January 2011, in the first hours of a crisp Australian morning, Scott O'Neill walks up to a yellow bungalow in a suburb of Cairns.[58] He sports glasses, a goatee, jeans, and an off-white shirt, with 'Eliminate Dengue' written over the breast pocket. That's both the name of the organisation that O'Neill created and its goal: eliminate dengue fever, from Cairns, from Australia, and perhaps eventually from the world. The tools with which he will accomplish this feat are sitting in the small plastic cup that he now holds in his hand. He carries it towards the house, past a fence, down a flower-lined patio, and up to a large palm tree. His pace is deliberate and a little self-conscious. This is a big moment, and around 20 people are watching, filming, joking. O'Neill stops and looks up. 'Are you ready?' he says. The crowd cheers. They have been waiting for this for a long time. O'Neill pulls the lid off the cup, and a few dozen mosquitoes fly out into the morning air. 'Go, babies, go!' says an onlooker.

These mosquitoes are *Aedes aegypti*, a black-and-white species that transmits the virus which causes dengue fever. Through its bites, it infects as many as 400 million people every year. O'Neill has never had dengue himself, but he has seen others go through it. He knows about the fevers, headaches, rashes, and the severe joint and muscle pains. He knows that there is no vaccine or effective treatment. The only real way of controlling dengue is through prevention. We can kill *Aedes* mosquitoes with insecticides. We can stop them from biting, using repellents or nets. We can remove the open, stagnant water in which the insects breed. But despite these strategies, dengue fever is still common, and increasingly so. A new solution is needed – and O'Neill has one. His plan, unorthodox though it sounds, is to beat the disease by releasing *even more* of the *Aedes* mosquitoes that carry it. But his insects are different from their wild counterparts. They have been loaded with a bacterium that you will already be familiar with – the ubersymbiont *Wolbachia*.[59]

O'Neill found that *Wolbachia* stops *Aedes* mosquitoes from carrying dengue viruses, turning them from vectors into dead ends. Of course, it would be impossible to collect every wild mosquito and shoot them up with a symbiont, but O'Neill doesn't have to. He just has to release a few *Wolbachia*-carrying insects into the wild, and wait. Remember that this bacterium is a master manipulator, with many tricks for spreading through a population of insects. The most common is cytoplasmic incompatibility, in which infected females, which pass the microbe to the next generation, are more likely to lay viable eggs than their uninfected peers. This advantage means that *Wolbachia* can quickly spread through an area – and its ascendancy implies dengue's downfall. O'Neill's plan is to release enough *Wolbachia*-laden mosquitoes into the wild to create an entirely dengue-resistant population. The ones he set loose in Cairns were the first. This was the culmination of decades of obsessive hard work and hair-pulling frustration. 'It seems like my whole life,' says O'Neill.

His quest to turn *Wolbachia* into a dengue fighter began in the 1980s, meandered through several wasted years, and hit many a dead end. It only started bearing fruit in 1997, when he read about an unusually virulent strain of *Wolbachia* that infects fruit flies. This strain, known as 'popcorn', would reproduce like mad in the muscles, eyes, and brains of adults, filling a fly's neurons so thoroughly that they become 'akin to a bag full of popcorn' – hence the name. These infections are so severe that they can halve a fly's lifespan. 'That was a light-bulb moment for me,' says O'Neill. He knew that dengue viruses take time to reproduce in mosquitoes, and even more time to reach the salivary glands where they can jump into a new host. This means that only old mosquitoes can transmit dengue. If O'Neill could halve the insects' lives, they would die before they got a chance to spread the virus. All he needed to do was to get popcorn into *Aedes*.

Wolbachia infects many mosquitoes – remember that it was originally discovered in a *Culex* before anyone realised how omnipresent it is. But as luck would have it, it doesn't touch either of the two groups that cause the most human suffering: *Anopheles*, which carries malaria, or *Aedes*, which spreads Chikungunya, yellow fever, and dengue. O'Neill was going to have to play matchmaker and create a new symbiosis from scratch. He couldn't just inject adults with *Wolbachia*, though; he needed to inject an egg, so that every part of the resulting insect would carry the microbe. He and his team would look down a microscope and, ever so delicately, try to lightly puncture a mosquito egg with a needle bearing *Wolbachia*. They did this hundreds of thousands of times, over many years. It never worked. 'I burnt the careers of all these students and I was so frustrated that I was ready to walk away,' says O'Neill. 'But I just had this sadistic streak in me. This particularly bright student came into the lab in 2004, and I couldn't help myself. I put the old project in front of him and he bit really hard. He was Conor McMeniman. He was one of the best students I ever had. He made it work.' It took thousands more attempts, but McMeniman finally managed to stably infect an egg in 2006, creating a line of *Aedes* that naturally carried *Wolbachia*. In the course of this story, we have seen alliances between animals and microbes that are millions of years old. Here's one that is, at the time of writing, ten years old.[60]

But after all that work, the team discovered a fatal flaw in their plans: the popcorn strain was *too* virulent. Besides killing females prematurely, it also reduced the number of eggs they laid, and the viability of those eggs, thus sabotaging its own chances of moving into the next generation of mosquitoes. Simulations revealed that if it was ever unleashed into the wild, it just wouldn't spread.[61] It was terrible news.

O'Neill soon learned that none of that mattered. In 2008, two groups of researchers independently discovered that *Wolbachia* made fruit flies resistant to the group of viruses responsible for dengue, yellow fever, West Nile fever and other diseases. When O'Neill saw that, he immediately asked his team to feed their *Wolbachia*-infected mosquitoes with blood that had been spiked with dengue virus. The virus utterly failed to take hold. Even when the team injected it straight into the insects' guts, *Wolbachia* stopped it from replicating. That changed everything. The team didn't need *Wolbachia* to shorten a mosquito's lifespan. Its mere presence would be enough to prevent the spread of dengue! Better still, the team didn't need popcorn any more. Other less virulent strains were similarly protective, and would spread far more easily. 'After years and years of banging our head against the wall, we suddenly realised that we didn't need to,' says O'Neill.[62]

The team switched to a different strain called wMel, which had a track record of spreading through wild insect populations, but was an altogether gentler companion than popcorn, with none of the same life-shortening, brain-destroying, egg-killing effects. Would it spread? To find out, O'Neill's team built two insect aviaries: giant, walk-in cages, which he filled with mosquitoes. For every one uninfected insect, they added two wMel carriers. They also included a makeshift porch for the mosquitoes to hide under and a pile of sweaty gym towels to attract them. And for fifteen minutes a day, they added some succulent team members to feed the *Wolbachia*-infected mosquitoes. Every few days, the team collected eggs from the cages and checked them for *Wolbachia*. They found that, within three months, every mosquito larva inside was infected with wMel.[63] Everything suggested that their big idea would work. All the signs were saying: Go.

So, they did. Since 2006, well before the team had a mosquito with *Wolbachia*, they had been talking to the residents of two Cairns suburbs – Yorkeys Knob and Gordonvale – about their plans.[64] Hi, they said, we have a plan to get rid of dengue fever. Yes, we know that you've always been told to kill mosquitoes because they make you sick, but now, we'd appreciate it if you let us release *more* mosquitoes. No, they're not genetically modified, but we have loaded them up with a microbe with a penchant for spreading rapidly. Also, *Aedes* mosquitoes don't migrate very far, so for this plan to work, we're going to have to do lots of releases, including on your property. Yes, they'll probably bite you. No, no one has ever done this before. Are you in?

Amazingly enough, they were. For two years, the Eliminate Dengue team ran focus groups, talks in town halls and local pubs, and a shopfront drop-in clinic where people could ask questions. They knocked on a lot of doors. 'The project requires a lot of trust, and we got it, but it didn't happen overnight,' says O'Neill. 'We were very authentic in how we listened to people. When they had concerns, we addressed them. We even did experiments.' For example, they showed that *Wolbachia* couldn't infect fish, spiders and other predators that bit the mosquitoes, or humans whom the mosquitoes bit. Slowly, even sceptics became supporters. 'This local volunteer group, who mobilise people to help the community if floods and cyclones happen, asked if they could go door-to-door on our behalf to get people to release mosquitoes from their houses,' says O'Neill. 'That was a real turning point for me.' By 2011, when the mosquitoes were ready, the project had the support of 87 per cent of the residents.

It began in earnest on that January morning, with the cup that O'Neill ceremoniously opened. 'We were all a bit giddy,' O'Neill recalls. 'We had been working on this thing for frigging decades. A whole bunch of us were there for that moment, people who had been on the journey for a long time.' The team marched through the streets, pausing at every fourth house to release a few dozen mosquitoes. Within two months, they had liberated some 300,000 of them, pausing only to duck an incoming cyclone. Every two weeks, the team would then collect mosquitoes from the suburbs using a grid of traps, and test the insects for *Wolbachia*. 'It actually worked better than expected,' says O'Neill. By May, *Wolbachia* was sitting happily in 80 per cent of the Gordonvale mosquitoes and 90 per cent of those in Yorkeys Knob.[65] In just four months, the dengue-proof insects had almost totally replaced the native ones. For the first time in history, scientists had transformed a population of wild insects to stop them from spreading human diseases. And they did it through symbiosis.

But O'Neill's organisation isn't called 'Transform Mosquitoes'. It's called 'Eliminate Dengue'. Have they done that? There certainly haven't been any new cases in the two suburbs since 2011 – an encouraging sign, if not a definitive one. Neither area was a dengue hotspot to begin with. Nor is Australia, for that matter. O'Neill will be able to declare victory only when his mosquitoes repress dengue in the countries where it's most prevalent, which is why he is now expanding his work to Brazil, Colombia, Indonesia, and Vietnam.[66] When he started Eliminate Dengue in 2004, it was just him and his lab members. Now, it's an international team of scientists and health workers.

Back in Australia, the team are starting to disperse their mosquitoes through the northern city of Townsville. With some 200,000 residents to address, the team can't go knocking on every door. Instead, they rely on media coverage, big public events, and citizen science initiatives, where local people – even schoolchildren – volunteer their time. It's also too cumbersome to release adult mosquitoes. Instead, the team hands containers with eggs, water, and food to homeowners, who let the mosquitoes grow up in their gardens. 'Ultimately, we want to go to tropical megacities,' says O'Neill.

Each new place presents its own challenges. For example, if a city is gratuitous in its use of insecticides, the resident mosquitoes will probably be partly resistant. Releasing naïve Australian-born mosquitoes into such an environment would be pointless: they would succumb to poison long before they passed on their symbionts. So, the *Wolbachia*-infused mosquitoes need to be at least as resistant as the local ones. Cross-breeding can help. At the Indonesian chapter of Eliminate Dengue, scientists breed the *Wolbachia*-carriers with local mosquitoes for several generations, so that the insects they release are as close to the indigenous ones as possible. That should help them to mate more successfully, too. 'Every location is unique,' says O'Neill, 'but we're seeing that *Wolbachia* works well in every setting. Everything suggests that it should be possible to roll it out globally. In two to three years, we should have good evidence showing its impact. In ten to fifteen years, we should be able to make a significant dent in dengue.'

Sceptics would argue that evolution produces a countermeasure to every measure, a parry to every thrust. Dengue viruses should eventually become resistant to the encroaching wave of *Wolbachia*, and start infecting mosquitoes again. (As British scientist Leslie Orgel once famously said: 'Evolution is cleverer than you are.') But Elizabeth McGraw, a long-standing member of the Eliminate Dengue team, is optimistic. Her team has shown that *Wolbachia* protects against viral infections in several ways. It boosts the mosquito's immune system. It also competes for nutrients like fatty acids and cholesterol, which dengue virus needs in order to reproduce.[67] 'The more mechanisms you have, the less likely you'll get resistance,' she says. 'For an evolutionary biologist, that's really heartening.'

O'Neill and McGraw also argue that the spectre of resistance haunts every possible control measure, such as insecticides and vaccines. Unlike these other solutions, *Wolbachia* is alive, and could counter-adapt to any viral adaptations. It is also safe and cost-effective. While insecticides are toxic and must be continuously resprayed, *Wolbachia*-carrying mosquitoes have no side effects and can sustain themselves when released. 'Once it's going, it's ongoing,' says O'Neill. 'We're trying to bring the cost in to two to three dollars per person.'

O'Neill marvels at how far the study of *Wolbachia* has come. 'We were a fairly innocent lab that studied symbiosis,' he says. 'It was an area of basic science, but something wonderful and applied will come out of it.' As well as thwarting dengue virus, *Wolbachia* stops mosquitoes from carrying the Chikungunya and Zika viruses or the *Plasmodium* parasites that cause malaria; a team of Chinese and American scientists has now successfully melded the microbe with the *Anopheles* mosquito that spreads malaria.[68] And yet more researchers are trying to use *Wolbachia* to control insect pests like tsetse flies, which spread sleeping sickness, and bed bugs, which spread sleepless nights. 'This is just part of the whole new way of thinking, about the microbial ecology of organisms and about how that relates to disease,' says O'Neill.

In 1916, a hundred years before this book first arrived on shelves, the tempestuous Russian scientist Elie Metchnikoff passed away, after decades spent imbibing the microbes in sour milk. Could he have imagined that the approach he pioneered would one day spawn a multibillion-dollar industry, whose products, even though their worth is still in doubt, would grace supermarket shelves around the world? In 1923, the American microbiologist Arthur Isaac Kendall published a new edition of his textbook on bacteriology, in which he predicted that 'the time is coming' when people would use the bacteria of the human gut to cure intestinal diseases. Could he have predicted that organisations would now be freezing human excrement and sending it out to hospitals to be transplanted into patients? In 1928, the British bacteriologist Frederick Griffith showed that bacteria could take on characteristics from their peers, transforming themselves through a factor later shown to be DNA. Could he have foreseen that scientists would be able to tweak the genetic material of microbes so precisely and routinely that they could program bacteria to hunt and destroy their own kind? And in 1936, the entomologist Marshall Hertig decided to name an obscure little bacterium after his friend Simeon Burt Wolbach, some twelve years after the duo first spotted the microbe in a Bostonian mosquito. Could either have known that *Wolbachia* would turn out to be one of the planet's most successful bacteria? Or that so many scientists would study it that they would organise a bi-annual, *Wolbachia*-devoted conference to share their results? Or that it might be the key to stopping nematode worms from afflicting 150 million people a year with blindness or disability? Or that scientists would one day implant the bacterium *into* mosquitoes, in a global effort to control dengue fever and other diseases?

Surely not. For most of human existence, microbes were hidden from sight, visible only through the illnesses they caused. Even after Leeuwenhoek first saw them 350 years ago, they loitered in obscurity. When they finally rose to prominence they were cast as rogues, sooner to be eradicated than embraced. Even when scientists noticed the bacteria that swarm in human guts, or those that nestle inside insect cells, the discoveries were questioned and dismissed. Only recently have they migrated from the neglected fringes of biology to its spotlight-hogging centre. Only recently have we learned enough about the microbial world to start manipulating it. Our attempts are still basic and stumbling, and our confidence is sometimes exaggerated, but the potential is enormous. We have finally started to use everything we have learned since Leeuwenhoek first thought to study pond water to improve our lives.

10. TOMORROW THE WORLD

The house I'm standing in is a Platonic vision of the all-American suburban idyll. Outside, there are white clapboards, a rocking chair on a porch, and kids riding around on bicycles. Inside, there's more space than Jack Gilbert and his wife Kat know what to do with. Like me, they're British, and are used to snugger spaces. They're also warm and good-humoured: Jack is a dervish of energy, while Kat is poised and grounded. One of their sons, Dylan, is watching cartoons. The other, Hayden, for reasons best known to him, is trying to punch me in the bum. I am protecting myself by backing up against the kitchen counter, and nursing a cup of tea. And as I do that, I'm also passively ejecting microbes all over the cup, the counter, and the rest of this beautifully furnished kitchen.

In fairness, so are the Gilberts. As we've seen, along with hyenas, elephants, and badgers, we humans release bacterial smells into the air around us. But we also release the bacteria themselves. All of us are constantly seeding the world with our microbes. Every time we touch an object, we leave a microbial imprint upon it. Every time we walk, talk, scratch, shuffle, or sneeze, we cast a personalised cloud of microbes into space.[1] Every person aerosolises around 37 million bacteria per hour. This means that our microbiome isn't confined to our bodies. It perpetually reaches out into our environment. When I sat in Gilbert's car on the drive over here, I bled microbes all over his seat. Now that I'm reclining on his kitchen counter, I'm autographing it with my bacteria. I contain multitudes, yes, but only some of them; the rest, I extend into the world like a living aura.

To analyse these auras, the Gilberts recently swabbed their light switches, doorknobs, kitchen counters, bedroom floors, and their own hands, feet, and noses.[2] They did this every day for six weeks. They also recruited and trained six other families, including singletons, couples, and families, to do the same. The results of this study – the Home Microbiome Project – showed that every home has a distinctive microbiome that largely comes from the people who live in it. Their hand bacteria coat the light switches and doorknobs. Their foot microbes cover the floors. Their skin bugs get on the kitchen surfaces. And all of this happens with astonishing speed. Three of the volunteers moved house over the course of the study and their new abodes quickly took on the microbial character of their old ones, even when, in one case, that old accommodation was a hotel room. Within 24 hours of moving into a new place we overwrite it with our own microbes, turning it into a reflection of ourselves. When people invite you to 'make yourself at home', you and they really have no choice in the matter.

We also change the microbes of our housemates. Gilbert's team found that room-mates share more microbes than people who live apart, and couples are even more microbially similar. ('All that I am I give to you and all that I have I share with you', as the marriage vows go.) And if there's a dog around, these connections become supercharged. 'Dogs bring in bacteria from the outside to the inside, and they increase the microbial traffic between people,' says Gilbert. On the basis of his results, and on Susan Lynch's work showing that dog dust contains allergy-suppressing microbes, the Gilberts got a dog of their own. He's a ginger-and-white mix of golden retriever, collie, and Great Pyrenees, who answers to Captain Beau Diggley. 'We saw the benefit in increasing the microbial diversity of the home, and we wanted to make sure that our kids had that capacity to train their immune systems,' says Gilbert. 'Hayden named him; where did the name come from, Hayden?' Hayden replies: 'From my head.'

Whether dog or human, all animals live in a world of microbes. And by moving through that world, we change the microbes in it. In travelling to Chicago to visit the Gilberts, I have left my skin microbes in their home, my hotel room, a few cafes, several taxis, and one aeroplane seat. The good Captain Diggley is a fuzzy conduit that shuttles microbes from the soil and water of Naperville into the Gilbert residence. A Hawaiian bobtail squid, come the dawn, flushes its luminous *Vibrio fischeri* partners into the surrounding water. Hyenas spray microbial graffiti onto stalks of grass. And all of us constantly welcome microbes onto and into our bodies, whether through inhalation or ingestion, touches or footfalls, injuries or bites. Our microbiomes have wide-reaching tendrils that root us in the wider world.

Gilbert wants to understand those connections. He wants to be an all-seeing border officer for the human body, who knows exactly which microbes are coming in (and their point of origin), and which ones are leaving (and their destination). But humans make his job very difficult. We interact with so many different objects, people, and places that it becomes a nightmare to trace the paths of any particular bacterium. 'I'm an ecologist; I want to treat the human being like an island,' he says. 'But I'm literally not allowed. I put in a proposal to take some people and lock them in a space for six weeks, and the institutional review board said no.'

That's why he turned to dolphins.

'How many samples would you like?' asks veterinarian Bernie Maciol.

'How many have you done?' says Gilbert.

'Three.'

'Can you do replicates of those? And maybe some from another skin site? What about the armpit? No, not armpit. Whatever that is. What do you call a dolphin's armpit?'[3]

We are in the Shedd Aquarium's dolphin exhibit – a large tank, overlooked by artificial rocks and trees. Jessica, a trainer in a black-and-blue wetsuit, sits in the water and slaps its surface with her hand. A Pacific white-sided dolphin named Sagu swims up. He's a beautiful animal, with skin like a laminated charcoal drawing. He's obedient, too: when Jessica holds her hands palms-down and waves them to the side, Sagu rolls over and exposes his milky-white stomach. Maciol reaches across, swabs Sagu's armpit with a cotton bud, seals it in a tube, and passes it back to Gilbert. She does the same for two other dolphins, Kri and Piquet, who are quietly mooching next to their respective trainers.

'We've been doing blowhole sampling, faecal sampling, and skin sampling,' Jessica tells me. 'For the blowhole, I'll rest their head in my hand, put an agar plate over the hole, and tap to make the dolphin do a forced exhale. For the faecal sample, I'll make them roll over, insert a small rubber catheter and pull it out. We're not short of poop around here.'

This Aquarium Microbiome Project offers Gilbert what he cannot get from his Naperville house or any of the other homes that he has sampled – a kind of omniscience. Here are animals whose environment is fully known. Everything about the water – temperature, salinity, chemical content – can be measured, and regularly is. Here, Gilbert can analyse the microbiome of the dolphins' bodies, water, food, tanks, trainers, handlers, and air, and he has done so once a day for six weeks. 'These are real animals with their own real microbiomes living in a real environment, and we've catalogued all of the microbial interactions they have with that environment,' he says. And that should give him an unprecedented view of the connections between the microbes in an animal's body and those in the surrounding world.

The aquarium is running several such projects to improve the lives of its charges.[4] Bill Van Bonn, the Shedd's vice-president for animal health, tells me that the entire 3-million-gallon water supply in the main oceanarium used to pass through a life-support loop that cleaned and filtered it every three hours. 'You know how much energy it takes to push that water? Why do we do it that often? Because we're going to make this water so clean that it'll be absolutely the best thing,' he says, putting on a mock gung-ho tone. 'But when we back it up and do it half as much, what happens? Nothing! The water chemistry and the animals' health actually improves!'

Van Bonn suspects that in shooting for sanitation their intense cleaning regimes had gone too far. They ended up stripping the microbes from the aquarium environment, preventing mature and diverse communities from establishing themselves, and creating opportunities for weedy and harmful species to exploit. Sound familiar? That's exactly what antibiotics do in the guts of hospital patients. They divest an ecosystem of its native microbes, and allow competing pathogens like C-diff to flourish in their stead. In both settings, sterility is a curse not a goal, and a diverse ecosystem is better than an impoverished one. These principles are the same whether we're talking about a human intestine or an aquarium tank – or even a hospital room.

'I'm Dr Jack Gilbert, and *that* is a hospital,' says Jack Gilbert, gesturing with his thumb at the massive hospital looming behind him.

We're now at the University of Chicago's Center for Care and Discovery, a shiny new building that looks like a giant opera gateau, with several grey, orange, and black layers. Gilbert stands in front of it, doing repeated takes for a promotional video. I'm not convinced that the cameraman's microphone is going to pick up any decent audio over the sound of Chicago's unforgiving wind. I'm more convinced that Gilbert is very cold. And I'm totally convinced that, yes, that is indeed a hospital.

Just before it opened in February 2013, Gilbert's student Simon Lax led a team of researchers through the eerily empty hallways, armed with bags of Q-tips and a plan. They swept through ten patient rooms and two nurse stations, spread over two floors: one for short-stay patients recovering from elective surgery, and another for long-term ones like cancer patients and transplant recipients. But none of the rooms were home to any humans yet. Their only residents were microbes, which Lax's team collected. They swabbed the pristine floors, the gleaming bedrails and taps, and the perfectly folded sheets.

They collected samples from light switches, door handles, air vents, phones, keyboards, and more. Finally, they fitted the rooms with data loggers that would measure light, temperature, humidity, and air pressure, carbon dioxide monitors that would automatically record if a room was occupied, and infrared sensors that could tell when people entered or left. After the grand opening, the team carried on their work, collecting more weekly samples from the rooms and the patients inside them.[5]

Just as others have catalogued the developing microbiome of a newborn baby, Gilbert has, for the first time, catalogued the developing microbiome of a newborn building. His team is busy analysing the data now, to work out how the presence of humans has changed the edifice's microbial character, and whether those environmental microbes have flowed back into the occupants. Nowhere are those questions more important than in a hospital. There, the flow of microbes can mean life or death – a *lot* of deaths. In the developing world, around 5 to 10 per cent of people who check into hospitals and other healthcare institutions pick up some kind of infection during their stay, falling ill in the very places that are meant to make them healthier. In the United States alone, this means around 1.7 million infections and 90,000 deaths a year. Where do the pathogens behind these infections come from? Water? The ventilation system? Contaminated equipment? Hospital staff? Gilbert plans to find out. Through the mammoth set of data that his team have amassed, he should be able to trace the movements of pathogens from, say, a light switch to a doctor's hand to a patient's bedrail. And he should be able to work out ways of curtailing that life-threatening traffic.

This isn't a new problem. Ever since the 1860s, when Joseph Lister instigated sterile techniques in his hospital, cleaning regimes have helped to curb the spread of pathogens. Simple measures like hand-washing have undoubtedly saved countless lives. But just as we have gone overboard in taking unnecessary antibiotics or lathering ourselves in antibacterial sanitisers, we have also gone too far in cleaning our buildings – even our hospitals. As an example, one US hospital recently spent around $700,000 to install flooring that had been impregnated with antibacterial substances, despite having no evidence that such measures work. They might even make things worse. As in the dolphin enclosure and the human gut, perhaps the quest to sterilise our hospitals has created dysbiosis in the microbiomes of our buildings. By removing harmless bacteria that would otherwise impede the growth of pathogens, perhaps we have inadvertently constructed a more dangerous ecosystem.

'You want to bring in microbes that are benign or aren't interacting very much, and just populating surfaces,' adds Sean Gibbons, another of Gilbert's students. 'Diversity is good.' And sanitation, when taken too far, can cause diversity to collapse. Gibbons showed this by studying public toilets.[6] He found that thoroughly scrubbed toilets are first colonised by faecal microbes, which are launched into the air by roiling, flushed water. Those species are eventually outcompeted by a diverse range of skin microbes, but once the toilet gets scrubbed again, the communities go back to square one. So, here's the irony: toilets that are cleaned too often are more likely to be covered in faecal bacteria.

Jessica Green, an Oregon-based engineer-turned-ecologist, found a similar pattern among the microbes that float inside air-conditioned hospital rooms.[7] 'I assumed that the microbial community of the indoor air would be a subset of that of outdoor air,' she says. 'It really surprised me that we saw little to no overlap between the two.' Outdoors, the air was full of harmless microbes from plants and soils. Indoors, it contained a disproportionate number of potential pathogens, which are normally rare or absent in the outside world, but had been launched from the mouths and skins of hospital residents. The patients were effectively stewing in their own microbial juices. And the best way of fixing that was remarkably simple: open a window.

The legendary life-saver Florence Nightingale advocated as much some 150 years earlier. She had no explicit knowledge of the microbiome but, during the Crimean War, she noticed that patients would recover from infections more readily if she opened a window. 'Always, air from the air without, and that, too, through those windows through which the air comes freshest,' she wrote. This makes perfect sense to an ecologist: fresh air brings in harmless environmental microbes that take up space and exclude pathogens. But the idea of deliberately inviting microbes into a room deeply contradicts our assumptions about how hospitals should work. 'The model that we're working with, in hospitals and also many different buildings, is to keep the outdoors out,' says Green. It's such an ingrained attitude that when she did her study, she had to convince the hospital to let her prise some windows open – they had been bolted shut.

Rather than trying to exclude microbes from our buildings and public spaces, perhaps it is time to lay the welcome mat out for them. We have already been doing so blindly and unintentionally. In 2014, Green's team visited a shiny new university building called Lillis Hall and collected dust samples from 300 classrooms, offices, toilets, and more. They showed that many features of the building's design influenced the microbes in the dust, including the size of the rooms, how connected they are to each other, how often they are occupied, and how they are ventilated. Almost every architectural design choice affects the microbial ecology of buildings, which could then affect the microbial ecology of us. Or, as Winston Churchill said, 'We shape our buildings, and afterwards our buildings shape us.' And we can control that process, Green says, through what she calls 'bioinformed design'. That is, we can shape our buildings to select for the microbes we live with. As always, there are parallels to the world we can see: by planting strips of wild flowers along the edges of their fields, farmers can boost the numbers of pollinating insects. Green hopes to devise similar architectural tricks that can boost the diversity of beneficial microbes. 'Within the decade, architects could implement our findings in their practice,' she says.[8]

Jack Gilbert agrees, and has even bigger plans: he wants to deliberately seed buildings with bacteria. The microbes won't be sprayed or plastered onto walls. Instead, they'll come caged within tiny plastic spheres, created by engineer Ramille Shah. She will use three-dimensional printers to fashion balls that contain a warren of microscopic nooks and crannies. Gilbert will then impregnate these with useful bacteria like the fibre-digesting and inflammation-quenching *Clostridia*, as well as nutrients that nourish those microbes. These bacteria should then jump over to anyone that interacts with the spheres. Gilbert is testing this with germ-free mice. He wants to see if the bacteria are stable in their cages, if they really do jump into rodents that play with the balls, if they last in their new hosts, and if they can cure the rodents of inflammatory diseases. If that works, Gilbert has visions of testing the microbial spheres in office blocks or hospital wards. He imagines adding them to the cots in neonatal intensive care units, so that the infants would 'be constantly exposed to a rich microbial ecosystem that we've designed to be beneficial'. He adds, 'I want to create 3-D printable teething toys, too. You can imagine children playing with these.'

These spheres are effectively a different take on probiotics – a way of delivering beneficial microbes not through yoghurts or FMTs, but via an animal's surroundings. 'I don't want to put the microbes in their food and shove it down their gullet,' he says. 'I want the microbes to interact with their nasal membranes, their mouths, and their hands. I want them to experience that microbiome in a more natural way.'

'I want to call them bioballs,' he adds. 'Or maybe microballs.'

I tell him that he cannot call them microballs. He sniggers, proving my point.

'With this hand here, I shook hands with the women's world squash champion yesterday. I took her microbiome and I'm giving it to you,' says Luke Leung, shaking Gilbert's hand.

'So now is my hand going to be really good at squash?' asks Gilbert.

'Just the right hand,' says Leung. 'If you're a lefty, I'm sorry.'

Leung is an architect whose impressive portfolio includes the world's tallest building – the Burj Khalifa in Dubai. Since meeting Gilbert, he has also become something of a microbiome fanatic. So has Karen Weigert, Chicago's chief sustainability officer. The four of us meet in a posh restaurant for lunch, surrounded by sharp-suited executives and a view of Lake Michigan. 'You don't think about this as being alive,' says Gilbert, waving his finger at the impeccably fashioned interiors, the vaulted ceiling, and the skyscrapers looming outside. 'But it is alive. It's a living, breathing organism. Bacteria are the main things here.'

Gilbert is here to talk to Leung and Weigert about implementing his ideas on a much larger scale. He wants to use the principles that he is learning through the home, aquarium, and hospital projects to shape the microbiomes of entire cities, starting with Chicago. Leung is an ideal partner. In several of his buildings, he has routed the ventilation system so that it flows through a wall of plants, which not only pleases the eye but also filters the air. To him, Gilbert's idea of lacing walls with microbial spheres – which I've suggested should be called Baccy Balls – makes perfect sense. Weigert is also excited about using bacteria in architecture, and she asks Gilbert if the Baccy Balls would work in low-income housing, as well as in impressive skyscrapers. Yes, he says. He wants to make them as cheaply as possible, and certainly more so than a dramatic wall of plants.

Reassured, Weigert switches the conversation to Chicago's perennial problem with flooding. The sewer system backs up a lot and will probably do so more and more as the global climate changes. 'Is there something we can do to manage flooding, or after-effects like mould?' she asks. 'There actually is,' says Gilbert. In a different project, he has been working with L'Oréal to identify bacteria that can prevent dandruff and dermatitis, by stopping fungi from germinating on the scalp. These microbes could form the basis of anti-dandruff probiotic shampoos. But they could also be used to create 'micro-wetlands' that stop flooded homes from becoming overrun by mould. If a home floods, fungi would get a bonanza of water, but also face a bloom of antifungal microbes. 'You'd get automatic built-in mould control,' says Gilbert.

'So how real is all of this? Where are you with it?' asks Weigert.

'We've got the fungal control agents, and we're trying to work out how to implant them into plastics,' says Gilbert. 'We're probably two or three years off from having something that we'd feel comfortable inserting in somebody's home – someone who wasn't a colleague. And it may be three or four years before we have something reliable we can roll out.'

I joke that scientists always optimistically predict that their work is five years away from a real application.

Gilbert laughs. 'Well, I said three or four, so I'm being even more optimistic.'

So is Leung. 'We've been getting pretty good at killing bacteria, but we want to revitalise that relationship,' he says. 'We want to understand how the bacteria can help us in the built environment.'

And as a designer, I ask him, how soon do you think we'll be able to actually create buildings with that in mind?

He pauses. 'Let's say five years?'

Manipulating the microbiomes of buildings and cities is just the start of Gilbert's ambitions. As well as the hospital and aquarium initiatives, he is also studying the microbiomes of a local gym and a college dorm. The Home Microbiome Project revealed that people can be tracked, to an extent, by the microbes they leave behind, so he and Rob Knight – the two are close friends – are looking into forensic applications. He is studying the microbiomes of a wastewater treatment plant, floodplains, oil-contaminated waters in the Gulf of Mexico, prairies, a neonatal intensive care unit, and Merlot grapes. He is looking for microbes that can prevent dandruff, those responsible for allergies to cow's milk, and those that might be involved in autism. He is searching for dust microbes that might explain why two different American religious sects – the Amish and the Hutterites – have such wildly different rates of asthma and allergies. He is studying how gut microbes change over the course of the day and whether that affects our risk of becoming fat. He is analysing samples from several dozen wild baboons to see if the females that are most successful at rearing young have anything distinctive in their microbiomes.

Finally, together with Knight and Janet Jansson, he is coordinating the Earth Microbiome Project – a breathtakingly ambitious plan to take full stock of the planet's microbes.[9] The team are making contact with people who work on oceans or grasslands or floodplains, and persuading them to share their samples and their data. Ultimately, they want to be able to predict the kinds of microbes that live in a given ecosystem by plugging in basic factors like temperature, vegetation, wind speed, or levels of sunlight. And they want to predict how those species would respond to environmental changes, like the flooding of a river, or the passage from night to day. As goals go, it is ludicrously ambitious; some would say, unachievably so. But Gilbert and his colleagues are undeterred. Recently they have even petitioned the White House to launch a Unified Microbiome Initiative—a coordinated drive to build better tools for studying the microbiome and spurring more cooperation between different camps of scientists.[10]

Now is a time for thinking big. It's a time when families can be persuaded to swab their houses, when aquarium managers are as concerned about the invisible life in their waters as they are about the charismatic dolphins, when hospitals are seriously considering *adding* microbes to walls rather than removing them, and when architects and civil servants can discuss faecal transplants over an expensive three-course meal. It's the start of a new era, when people are finally ready to embrace the microbial world.

When I walked through San Diego Zoo with Rob Knight at the start of this book, I was struck by how different everything seemed with microbes in mind. Every visitor, keeper, and animal looked like a world on legs – a mobile ecosystem that interacted with others, largely oblivious to their inner multitudes. When I drive through Chicago with Jack Gilbert, I experience the same dizzying shift in perspective. I see the city's microbial underbelly – the rich seam of life that coats it, and moves through it on gusts of wind and currents of water and mobile bags of flesh. I see friends shaking hands, saying, 'How do you do?', and exchanging living organisms. I see people walking down the street, ejecting clouds of themselves in their wake. I see the decisions through which we have inadvertently shaped the microbial world around us: the choice to build with concrete versus brick, the opening of a window, and the daily schedule to which a janitor now mops the floor. And I see, in the driver's seat, a guy who notices those rivers of microscopic life and is enthralled rather than repelled by them. He knows that microbes are mostly not to be feared or destroyed, but to be cherished, admired, and studied.

This is the viewpoint from which all the stories in this book are told, from the decades-long project to fatally remove *Wolbachia* from nematode worms to the continuing quest to understand how milk nourishes a baby's bacteria; from intrepid expeditions into belching vents of the deep oceans to quieter attempts to uncover the symbiotic secrets of humble aphids. All of these endeavours were propelled by curiosity, awe, and the exhilaration of exploration. It was the unquenchable, voracious urge to know more about nature and our place in it that drove van Leeuwenhoek to peer at some water through his magnificent hand-crafted microscopes, and open up a world that no one knew existed. And that same urge – that spirit of discovery – is very much alive today.

While writing this chapter, I attended a conference about animal-microbe symbioses, featuring many of the people who have appeared in these pages. During a lunch break, the Japanese symbiont king Takema Fukatsu disappeared off into the surrounding forest and came back with several golden tortoise beetles – gorgeous little baubles with metallic gold shells. Later that night, beewolf whisperer Martin Kaltenpoth told me excitedly about how he had watched one of Fukatsu's beetles change colour from gold to red in front of his eyes. Who knows what symbionts they carry, or how the bacteria and the beetles have changed each other's lives? And on the final day, while everyone waited for a coach, aphid expert Lee Henry ducked away from the main group. He returned five minutes later with a tube full of aphids, which he had plucked from a bush growing next to the conference centre. That particular species, he told me, has fully domesticated *Hamiltonella*, the part-time associate that occasionally protects aphids from parasitic wasps. How? When? Why? Henry was excited to find out.

To peer into this world is to peer into William Blake's grain of sand. When we begin to understand our microbiomes, our symbionts, our inner ecosystems, our staggering multitudes, every walk bristles with opportunity for discovery. Every innocuous bush sings with incredible stories. Every part of the world is full of partnerships that have been playing themselves out for hundreds of millions of years, and that have affected all the flora and fauna we know.

We see how ubiquitous and vital microbes are. We see how they sculpt our organs, protect us from poisons and disease, break down our food, uphold our health, calibrate our immune system, guide our behaviour, and bombard our genomes with their genes. We see the lengths to which animals must go to keep their multitudes in check, from the ecosystem managers of the immune system to the bacteria-feeding sugars in breast milk. We see what happens when those measures break: bleached reefs, inflamed guts, and obese bodies. We see, conversely, the rewards of a harmonious relationship: the ecological opportunities that open up to us, and the accelerated pace with which we can grasp them. We see how we might start to control these multitudes for our own benefit, transplanting entire communities from one individual to another, forging and breaking symbioses at will, or even engineering new kinds of microbes. And we learn the secret, invisible, and wondrous biology behind the gutless worms that thrive in an abyssal Eden, the mealybugs that suck the juices of plants, the corals that construct mighty reefs, the small stinging hydras that cling to pondweed, the beetles that bring down forests, the adorable squid that create their own light shows, the pangolin curled around a zookeeper's waist, and the disease-fighting mosquitoes flying off into a bright Australian dawn.

ACKNOWLEDGEMENTS

This is not the part where I thank my microbes. This is the part where we ignore the little critters for a spell and focus entirely on the hosts.

Every book is the product of more than one mind, and a book on symbiosis and partnerships must be especially so. Foremost among those minds are Stuart Williams from Bodley Head and Hilary Redmon formerly at Ecco. I'd call them editors, but they feel more like co-conspirators. Right from the start, both of them immediately grasped the book I was trying to write: a story of the microbiome that would span the entire animal kingdom, without focusing narrowly on humans, health or diet. They nourished that idea, often understanding it better than I did myself. They championed it tirelessly, provided incisive, insightful, and invaluable edits, and were never less than total joys to work with. Thanks also to PJ Mark who escorted the book into American shores, and Denise Oswald who took up the editorial baton from Hilary at Ecco.

David Quammen was the first person I told about the idea for this book and he has been an astoundingly gracious supporter from the start. His masterwork, *Song of the Dodo*, helped to erode an early writing block, as at various points did Helen Macdonald's *H is for Hawk*, David George Haskell's *The Forest Unseen* and Kathryn Schulz's *Being Wrong*. Their works sat on my shelf as reminders of the quality I aspired to reach.

Several other people shaped the environment in which writing this book was possible. Alice Trouncer gave me a dozen years of love and adventure, and held me above water as I built a career as a writer; wife, friend, confidante, dance partner and all-round wonderful human, I'll always be grateful to her. Alice See, my mother, has never once flinched in her faith and support; she is a rock. Carl Zimmer has been a friend, mentor and inspiration, his skill as a writer matched only by his generosity as a person. Virginia Hughes read the first complete chapter and provided invaluable feedback. Meehan Crist, David Dobbs, Nadia Drake, Rose Eveleth, Nikki Greenwood, Sara Hiom, Alok Jha, Maria Konnikova, Ben Lillie, Kim Macdonald, Maryn McKenna, Hazel Nunn, Helen Pearson, Adam Rutherford, Kathryn Schulz and Beck Smith all helped to ground a turbulent year. And Liz Neeley, an indefatigable whirlwind of joy, wit, and optimism, has transformed and enriched my life in ways that continue to amaze me; she makes a secret cameo in an early chapter.

In writing this book, and in reporting on microbes for a decade, I've interviewed hundreds of researchers, who have been unflaggingly generous with their time and knowledge. That's a quality I regularly find among scientists, but especially so among those who study symbiosis, partnerships, and cooperation – you are, it seems, what you research. There are too many to list here, but I especially want to single out Jonathan Eisen, Jack Gilbert, Rob Knight, John McCutcheon and Margaret McFall-Ngai for supporting the project, acting as intellectual sounding boards and providing their views on the finished manuscript. Eisen, in particular, has always championed a measured and critical take on microbiome science that has informed my writing for years; hopefully I'll have avoided an Overselling the Microbiome Award with this book. I'm also grateful to Knight for organising the trip to the zoo that featured in the opening pages, and to Gilbert for taking me on a manic tour of Chicago.

Thanks also to: Martin Blaser, Seth Bordenstein, Thomas Bosch, John Cryan, Angela Douglas, Jeff Gordon, Greg Hurst, Nicole King, Nick Lane, Ruth Ley, David Mills, Nancy Moran, Forest Rohwer, Mark Taylor and Mark Underwood for either showing me round their labs or offering especially detailed and enlightening discussions; Nell Bekiares for introducing me to some squid; Dave O'Donnell, Maria Karlsson and Justin Serugo for letting me hold some germ-free mice; Bill Van Bonn for showing me round the Shedd Aquarium; Elizabeth

Bik, whose Microbiome Digest newsletter was the single best way of keeping up with an ever-proliferating literature; historians Jan Sapp and Funke Sangodeyi, whose books and thesis respectively provided critical insights into the rich history of this field; the lively community of geneticists and microbiologists on Twitter, whose critical eyes and open discussions have informed my views and kept me honest; and Nicole Dubilier and Ned Ruby for letting a journalist – boo, hiss! – crash the esteemed Gordon Research Conference on Animal-Microbe Symbiosis, a sparkling and vibrant week of science, hiking and, regrettably, cornhole.

Sadly, I talked to many people whose work or names could not be represented in these pages; the field is vast and a book cannot be a comprehensive review of it. I also note that many students, postdocs and collaborators contributed to studies described in these pages that, by necessity, are associated only with one or two key names. I have tried to make amends in the endnotes, but regardless, I offer sincere thanks, sad condolences and a note that this is far from the last time I will be writing about these topics.

Finally, my most heartfelt thanks must go to my agent, Will Francis. Early on, a friend of mine told me that a good agent could help you shape your ideas, sell your book ferociously, or help with promotion and publicity, but that no agent would be strong in all three. Will was. He pestered me for years about writing a book, graciously ignored the email that I wrote in January 2014 boldly stating that no such plans would ever unfold and please could he stop pestering me, graciously accepted the email I sent three weeks later frantically backtracking on my earlier statement, and helped me to shape my nebulous idea into a solid proposal. He's a friend – a symbiont, perhaps – and these pages are marbled with his influence.

LIST OF ILLUSTRATIONS

NOTES

PROLOGUE: A TRIP TO THE ZOO

1. In this book, I use the terms 'microbiota' and 'microbiome' interchangeably. Some scientists will argue that microbiota means the organisms themselves, while microbiome refers to their collective genes. But one of the very first uses of microbiome, back in 1988, used the term to talk about a group of *microbes* living in a given place. That definition persists today – it emphasises the 'biome' bit, which refers to a community, rather than the 'ome' bit, which refers to the world of genomes.
2. This imagery was first used by the ecologist Clair Folsome (Folsome, 1985).
3. Sponges: Thacker and Freeman, 2012; placozoans: personal communication from Nicole Dubilier and Margaret McFall-Ngai.
4. Costello et al., 2009.
5. There are plenty of good general reviews about the importance of microbes to animal lives, but 'Animals in a bacterial world, a new imperative for the life sciences' stands out as one of the best (McFall-Ngai et al., 2013).

1. LIVING ISLANDS

1. When I was a kid, I saw Sir David Attenborough use this framing device in his seminal series *Life on Earth* and it has stuck with me ever since.
2. The other half comes from land plants, which conduct photosynthesis using domesticated bacteria – chloroplasts. So technically, *all* the oxygen you breathe comes from bacteria.
3. It's estimated that every human contains 100 trillion microbes, most of which live in our guts. By comparison, the Milky Way contains between 100 million and 400 million stars.
4. McMaster, 2004.

5. It is clear that mitochondria did evolve from an ancient bacterium that fused with a host cell, but whether this event was itself the origin of eukaryotes or just one of many milestones in their evolution is still hotly disputed among scientists. To my mind, the proponents of the former idea have assembled a strong set of evidence for their claims. I wrote about their arguments in more detail for the online magazine *Nautilus* (Yong, 2014a), and you can read an even more detailed account in Nick Lane's book, *The Vital Question* (Lane, 2015a).
6. Size isn't a strict prerequisite for having a microbiome: some single-cell eukaryotes also carry bacteria in and on their cells, although their communities are understandably smaller than ours.
7. Judah Rosner calls the 10-to-1 ratio a 'fake fact', which he traced back to a microbiologist named Thomas Luckey (Rosner, 2014). In 1972, Luckey estimated, with little in the way of evidence, that there are 100 billion microbes in a gram of intestinal contents (fluid or faeçes), and 1,000 grams of such contents in an average adult – giving a total of 100 trillion microbes. Eminent microbiologist Dwayne Savage then took this figure and contrasted it with the 10 trillion human cells in our bodies – a figure pulled from a textbook that, again, cited no supporting evidence.
8. McFall-Ngai, 2007.
9. Li et al., 2014.
10. Hoopoes: Soler et al., 2008; leafcutter ants: Cafaro et al., 2011; Colorado potato beetle: Chau et al., 2011; pufferfish: Chung et al., 2013; cardinalfish: Dunlap and Nakamura, 2011; ant lion: Yoshida et al., 2001; nematodes: Herbert and Goodrich-Blair, 2007.
11. These same glowing microbes got into the wounds of soldiers during the American Civil War and disinfected them; the troops called the mysterious protective light the 'Angel's Glow'.
12. Gilbert and Neufeld, 2014.
13. See http://wallacefund.info/ for more on Wallace's life.
14. *The Song of the Dodo* masterfully recounts the adventures of both Wallace and Darwin (Quammen, 1997).
15. Wallace, 1855.
16. O'Malley, 2009.
17. This concept, and the ecological nature of the microbiome, are well explained in these papers: Dethlefsen et al., 2007; Ley et al., 2006; Relman, 2012.
18. Huttenhower et al., 2012.
19. Fierer et al., 2008.
20. Several researchers have looked at the changing microbiomes of infants, including their own; Fredrik Bäckhed did so most recently (and most thoroughly) by analysing samples from 98 infants during their first year of life (Bäckhed

et al., 2015). Tanya Yatsunenko and Jeff Gordon also did a seminal study in three separate countries, in which they showed how a child's microbes change over its first three years of life (Yatsunenko et al., 2012).

21. Jeremiah Faith and Jeff Gordon showed that most strains in the gut stay there for decades: rising, falling, but always keeping a presence (Faith et al., 2013). Other teams have shown that the microbiome is incredibly dynamic over shorter timescales (Caporaso et al., 2011; David et al., 2013; Thaiss et al., 2014).

22. Quammen, 1997, p. 29.

23. This work was done together with Peter Dorrestein (Bouslimani et al., 2015).

24. Frederic Delsuc led this study (Delsuc et al., 2014).

25. Scott Gilbert, a developmental biologist, has wrestled with this seemingly trivial problem for years (Gilbert et al., 2012).

26. Relman, 2008.

2. THE PEOPLE WHO THOUGHT TO LOOK

1. Details of Leeuwenhoek's life can be found in Douglas Anderson's website Lens on Leeuwenhoek (http://lensonleeuwenhoek.net/) and two biographies: *Antony Van Leeuwenhoek and His 'Little Animals'* (Dobell, 1932), *The Cleere Observer*. (Payne, 1970). His influence is also discussed in papers by Douglas Anderson (Anderson, 2014) and Nick Lane (Lane, 2015b), both of which I have quoted from. There is no standardised spelling for the man's name, and I am using the same one that Dobell chose.

2. Leeuwenhook, 1674.

3. He meant cheese mites – the smallest creatures then known.

4. There is some dispute about this. In the 1750s, two decades before Leeuwenhoek looked at water, the German scholar Anthanasius Kircher studied the blood of plague victims and described 'poisonous corpuscles', each of which changed 'into a little invisible worm'. His descriptions are vague, but it seems more likely that he was describing red blood cells or bits of dead tissue rather than the plague bacterium *Yersinia pestis*.

5. Leeuwenhoeck, 1677.

6. Dobell, 1932, p. 325.

7. Alexander Abbott wrote that, 'Throughout all of Leeuwenhoek's work, there is a conspicuous absence of the speculative. His contributions are remarkable for their purely objective nature' (Abbott, 1894, p. 15).

8. The stories of Pasteur, Koch, and their contemporaries are lucidly told in *Microbe Hunters* (Kruif, 2002).

9. Dubos, 1987, p. 64.

10. Chung and Ferris, 1996.

11. Hiss and Zinsser, 1910.

12. Sapp, 1994, pp. 3–14. Sapp's book, *Evolution by Association*, is the most comprehensive history of symbiosis research yet published – a landmark historical work.

13. Ibid., pp. 6–9. Albert Frank coined it first in 1877; Anton de Bary is arguably more famous for it, even though he did not use it until a year later.

14. Buchner, 1965, pp. 23–24.

15. Kendall, 1923.

16. Quoted in Zimmer, 2012.

17. Many of their observations were accurate; others less so, including the claim that Arctic mammals are sterile (Kendall, 1923).

18. Kendall, 1909.

19. Kendall, 1921.

20. Metchnikoff talked about his ideas in a public lecture (see The Wilde Lecture, 1901); his Dostoevsky-esque nature is noted in Kruif, 2002, and his influence in Dubos, 1965, pp. 120–121.

21. Bulloch, 1938.

22. Funke Sangodeyi is one of the few historians to catalogue this phase in the history of microbial ecology, and her thesis (Sangodeyi, 2014) is well worth reading for that reason.

23. Robert Hungate, a fourth-generation descendant of the Delft School, was intrigued by the gut microbes of plant-eaters like termites and cattle. He developed a way of coating the inside of a test tube with agar, while flushing out any oxygen using carbon dioxide. Using this 'roll tube method', bacteriologists could finally grow the oxygen-hating microbes that dominated animal guts, including our own (Chung and Bryant, 1997).

24. Following the example set by Leeuwenhoek, American dentist Joseph Appleton looked at bacteria in the mouth. Between the 1920s and 1950s, he and others examined how these communities changed during oral diseases, and how they were influenced by saliva, food, age, or seasons. Mouth microbes proved to be more amenable subjects than their gut counterparts: they were easier to collect with swabs, and they tolerated oxygen. In studying them, Appleton helped to turn dentistry – itself a marginalised part of medicine – into a true science rather than just a technical profession (Sangodeyi, 2014, pp. 88–103).

25. Rosebury, 1962.

26. Rosebury also wrote the first popular science book about the human microbiota – the bestseller, *Life on Man*, published in 1976.

27. Dwayne Savage gives an excellent account of all the work that followed (Savage, 2001).

28. Moberg's excellent biography of René Dubos provides many rich details about his life (Moberg, 2005).
29. Dubos, 1987, p. 62.
30. Dubos, 1965, pp. 110–146.
31. The quote comes from a *New York Times* interview (Blakeslee, 1996). For excellent accounts of Woese's groundbreaking work, see John Archibald's *One Plus One Equals One* (Archibald, 2014) and Jan Sapp's *The New Foundations of Evolution* (Sapp, 2009).
32. Woese did not come up with this idea. Francis Crick, one of the co-discoverers of the DNA double helix, had proposed a similar strategy in 1958, while Linus Pauling and Emil Zuckerkandl proposed using molecules as 'documents of evolutionary history' in 1965.
33. Postdoc George Fox was Woese's collaborator and the co-author of his iconic paper (Woese and Fox, 1977).
34. Morell, 1997.
35. Right across the tree of life, this approach, known as molecular phylogenetics, has splintered many groups that were united on the basis of misleading physical traits, and united organisms that are actually similar despite all appearances. It also proved, beyond a shadow of a doubt, that mitochondria – those bean-shaped power plants found in all complex cells – were formerly bacteria. These structures had their own genes, which were remarkably similar to bacterial genes. The same was true of the chloroplasts, which allow plants to harness the sun's energy in photosynthesis.
36. The Yellowstone study: Stahl et al., 1985. Pace had applied the same technique to the bacteria inside deep-sea worms; those results were published a year earlier, but didn't uncover any new species.
37. Pace's Pacific Ocean study: Schmidt et al., 1991; the recent survey of a Colorado aquifer: Brown et al., 2015.
38. Pace et al., 1986.
39. Handelsman, 2007; National Research Council (US) Committee on Metagenomics, 2007.
40. Kroes et al., 1999.
41. Eckburg, 2005.
42. Critical early studies from Jeff Gordon's lab included Bäckhed et al., 2004; Stappenbeck et al., 2002; Turnbaugh et al., 2006.
43. In December 2007, the US National Institutes for Health launched the Human Microbiome Project – a five-year initiative that would characterise the nose, mouth, skin, gut, and genital microbiomes of 242 healthy volunteers. With US $115 million behind it, the project consumed the time of around 200 scientists, and produced 'the most extensive catalogue yet of organisms and genes

pertaining to our microbiomes'. A year later, a similar programme called MetaHIT was launched in Europe, focused on the gut and funded to the tune of 22 million euros. Other ventures were launched in China, Japan, Australia, and Singapore. These projects are documented in Mullard, 2008.

44. I wrote about my visit to Micropia for the *New Yorker* (Yong, 2015a).

3. BODY BUILDERS

1. This scene appears in a profile I wrote about McFall-Ngai for *Nature* (Yong, 2015b).

2. McFall-Ngai's work with the bobtail squid: McFall-Ngai, 2014. The role of cilia in recruiting *V. fischeri* is unpublished at the time of writing. The terraforming that occurs when *V. fischeri* touches the squid was revealed by postdoc Natacha Kremer in 2013 (Kremer et al., 2013). The events that happen after *V. fischeri* reaches the crypts were detailed by McFall-Ngai and Ruby themselves in 1991 (McFall-Ngai and Ruby, 1991). McFall-Ngai first stated that *V. fischeri* affects the squid's development in 1994 (Montgomery and McFall-Ngai, 1994). The MAMPs were identified by Tanya Koropatnick and others in 2004 (Koropatnick et al., 2004).

3. Karen Guillemin showed that the guts of zebrafish mature properly only when they are exposed to microbes and the LPS molecules on their surface (Bates et al., 2006). And Gerard Eberl found that PGN similarly affects the development of a mouse's gut (Bouskra et al., 2008). The influence of microbes on animal development is discussed in Cheesman and Guillemin, 2007; Fraune and Bosch, 2010.

4. Coon et al., 2014.

5. Rosebury, 1969, p. 66.

6. Fraune and Bosch, 2010; Sommer and Bäckhed, 2013; Stappenbeck et al., 2002.

7. Hooper, 2001.

8. Hooper's work inspired John Rawls to carry out the same experiment in germ-free zebrafish, in which he found a largely overlapping set of microbe-activated genes (Rawls et al., 2004).

9. Gilbert et al., 2012.

10. Most bacteria consist of single cells, but, as always in biology, there are exceptions. Under some conditions, *Myxococcus xanthus* forms cooperative predatory colonies consisting of millions of cells that move, develop, and hunt as one.

11. Alegado and King, 2014.

12. The great German biologist Ernst Haeckel imagined the earliest animals as hollow spheres of bacteria-eating cells. He named this hypothetical colony

Blastaea and, as was his wont, he drew it. His sketch looks uncannily similar to the choano rosettes that King's son doodled on his pad.

13. As described in Alegado et al., 2012; the name means 'cold eater from Machipongo'.

14. As reviewed in Hadfield, 2011.

15. Leroi, 2014, p. 227.

16. It took almost a decade for Hadfield to find out *how* the bacteria trigger the worms' transformation. The answer turns out to be surprisingly violent. Together with Nick Shikuma at the California Institute of Technology, Hadfield found that P-luteo produces toxins called bacteriocins, which are used to wage war against other microbes (Shikuma et al., 2014). Each one is a microscopic, spring-loaded machine that punches holes into other cells to create fatal leaks. A hundred of them will merge into a large dome-shaped cluster with all the dangerous ends pointing outwards. These domes litter P-luteo's biofilms like landmines. Hadfield thinks that when a larval worm touches one of these mines – wham! – suddenly 'one of its cells gets a whole lot of holes punctured in it'. That might be enough to trigger a nervous signal that tells the worm it's time to grow up.

17. Hadfield, 2011; Sneed et al., 2014; Wahl et al., 2012.

18. Gruber-Vodicka et al., 2011; the regeneration results are as yet unpublished.

19. Sacks, 2015.

20. Several studies have shown that microbes affect fat (Bäckhed et al., 2004), the blood–brain barrier (Braniste et al., 2014), and bone (Sjögren et al., 2012); other relevant research is reviewed in Fraune and Bosch, 2010.

21. Rosebury, 1969, p. 67.

22. And not just any old microbiome, either. Dennis Kasper showed that a germ-free mouse will develop a hearty, vigorous immune system if it receives a normal set of mouse microbes, but not if it gets an ensemble from a human, or even from a rat (Chung et al., 2012). This suggests that specific sets of microbes have co-evolved with their hosts to promote good health by generating robust immune systems. Even viruses play a role. When Ken Cadwell infected germ-free mice with a strain of norovirus related to the one that frequently blights passengers in cruise ships with bouts of vomiting, he saw that the rodents developed more white blood cells of various types. The virus was behaving like a microbiome rich in bacteria (Kernbauer et al., 2014).

23. The connections between the immune system and the microbiome have been thoroughly reviewed in Belkaid and Hand, 2014; Hooper et al., 2012; Lee and Mazmanian, 2010; Selosse et al., 2014. The importance of microbes in early life was demonstrated in Olszak et al., 2012.

24. Dan Littman and Kenya Honda showed that segmented filamentous bacteria (SFB) can induce pro-inflammatory immune cells (Ivanov et al., 2009). Honda also showed that *Clostridia* bacteria can induce anti-inflammatory cells (Atarashi et al., 2011).

25. To understand how important these are, just look at HIV: the virus is so greatly feared precisely because it destroys helper T cells, leaving people unable to mount an immune response against even weak pathogens.

26. Mazmanian's original study on B-frag and PSA: Mazmanian et al., 2005; former lab member June Round was critically involved in the later work: Mazmanian et al., 2008; Round and Mazmanian, 2010.

27. B-frag isn't found in every gut. Thankfully, it's just one of a legion of microbes with similar properties. Wendy Garrett showed that many of these work by producing the same chemicals, such as short-chain fatty acids (SCFAs), which can stimulate the anti-inflammatory branches of the immune system (Smith et al., 2013b).

28. Theoretically speaking. In reality, we still don't know what most of those genes do, but these gaps in our knowledge will eventually be filled.

29. The importance of microbial metabolites is reviewed in Dorrestein et al., 2014, Nicholson et al., 2012, and Sharon et al., 2014.

30. Leopard urine also smells of popcorn. If you're driving through the African savannah and you smell the redolent twang of buttery corn, beware.

31. Theis et al., 2013.

32. Scent gland research: Archie and Theis, 2011; Ezenwa and Williams, 2014; the smell of identical twins: Roberts et al., 2005; locust, cockroach, and mesquite bug studies: Becerra et al., 2015; Dillon et al., 2000; Wada-Katsumata et al., 2015.

33. Lee et al., 2015; Malkova et al., 2012.

34. Willingham, 2012.

35. Mazmanian presented this work, done by postdoc Gil Sharon, at a recent conference; it is unpublished at the time of writing.

36. This story is recounted in Beaumont's own words (Beaumont, 1838), and in a later biography (Roberts, 1990).

37. Despite his injury, St Martin outlived Beaumont by 27 years; the latter died after slipping on ice.

38. There are plenty of reviews on this topic, more so than actual research papers; here's a selection: Collins et al., 2012; Cryan and Dinan, 2012; Mayer et al., 2015; Stilling et al., 2015. One of the seminal studies was done in 1998, when Mark Lyte infected mice with *Campylobacter jejuni* – a bacterium that causes food poisoning. He used such a low dose that the mice didn't even mount an immune response, much less get sick – but they did behave more anxiously (Lyte et al., 1998). In

2004, another team from Japan showed that germ-free rodents respond more strongly to stressful situations (Sudo et al., 2004).

39. The flood of papers in 2011 included work by Jane Foster (Neufeld et al., 2011); Sven Petterson (Heijtz et al., 2011); Stephen Collins (Bercik et al., 2011); and John Cryan, Ted Dinan, and John Bienenstock (Bravo et al., 2011).

40. Bravo et al., 2011.

41. John Bienenstock led this work. The JB–1 strain of *L. rhamnosus* originally came from his lab – hence the name – and he gave his Irish colleagues confidence by repeating all their experiments in Canada, using a different group of mice and slightly different techniques. And he still got the same results. That's when the team knew they were really on to something. 'We said: Christ, this is excellent,' he tells me. 'Most of these bloody things are so un-robust when you go from lab to lab.'

42. Some microbes can make neurotransmitters directly, and others persuade our gut cells to churn them out. People often think of these substances as *brain* chemicals, but at least half of the dopamine in our bodies exists in the gut, as does 90 per cent of our serotonin (Asano et al., 2012).

43. Tillisch et al., 2013.

44. Results are unpublished at the time of writing.

45. One American team took the microbes of mice that ate a high-fat diet and implanted them in the guts of mice that were raised on normal chow. The recipients became more anxious and had poorer memories (Bruce-Keller et al., 2015).

46. Joe Alcock proposed this idea (Alcock et al., 2014).

47. I talked about these mind-controlling parasites in my TED talk (Yong, 2014b).

48. *T. gondii* might also affect human behaviour: some scientists have suggested that infected people show personality differences, run a higher risk of car accidents, and are more likely to develop schizophrenia.

4. TERMS AND CONDITIONS APPLY

1. The history of Wolbach and Hertig's work is detailed in Kozek and Rao, 2007.

2. Stouthamer's wasps: Schilthuizen and Stouthamer, 1997; Rigaud's woodlice: Rigaud and Juchault, 1992; Hurst's butterflies: Hornett et al., 2009; reviews of all the above: Werren et al., 2008 and LePage and Bordenstein, 2013.

3. An earlier study put the figure at 66 per cent (Hilgenboecker et al., 2008) but a more recent one proposed a more modest 40 percent (Zug and Hammerstein, 2012).

4. There are probably oceanic bacteria that are more common. One of them – *Prochlorococcus* – is so common that a millilitre of water scooped from the ocean's surface probably contains 100,000 of them. Together, they produce around 20 per cent of the oxygen in the air. Take five breaths; the oxygen in one of them came from these bacteria. But theirs is a story for another book.

5. The nematodes: Taylor et al., 2013; flies and mosquitoes: Moreira et al., 2009; bed bugs: Hosokawa et al., 2010; the leaf miner: Kaiser et al., 2010; the wasp: Pannebakker et al., 2007. The reason behind the wasp's dependency is perverse. Like all animals, the wasp has self-destruct programs that kill its own cells if they become damaged or cancerous. *Wolbachia* tamps down these programs, so the wasp has compensated by making them unusually sensitive. Now, if you remove *Wolbachia*, the wasp mistakenly destroys the tissues that support its own eggs. It has been struggling against the microbe for so long that it has come to rely on it. *Wolbachia* doesn't really provide it with any benefits, but the two are stuck with each other nonetheless.

6. As reviewed in Dale and Moran, 2006; Douglas, 2008; Kiers and West, 2015; McFall-Ngai, 1998.

7. Blaser, 2010.

8. Broderick et al., 2006.

9. Theodor Rosebury hated the term 'opportunistic'. 'The name bespeaks analogy again – microbes sharing human vices,' he wrote. 'All microbes, all living things, respond in some way to changes in their situation. All possible degrees and kinds of opportunity change harmless microbes to harmful ones.' He coined a different term – *amphibiosis* – for natural partnerships that are helpful in some contexts and harmful in others. It's a fine term – beautiful, even – but perhaps unnecessary since many (if not most) partnerships are like this.

10. Zhang et al., 2010.

11. Marmalade hoverfly: Leroy et al., 2011; mosquitoes: Verhulst et al., 2011.

12. Polio: Kuss et al., 2011. Another virus called MMTV, which causes breast cancer in mice, uses bacterial molecules like fake ID cards, displaying them to the immune system to garner safe passage into the gut (Kane et al., 2011).

13. Wells et al., 1930.

14. Oxpeckers: Weeks, 2000; cleaner fish: Bshary, 2002; ants and acacias: Heil et al., 2014.

15. Kiers said this at a conference; her views are also recounted in West et al., 2015.

16. McFall-Ngai tells me that the squid are exceptionally good at weeding out dark symbionts, and can somehow detect even a few of these lightless mutants in their crypts – and evict them.

17. Reviewed in Bevins and Salzman, 2011.

18. Stomach acid: Beasley et al., 2015; Ants and formic acid: interview with Heike Feldhaar.

19. Stinkbugs: Ohbayashi et al., 2015; bacteriocytes: Stoll et al., 2010.
20. This happens in weevils, which use an antimicrobial chemical to stop the bacteria in their cells from reproducing; if you stop them from making that chemical, the bacteria multiply, escape, and run amok throughout the insects' bodies (Login and Heddi, 2013).
21. Abdelaziz Heddi discovered the weevil's ability: Vigneron et al., 2014. Many other animals, including other insects, clams, worms, and grazing mammals, can digest their microbes for extra nutrition. This side of symbiosis is rather neglected. Scientists often assume that microbes get something out of their relationships with animals, whether it's nutrients, protection, or stable environments – but such benefits are rarely proven. In a provocative paper called 'The symbiont side of symbiosis: do microbes really benefit?', Justine Garcia and Nicole Gerardo write, 'In cases where there is no evidence of a symbiont benefit, symbionts may instead be more akin to prisoners or farmed crops than equal partners' (Garcia and Gerardo, 2014).
22. Interview with Rohwer.
23. Barr et al., 2013.
24. I should note that this is but one of *many* theories about the origins of the immune system.
25. Vaishnava et al., 2008.
26. The most important of these is an antibody called immunoglobulin A, or IgA. The gut makes an absurd amount of it – around a teaspoon every day. But there isn't just one mass-produced version of IgA. Instead, it's something of an artisanal molecule, which comes in an endless variety of subtly different shapes, each one designed to recognise and neutralise a different microbe. By sampling the microbes in the demilitarised zone, the immune cells within the gut can make a wide range of bespoke IgAs that target the most common species. They then release these antibodies into the mucus, which pile onto local microbes, creating an immobilising coat. This system is so effective that around half of the bacteria in our gut are restrained by IgA straitjackets. As the community of microbes changes, so too does the array of IgAs that are sent out to detain them. It is a wonderfully flexible and adaptive system.
27. As reviewed in Belkaid and Hand, 2014; Hooper et al., 2012; Maynard et al., 2012.
28. Hooper et al., 2003.
29. This hypothesis was first stated in McFall-Ngai, 2007. There are some holes in it; for example, if the vertebrate immune system is so important for controlling our complex microbiome, how do corals and sponges harbour extensive communities with much simpler immune systems?
30. Elahi et al., 2013.

31. Rogier et al., 2014.
32. As reviewed in Bode, 2012; Chichlowski et al., 2011; Sela and Mills, 2014.
33. Kunz, 2012.
34. The team includes German himself, microbiologist David Mills, chemist Carlito Lebrilla, and food scientist Daniela Barile.
35. Robert Ward led this work (Ward et al., 2006). David Sela led the genome sequencing (Sela et al., 2008).
36. This can have dramatic effects: in a study in Bangladesh, Mills's team found that infants who are colonised by lots of *B. infantis* have better responses to polio and tetanus vaccines.
37. Mills tells me that *B. infantis* isn't always *B. infantis*. People often misidentify it, and apply the name to other very different microbes. One of these '*B. infantis*' strains can be found in popular yoghurts, but Mills uses *that* strain as a negative control in his experiments. It behaves very little like the milk-specialist that he studies.
38. David Newburg has led most of this work (Newburg et al., 2005); Lars Bode led the HIV study (Bode et al., 2012).
39. It could also be a way for mothers to *manipulate* their babies. It's in a baby's interest to monopolise as much of its mother's attention as possible, and evolution has given babies many ways of doing so: crying, nuzzling, and sheer adorableness. But a mother has to divide her care between many children, present and future. If she spends too much effort on any one, she might not have enough energy to raise more. So evolution ought to equip mothers with countermeasures – and evolutionary biologist Katie Hinde suspects that milk is one of them. It nourishes specific microbes, and, as we saw last chapter, some microbes can affect the behaviour of their host. By altering the HMO content of her breast milk, perhaps a mother can (inadvertently) elect for mind-manipulating microbes that influence her baby in ways that benefit her. For example, if the infant is less anxious, it might become independent sooner, leaving mum to focus on other kids.
40. Importance of glycans: Marcobal et al., 2011; Martens et al., 2014; fucose and sick mice: Pickard et al., 2014.
41. Reviewed in Fischbach and Sonnenburg, 2011; Koropatkin et al., 2012; Schluter and Foster, 2012.
42. Reviewed in Kiers and West, 2015; Wernegreen, 2004.
43. Genes that allow their owners to sense and adapt to changing environments are quick to go. After all, these microbes no longer have to cope with the vagaries of weather, temperature, or food supplies. In the cushy confines of an insect's cells, they can settle down into millions of years of constancy. They also tend to lose genes for repairing or reshuffling their DNA, which stops them from fixing problems in their remaining sequences.

44. Reviewed in McCutcheon and Moran, 2011; Russell et al., 2012; Bennett and Moran, 2013.
45. It's debatable whether these count as separate species or not; it's such a weird set-up that the traditional definitions don't quite apply.
46. Matthew Campbell, James van Leuven, and Piotr Lukasik have led this work (Campbell et al., 2015; Van Leuven et al., 2014); the Chilean results are as yet unpublished.
47. Bennett and Moran, 2015.

5. IN SICKNESS AND IN HEALTH

1. Rohwer wrote about the Line Islands expedition in *Coral Reefs in the Microbial Seas* (Rohwer and Youle, 2010), a richly detailed and often hilarious read. Except for the experiments referenced below, other details from this section can be found in *Coral Reefs in the Microbial Seas.*
2. Rohwer's model of coral reef death is described in Barott and Rohwer, 2012; Lisa Dinsdale's work on coral microbes is published in Dinsdale et al., 2008; Jennifer Smith's experiment with fleshy algae is detailed in Smith et al., 2006; Rebecca Vega Thurber led the study of coral viruses: Thurber et al., 2008, 2009; Linda Kelly led the study of the black reefs: Kelly et al., 2012; Tracy McDole led the development of the microbialisation score: McDole et al., 2012.
3. When US satirist Stephen Colbert covered this virus experiment on his show, he asked, 'Who has been fucking the corals?'
4. There are coral diseases caused by single microbes; for example, white pox disease is the work of *Serratia marascens,* a bacterium found in soil and sewage. But such examples are the exception rather than the rule.
5. The concept of dysbiosis is reviewed in Bäckhed et al., 2012; Blumberg and Powrie, 2012; Cho and Blaser, 2012; Dethlefsen et al., 2007; Ley et al., 2006. The term is often wrongly credited to that eccentric Russian Elie Metchnikoff, but it was already in use decades earlier.
6. The alumni on Jeff Gordon's star-studded roster include many of the people we meet elsewhere in the book, including Justin Sonnenburg, Ruth Ley, Lora Hooper, and John Rawls. Rob Knight is a long-time collaborator. Sarkis Mazmanian says that he entered the field thanks to an opinion piece that Gordon wrote in 2001, 'before the microbiome was the microbiome'.
7. The facility is run by David O'Donnell and Maria Karlsson, who have been with Gordon since 1989, and Justin Serugo, a refugee from the Democratic Republic of Congo who worked as a janitor at the university before becoming part of the team. I'm grateful to them for showing me around.

8. In the 1940s, microbiologist James Reyniers and engineer Philip Trexler developed ways of mass-producing germ-free rodents (Kirk, 2012). They would remove uteruses from pregnant females, bathe them in disinfectants, transfer them into isolators, cut the foetuses out, and then rear them by hand. In this way, they bred germ-free mice, rats, and guinea pigs, before moving on to pigs, cats, dogs and even monkeys. The technique was clearly successful but those early isolators, with their cold steel, chunky gauntlets and small viewing windows, were inconvenient and prohibitively expensive. By 1957, Trexler had devised a plastic version with rubber gloves similar to those in Gordon's lab. It was easier to use and cost a tenth as much to make.
9. Fred Bäckhed led this study (Bäckhed et al., 2004).
10. The links between the microbiome and obesity are reviewed in Zhao, 2013 and Harley and Karp, 2012. The first study showing that obese people and mice have different gut communities was led by Ruth Ley (Ley et al., 2005), while Peter Turnbaugh did the experiments transplanting microbes from obese people into germ-free mice (Turnbaugh et al., 2006).
11. Patrice Cani led the work on *Akkermansia*, together with Willem de Vos who discovered it (Everard et al., 2013); Lee Kaplan led the study on gastric bypass surgery (Liou et al., 2013).
12. Ridaura et al., 2013.
13. Michelle Smith and Tanya Yatsunenko led this study; Mark Manary and Indi Trehan were also involved (Smith et al., 2013a).
14. As the great ecologist Bob Paine once said, 'compounded perturbations yield ecological surprises'. He was talking about national parks, islands, and estuaries. He could easily have been talking about our bodies (Paine et al., 1998).
15. The interplay between the microbiome and the immune system is reviewed in Belkaid and Hand, 2014; Honda and Littman, 2012; Round and Mazmanian, 2009.
16. There are hundreds of papers on inflammatory bowel disease and the microbiome, but I recommend the following reviews by leaders in the field: Dalal and Chang, 2014; Huttenhower et al., 2014; Manichanh et al., 2012; Shanahan, 2012; Wlodarska et al., 2015. See also Wendy Garrett's studies on how immunity affects the microbiome (Garrett et al., 2007, 2010), and these papers on the microbiome changes that accompany IBD: Morgan et al., 2012; Ott et al., 2004; Sokol et al., 2008.
17. Dirk Gevers led this study, which is one of the largest to look at connections between microbiome and IBD (Gevers et al., 2014).
18. Cadwell et al., 2010.
19. As reviewed in Berer et al., 2011; Blumberg and Powrie, 2012; Fujimura and Lynch, 2015; Kostic et al., 2015; Wu et al., 2015.

20. Gerrard's paper: Gerrard et al., 1976; Strachan's follow-up: Strachan, 1989; Strachan is sometimes wrongly credited as the father of the hygiene hypothesis, although in Strachan, 2015 he himself denies paternity for that *enfant terrible*, citing many thinkers who preceded him and claiming that his word choice 'owed more to an alliterative tendency than to my aspiration to claim a new scientific paradigm'.

21. As reviewed in Arrieta et al., 2015; Brown et al., 2013; Stefka et al., 2014.

22. Graham Rook coined the term 'old friends' (Rook et al., 2013).

23. Fujimura et al., 2014; the difference in richness might be because dogs are larger than cats, and spend more time outside.

24. Dominguez-Bello's study: Dominguez-Bello et al., 2010; epidemiological studies showing links between C-sections and later diseases: Darmasseelane et al., 2014; Huang et al., 2015.

25. Eugene Chang showed the influence of saturated fats (Devkota et al., 2012); Andrew Gewirtz studied the two additives (Chassaing et al., 2015).

26. Burkitt's adventures are reported in Altman, 1993; his views on fibre are quoted in Sonnenburg and Sonnenburg, 2015, p. 119.

27. Wendy Garrett and others showed that fibre-digesting bacteria produce SCFAs (Furusawa et al., 2013; Smith et al., 2013b); Mahesh Desai showed that without fibre, gut bacteria devour the mucus layer, and presented this unpublished work at a conference.

28. Justin and Erica Sonnenburg showed that a lack of fibre causes extinctions in the gut (Sonnenburg et al., 2016), and reviewed the benefits of fibre (Sonnenburg and Sonnenburg, 2014).

29. Several microbiome studies have looked at rural populations, including seminal papers by Carlotta de Filippo and Tanya Yatsunenko (De Filippo et al., 2010; Yatsunenko et al., 2012).

30. American Chemical Society, 1999.

31. The effects of antibiotics on the microbiome are reviewed in Cox and Blaser, 2014, which also contains estimates for antibiotic doses taken by children; studies showing how antibiotics affect the microbiome include Dethlefsen and Relman, 2011; Dethlefsen et al., 2008; Jakobsson et al., 2010; Jernberg et al., 2010; Schubert et al., 2015.

32. This was discovered in the 1960s, when scientists showed that faeces from mice could stop *Salmonella* from growing, but not if they were first treated with antibiotics (Bohnhoff et al., 1964).

33. Katherine Lemon used this analogy in Lemon et al., 2012.

34. Blaser's first experiments on antibiotics and obesity were done with colleague Ilseung Cho (Cho et al., 2012); the second study was led by Laura Cox (Cox et al., 2014); his epidemiological work was led by Leonardo Trasande (Trasande et al., 2013).

35. He said this on Twitter. Marshall is the man who confirmed that *H. pylori* causes gastritis by swallowing the bacterium himself.

36. Maryn McKenna's story about the post-antibiotic future (McKenna, 2013), and her book *Superbug* (McKenna, 2010), are required reading on this topic.

37. Rosebury, 1969, p. 11.

38. Blaser's studies of *H. pylori*: Blaser, 2005; his concern about its disappearance: Blaser, 2010 and Blaser and Falkow, 2009; the long history of *H. pylori* and humanity: Linz et al., 2007; the *Lancet* opinion piece: Graham, 1997; *H. pylori* does not affect overall mortality: Chen et al., 2013.

39. Zack Lewis led this work.

40. Studies of rural and hunter-gatherer microbiomes: Clemente et al., 2015; Gomez et al., 2015; Martínez et al., 2015; Obregon-Tito et al., 2015; Schnorr et al., 2014); a study of microbes in fossilised faeces: Tito et al., 2012.

41. Le Chatelier et al., 2013.

42. In Cameroon, people who are infected with a parasitic amoeba called *Entamoeba* have a wider variety of gut bacteria, and especially so if they also carry parasitic worms. The bacteria might be creating openings for the parasites, or the parasites might somehow boost the range of bacteria; either way, here's a case where the supposedly desirable diversity of rural people indicates the presence of something undesirable (Gomez et al., 2015).

43. Moeller et al., 2014.

44. Blaser, 2014, p. 6.

45. Eisen, 2014.

46. Mukherjee, 2011, pp. 349–356.

47. There are so many scientific review papers that link the gut microbiome to one disease or another that Elizabeth Bik, a tireless chronicler of new microbiome research, started a spoof hashtag on Twitter called #gutmicrobiomeandrandomthing. Entries included 'Gut Microbiome and Always Ending up Standing in the Slowest Line at the Cash Register', 'Gut Microbiota and the Art of Motorcycle Maintenance', and 'Gut Microbiome and the Prisoner of Azkaban'.

48. *The Allium*, 2014.

49. On dysbiosis, Fergus Shanahan cautions his fellow scientists to 'mind George Orwell's refrain, "the slovenliness of our language makes it easier to have foolish thoughts." Inaccurate thinking can arise when clinicians become captive to errors in nomenclature and imprecise terminology. Neologisms should be used with caution; they often are unnecessary or imply an understanding where none exists' (Shanahan and Quigley, 2014).

50. This argument appeared in a piece I wrote for the *New York Times*, about the contextual nature of the microbiome (Yong, 2014c).
51. Ruth Ley and Omry Koren did this work (Koren et al., 2012).
52. The vaginal studies were led by Larry Forney and Jacques Ravel (Gajer et al., 2012; Ma et al., 2012); the other body parts were analysed by Pat Schloss (Ding and Schloss, 2014).
53. Katherine Pollard led one study, and Rob Knight led the other (Finucane et al., 2014; Walters et al., 2014).
54. Susannah Salter and Alan Walker showed that extraction kits, which pull DNA out of swabs and samples and prepare them for sequencing, are almost always contaminated by low levels of microbial DNA (Salter et al., 2014).
55. For example, Pat Schloss created a program that could look at a given microbiome and predict how vulnerable it was to colonisation by C-diff: (Schubert et al., 2015).
56. A few scientists have tried to answer these questions by tracking their own microbiomes. Eric Alm and Lawrence David from MIT did so every day for a year. When David picked up traveller's diarrhoea in Bangkok, he could see his gut community go through a period of upheaval before bouncing back to normal. When Alm picked up *Salmonella* after an unlucky restaurant visit, he saw how the bug quickly dominated his gut, and how the community switched to a different state when he returned to health (David et al., 2014).
57. Sathish Subramanian led this work (Subramanian et al., 2014).
58. Andrew Kau also led this study, together with Planer (Kau et al., 2015).
59. Redford et al., 2012.

6. THE LONG WALTZ

1. Fritz's story: University of Utah, 2012; the initial characterisation of HS, led by Adam Clayton: Clayton et al., 2012; the second case hasn't been published yet.
2. Unlike Fritz's crab apple tree, the one that impaled the kid is still alive, so Dale is planning to visit it and grab a sample of strain HS from the wild. Then he can try the 'high-risk, high-reward experiment': to inject HS into insects and see if he can artificially set up a new symbiosis.
3. Dale can tell because these versions – say, the one in tsetse flies and the one in weevils – have both lost *different* genes from the full set possessed by HS. They evolved from ancestral HS-like microbes that became independently domesticated.

4. Aphids and sexual transmission: Moran and Dunbar, 2006; cannibalistic woodlice: Le Clec'h et al., 2013; bugs and backwash: Caspi-Fluger et al., 2012; human swallowing: Lang et al., 2014; wasps as dirty needles: Gehrer and Vorburger, 2012.
5. John Jaenike took mites that were sucking the blood of one species of fruit fly, and placed them onto a second species. Sure enough, the latter flies picked up microbes that were only found in the former (Jaenike et al., 2007).
6. The origins of new symbioses are discussed in Sachs et al., 2011 and Walter and Ley, 2011.
7. Kaltenpoth et al., 2005.
8. As reviewed in Funkhouser and Bordenstein, 2013; Zilber-Rosenberg and Rosenberg, 2008.
9. I once asked Fukatsu how he chose what to work on. He paused, pointed at some imaginary speck in the air, and said 'Oh! Interesting!' Then, he just smiled at me. When I put the same question to Martin Kaltenpoth, he said: 'I find a species that Takema isn't studying and then tell him that I'm working on it.' His stinkbug papers include Hosokawa et al., 2008, Kaiwa et al., 2014, and Hosokawa et al., 2012.
10. Pais et al., 2008.
11. Osawa et al., 1993.
12. To say the least. When I asked a broad range of microbiome scientists which findings they were most sceptical about, many pointed to these particular results.
13. Many underwater animals release their symbionts into the surrounding water, so that larvae and hatchlings have a decent supply around them. The bobtail squid does this every morning, at dawn. The medicinal leech sheds microbe-rich mucus from its gut every few days, and it's attracted to the leftover mucus of other leeches (Ott et al., 2015). Some nematode worms kill insects by vomiting hordes of toxic bacteria into their bloodstream; their larvae, which develop inside the dead insects, then suck up the killer symbionts for their own use (Herbert and Goodrich-Blair, 2007).
14. House-sharing humans: Lax et al., 2014; sociable baboons: Tung et al., 2015; roller derby players: Meadow et al., 2013.
15. Lombardo's idea, outlined in Lombardo, 2008, is just a hypothesis, but one that makes testable predictions. If he's right, animals that get microbes from the environment (like squid), or automatically inherit them (like aphids), are more likely to be solitary. Those that get microbes from their peers, as termites do, are more likely to have a more complex social system that regularly brings them into close contact with their contemporaries. To test this, scientists would need to draw family trees for different animal groups with both sociable and solitary members, like Fukatsu's stinkbugs, and to see if the evolution of microbial

partnerships consistently precedes the evolution of large groups. To my knowledge, no one has done that yet.

16. Fraune's first experiment: Fraune and Bosch, 2007; later studies showing how hydra select for the right microbes: Franzenburg et al., 2013; Fraune et al., 2009, 2010; a review of Bosch's work on hydra: Bosch, 2012.

17. As reviewed in Bevins and Salzman, 2011; Ley et al., 2006; Spor et al., 2011.

18. Whales and dolphins: interview with Amy Apprill; beewolves: Kaltenpoth et al., 2014.

19. Bee symbionts: Kwong and Moran, 2015; *Lactobacillus reuteri*: Frese et al., 2011; Rawls's swapping experiments: Rawls et al., 2006.

20. For example, Andrew Benson identified 18 regions within the mouse genome that affect the abundance of the most common gut microbes. Some of these regions influence the levels of individual microbe species, while others control entire groups (Benson et al., 2010).

21. Published under her married name, Lynn Sagan (Sagan, 1967).

22. Margulis and Fester, 1991.

23. The hologenome concept was first conceived by a biotechnologist named Richard Jefferson back in the 1980s, although he never got round to publishing it (Jefferson, 2010). He did present the theory at a conference in 1994, some thirteen years before the Rosenbergs independently came up with the same idea and name.

24. Hird et al., 2014.

25. As an example, Ruth Ley showed that human genes don't dictate the overall composition of our microbiome, but they do strongly influence the presence of specific groups. The most heritable bacterium in our bodies is a recently discovered and little-known species called *Christensenella* (Goodrich et al., 2014). Some people have it and others don't, and around 40 per cent of that variation is down to our genes. This enigmatic species is common during childhood, more prevalent among people with a healthy body weight, and often found together with a large network of other microbes. It might be a keystone species: one that's relatively rare but ecologically powerful.

26. The Rosenbergs propose the hologenome idea: Rosenberg et al., 2009; Zilber-Rosenberg and Rosenberg, 2008; Seth Bordenstein and Kevin Theis expand upon it: Bordenstein and Theis, 2015; Nancy Moran and David Sloan argue against it: Moran and Sloan, 2015.

27. Diane Dodd's experiment: Dodd, 1989; the Rosenberg's follow-up, led by Gil Sharon: Sharon et al., 2010.

28. Wallin: Wallin, 1927; Margulis and Sagan: Margulis and Sagan, 2002.

29. The first experiment with Werren: Bordenstein et al., 2001; the second one, done together with Robert Brucker: Brucker and Bordenstein, 2013.
30. Brucker and Bordenstein, 2014; Chandler and Turelli, 2014.

7. MUTUALLY ASSURED SUCCESS

1. Sapp, 2002.
2. René Dubos, the antibiotic-discovering bacteriologist whom we met in Chapter 2, brought Buchner's book to the attention of American publishers. This was one of the few historical moments when the study of insect symbionts entwines with the study of human microbes.
3. Moran's first study on *Buchnera* was done together with bacteriologist Paul Baumann (Baumann et al., 1995). Both now have symbionts named after them. *Baumannia* is found in the glassy-winged sharpshooter, and *Moranella* is found in the citrus mealybug, of which more later.
4. Nováková et al., 2013.
5. As reviewed in Douglas, 2006; Feldhaar, 2011.
6. For example, *Buchnera* can carry out every chemical reaction necessary for making the amino acids isoleucine or methionine – except the final one. It falls to the aphid to carry them over the finish line. Angela Douglas, Nancy Moran and others have sketched out these pathways in exquisite detail (Russell et al., 2013a; Wilson et al., 2010).
7. The funny thing is that different lineages of hemipterans have independently evolved the ability to drink phloem sap. Other insects haven't, even though they also have symbionts that could double as dietary supplements. So, why the hemipterans? Or rather, why not anything else? That is a mystery.
8. As reviewed in Wernegreen, 2004.
9. *Blochmannia* is closely related to *Buchnera* and that may not be a coincidence. Many carpenters farm aphids, like human farmers herding their livestock, and protect them from predators. In return, the aphids feed the ants with a sugary waste fluid called honeydew. And with that honeydew came aphid symbionts. Jennifer Wernegreen thinks that *Blochmannia* is the descendant of some symbiont that travelled out of an aphid's backside, ended up in an ant farmer, and stayed there (Wernegreen et al., 2009).
10. The history of the Galapagos Rift discovery is detailed in Smithsonian National Museum of Natural History, 2010 and especially in Robert Kunzig's *Mapping the Deep* (Kunzig, 2000), which also details the work on *Riftia* by Jones and Cavanaugh.
11. Cavanaugh published her ideas in 1981 (Cavanaugh et al., 1981), but it then took years of work to confirm that the bacteria work as she envisaged. Other

scientists had speculated about chemosynthetic microbes, but Cavanaugh was the first to show they exist, and that they form partnerships with animals. As a graduate student, she had discovered an entirely new way of life – and a surprisingly common one. Her work on *Riftia* is reviewed in Stewart and Cavanaugh, 2006.

12. As reviewed in Dubilier et al., 2008.

13. Dubilier discovers the two *Olavius* symbionts: Dubilier et al., 2001; she then discovers three more: Blazejak et al., 2005.

14. Ley et al., 2008a.

15. Another exception: the Iberian lynx – a tufty-eared European cat and a dedicated carnivore, which has an unexpected number of plant-digesting genes in its gut. It's possible that its microbiome has adapted to digest not just the rabbits that it hunts, but the plant matter in the guts of those rabbits (Alcaide et al., 2012).

16. On the proportion of a mammal's energy intake that comes from microbes: Bergman, 1990; reviews on the digestive systems of mammals: Karasov et al., 2011; Stevens and Hume, 1998.

17. Whales are interesting outliers. They're meat-eaters, feasting on tiny crustaceans, fish, or even other mammals. However, they evolved from plant-eating, deer-like animals, and have retained the large, multi-chambered foregut of their ancestors. They now use this foregut fermenter to deal with animal tissues, which, as Jon Sanders found, leaves them with a gut microbiome that's unlike anything on land, whether carnivore or herbivore (Sanders et al., 2015).

18. The hoatzin, a chicken-sized South American bird with a blue face, red eyes, orange feathers, and a punk crest, is also a foregut fermenter. It feeds largely on leaves, which it digests in its crop – an enlarged part of its gullet. Maria Gloria Dominguez-Bello showed that the bacteria in the crop are more similar to those in a cow's stomach than to those further down the bird's own gut (Godoy-Vitorino et al., 2012). It's unsurprising, then, that hoatzins stink like cow manure.

19. Ley et al., 2008b.

20. The three-toed sloth is an exception that proves the rule: it mostly eats the leaves of one particular tree, and so has a very restricted gut microbiome for a plant-eater (Dill-McFarland et al., 2015).

21. Hongoh, 2011.

22. This difference fooled some early biologists. Alfred E. Emerson saw that the most advanced termites lacked the protists that thrived in the lower termites, and so deduced that symbiotic microbes *prevented* animals from evolving 'higher social functions'. If he had known about the bacteria, he might have changed his tune.

23. Michael Poulsen led this work (Poulsen et al., 2014).

24. Amato et al., 2015.
25. David et al., 2013.
26. Chu et al., 2013.
27. W. J. Freeland and Daniel Janzen said, 'Presumably, when small amounts of a toxic food are given . . . there is selection for species or strains of bacteria capable of living with and degrading the toxin' (Freeland and Janzen, 1974).
28. Kohl et al., 2014.
29. This seems to be something that woodrats are very good at. Denise Dearing, who led Kohl's work, uncovered a similar story in another species (the white-throated woodrat) that lives in a different desert (the Lower Sonoran), specialises on a different plant (cactus), and tolerates a different toxin (oxalate). Detoxifying microbes were the heroes of this tale, too, and by transplanting them into naïve lab rats, Dearing could also turn them into successful cactus-eaters (Miller et al., 2014).
30. Reindeer and lichen: Sundset et al., 2010; Tannin-degraders: Osawa et al., 1993; coffee berry borer beetle: Ceja-Navarro et al., 2015.
31. Six, 2013.
32. Adams et al., 2013; Boone et al., 2013.
33. In contrast to the examples in this chapter, microbes can constrain their hosts, too. Insect symbionts tend to be more sensitive to high temperatures than their hosts, so their numbers plummet in hot weather – even Buchner saw that. That restricts the places where their hosts can thrive and might lead to 'mutualistic meltdown' in a warming world (Wernegreen, 2012). Brine shrimps – the sea monkeys of children's aquaria – have gut bacteria that help them to digest algae, but since these particular microbes love salty conditions, the shrimps are forced to live in saltier waters than they would normally prefer (Nougué et al., 2015). Microbes can also set dietary restrictions. Imagine that an insect starts eating a plant that makes ample amounts of a certain essential nutrient. Its symbiont no longer needs to supply that nutrient, so it quickly loses the genes for doing so. The host doesn't need to compensate for those losses, because of the plant. Everything's hunky-dory. Then, the plant starts dying out. Now, the insect has only two options: it can find another plant that makes the same nutrient, or it can pick up a new microbe as a supplement. If it can't do either, it's in trouble.
34. Wybouw et al., 2014.

8. ALLEGRO IN E MAJOR

1. Ochman et al., 2000.
2. This classic experiment was done in 1928 by British bacteriologist Frederick Griffith.

3. Avery's discovery was one of the most important in modern genetics because it suggested, against conventional wisdom, that DNA was the stuff of genes. At the time, most scientists believed that genes were made of proteins, which have infinitely varied shapes, and that DNA, with its four repeating building blocks, was boring and unworthy of attention. Avery showed otherwise. In many ways, he set the groundwork for later discoveries that would cement DNA's status as the all-important molecule of life (Cobb, 2013).

4. This was a monumental discovery – one that earned Lederberg a Nobel Prize in 1958, at the prodigal age of 33.

5. As reviewed in Boto, 2014; Keeling and Palmer, 2008.

6. Hehemann et al., 2010; *Zobellia*, incidentally, is named after marine microbiologist Claude E. ZoBell.

7. Paul Portier, a much-maligned symbiosis advocate of the early twentieth century, argued that we swallowed fresh mitochondria and other symbionts in our food, which revitalised the old ones inside our bodies by fusing with them. Not quite, but so close!

8. Unpublished data.

9. Smillie et al., 2011.

10. I'm excluding mitochondria from this; they'd stopped being free-living bacteria billions of years before animals evolved.

11. The Human Genome Project paper: Lander et al., 2001; the rebuttal, led by Jonathan Eisen and Steven Salzberg: Salzberg, 2001.

12. *Wolbachia* DNA in *Drosophila*: Salzberg et al., 2005; *Wolbachia* DNA in other animals: Hotopp et al., 2007; complete *Wolbachia* genome in *D. ananassae:* Hotopp et al., 2007.

13. This message still falls on deaf ears. When scientists sequence animal genomes they deliberately purge their results of anything bacterial, on the assumption that such sequences are contaminants. The pea aphid genome contains horizontally transferred *Buchnera* genes, but these have been stripped from the version that was uploaded into online databases. The fly *D. ananassae* has an entire *Wolbachia* genome in it, but you'd never be able to tell by looking at the publicly available genome – those sequences have been removed. This unforgiving approach makes sense because contamination *is* a genuine problem. But it also feeds the pernicious view that bacterial sequences are necessarily foreign, and must be discarded lest they contaminate the purity of an animal's genome. 'A circular argument ensues where genome sequencing projects remove all bacterial sequences because animals do not have HGT from bacteria, and the examination of those same genomes for HGT reinforces the notion that HGT from bacteria to animals does not occur,' wrote Dunning-Hotopp (Dunning-Hotopp et al., 2011).

14. A bacterium in your gut might be able to transfer its genes into one of your intestinal cells, but once that cell dies, the bacterial DNA goes with it. The gene might become part of *a* human genome, but never *the* human genome. In 2013, Dunning-Hotopp showed that these short-lived unions are surprisingly common (Riley et al., 2013). She analysed hundreds of human genomes that had been sequenced from body cells – the ones from kidneys or skin or livers, none of which get passed on to offspring. She found traces of bacterial DNA in around a third of them. They were especially common in cancer cells; an intriguing result with unclear implications. It might be that tumours are especially prone to genetic intrusions, or that bacterial genes help to transform healthy cells into cancerous ones.
15. Etienne Danchin has done much of this work (Danchin and Rosso, 2012; Danchin et al., 2010).
16. Acuna et al., 2012.
17. Several scientists have contributed to this work including Jean-Michel Drezen, Michael Strand, and Gaelen Burke: Bezier et al., 2009; Herniou et al., 2013; Strand and Burke, 2012.
18. This actually happened twice. A different lineage of wasps – the ichneumons – independently domesticated a different lineage of viruses and now use them in a similar way to the bracoviruses (Strand and Burke, 2012).
19. In parallel with the *tae* gene example (Chou et al., 2014), Seth Bordenstein revealed a similar story involving another kingdom-hopping antibiotic gene (Metcalf et al., 2014).
20. There is one other example of such a set-up: one bacterium has managed to find its way *inside* the mitochondria of ticks, where it now resides; it has been named *Midichloria*, after the much-maligned symbionts from the *Star Wars* universe that connect their owners with the Force.
21. McCutcheon bills these diminished microbes as 'conundrums of biological classification' (McCutcheon, 2013). They are obviously bacteria and they still possess their own distinct genomes. But they cannot survive alone and some of them (like *Moranella*) cannot even define their own boundaries. They're almost like mitochondria or chloroplasts. Those structures are called organelles, but to McCutcheon, organelles are just symbionts *in extremis* – the culmination of a long process of genetic loss and relocation, which irrevocably binds animals and bacteria together.
22. Graduate student Filip Husnik led this work (Husnik et al., 2013).
23. You might remember peptidoglycan as one of the MAMPs that control the development of Margaret McFall-Ngai's squid.
24. It gets weirder! In other species of mealybug, *Moranella* has been replaced by other symbionts. All of these, like *Moranella*, are related to HS, the bacterium that entered the hand of Thomas Fritz and that Colin Dale later identified.

25. They also worked with parasitoid expert Molly Hunter.
26. *Hamiltonella* is named after Bill Hamilton, the legendary evolutionary biologist who trained Moran.
27. The discovery of *Hamiltonella*: Oliver et al., 2005; discovery of *Hamiltonella*'s phage: Moran et al., 2005; the flexible nature of the aphid/*Hamiltonella symbiosis:* Oliver et al., 2008.
28. Moran and Dunbar, 2006.
29. As reviewed in Jiggins and Hurst, 2011.
30. On the Japanese bean bug, a study led by symbiont guru Takema Fukatsu: Kikuchi et al., 2012; on the many secondary symbionts of aphids: Russell et al., 2013b; on secondary symbionts and aphid success: Henry et al., 2013.
31. Jaenike identifies *Spiroplasma* as the secret of the flies' success: Jaenike et al., 2010; the symbiont's fast spread: Cockburn et al., 2013.
32. Molly Hunter discovered this sweep: Himler et al., 2011.
33. They might even be able to predict tomorrow's partnerships. A few years ago, Jaenike showed that *Spiroplasma* could protect other species of fruit fly, beyond the one he studied. One of these doesn't have any bacterial defenders yet, but is also targeted by sterilising nematodes. When Jaenike artificially united this fly with *Spiroplasma* in his lab, he saw that it could reproduce again (Haselkorn et al., 2013). In the wild, this alliance hasn't happened yet, for whatever reason, but it's so beneficial to the fly that it is surely bound to. And once it does, it will assuredly take off.

9. MICROBES A LA CARTE

1. Filarial diseases, and the *Wolbachia*-carrying nematodes that cause them, are reviewed in Taylor et al., 2010 and Slatko et al., 2010.
2. Bacteria-like structures seen in filarial nematodes: Kozek, 1977; Mclaren et al., 1975; the bacteria are identified as *Wolbachia:* Taylor and Hoerauf, 1999.
3. Taylor's colleague Achim Hoerauf co-led these trials (Hoerauf et al., 2000, 2001; Taylor et al., 2005).
4. Doxycycline had other benefits too. In parts of central Africa, it is incredibly hard to treat people with river blindness because they also carry a second filarial nematode called *Loa loa* – the so-called 'eyeworm'. If you kill the species that causes river blindness, the eyeworms die too, and *their* larvae are so large that they can block blood vessels and cause brain damage. But since the eyeworm doesn't have *Wolbachia*, doxycycline won't harm it. This drug can attack the parasites behind river blindness without inflicting heavy collateral damage.
5. The A•WOL consortium's strategy: Johnston et al., 2014; Taylor et al., 2014; the minocycline results are unpublished.

6. Voronin et al., 2012.
7. Rosebury, 1962, p. 352.
8. On the decline of amphibians: Hof et al., 2011; on Bd: Kilpatrick et al., 2010; Amphibian Ark, 2012.
9. Eskew and Todd, 2013; Martel et al., 2013.
10. Harris et al., 2006.
11. The discovery of the Conness frog population: Woodhams et al., 2007; J-liv protects against Bd in the lab: Harris et al., 2009. The field trial results with J-liv are unpublished at time of writing
12. Becker's work on golden frogs: Becker et al., 2015; the diversity of bacteria on frog skin: Walke et al., 2014; the Madagascar project, done with Molly Bletz: Bletz et al., 2013; how metamorphosis changes the microbiome, led by Valerie McKenzie: Kueneman et al., 2014.
13. Valerie McKenzie and Rob Knight have developed a method of predicting a frog's resilience to Bd based on its immune system, the mucus layer on its skin, and its microbiome (Woodhams et al., 2014).
14. Kendall, 1923, p. 167.
15. A history of probiotics research: Anukam and Reid, 2007.
16. The fate of ingested microbes: Derrien and van Hylckama Vlieg, 2015; Jeff Gordon's work on Activia yoghurt, led by Nathan McNulty: McNulty et al., 2011; Wendy Garrett's experiments: Ballal et al., 2015.
17. On the definition of probiotics: Hill et al., 2014; reviews of research on probiotics: Slashinski et al., 2012 and McFarland, 2014; Cochrane reviews: AlFaleh and Anabrees, 2014; Allen et al., 2010; Goldenberg et al., 2013.
18. Katan, 2012; *Nature*, 2013; Reid, 2011.
19. As reviewed in: Ciorba, 2012; Gareau et al., 2010; Gerritsen et al., 2011; Petschow et al., 2013; Shanahan, 2010.
20. Most of the research on probiotics focused on the gut, but the term could also refer to any product that contains beneficial microbes, including skin creams, shampoos, or mouthwashes. All of these products are being actively developed.
21. Excellent, but not entirely unblemished. Benign groups like *Lactobacillus* and *Bifidobacterium* have caused rare cases of blood poisoning. And in one infamous Dutch clinical trial, patients with acute pancreatitis were *more* likely to die when they took probiotics than a placebo (Gareau et al., 2010). Broadly speaking, these products are safe, but doctors might think twice about giving them to people who are critically ill or have immune deficiencies.
22. The story of Raymond Jones and *Synergistes*: Aung, 2007; CSIROpedia; *New York Times*, 1985; Jones does his first rumen transplant: Jones and Megarrity, 1986; *Synergistes jonesii* is described and named: Allison et al., 1992.
23. Ellis et al., 2015.

24. Interview with Denise Dearing; the white-throated woodrat from her experiments also uses *Oxalobacter* to detoxify the oxalate in the cactus it eats.
25. As reviewed in: Bindels et al., 2015; Delzenne et al., 2013.
26. Underwood et al., 2009.
27. Kenya Honda's work: Atarashi et al., 2013; the path to the clinic: Schmidt, 2013.
28. Reviews on FMT include Aroniadis and Brandt, 2014; Khoruts, 2013; Petrof and Khoruts, 2014; popular pieces include Nelson, 2014.
29. Petrof's team now use an entirely disposable system involving a 'stool hat' – essentially, a Tupperware container that attaches to a toilet seat – and some coffee filters.
30. Koch and Schmid-Hempel, 2011.
31. Hamilton et al., 2013.
32. Zhang et al., 2012.
33. Van Nood et al., 2013.
34. FMT and IBD: Anderson et al., 2012; the FMT/obesity trial: Vrieze et al., 2012.
35. This result was predictable. Remember Vanessa Ridaura and Jeff Gordon's experiment, in which gut microbes from lean mice were transplanted into fat ones, and the recipients only lost weight when they also ate a healthy diet.
36. Petrof and Khoruts, 2014.
37. Frozen stool samples work just as well as fresh ones: Youngster et al., 2014; OpenBiome's work is described in Eakin, 2014.
38. Microbiologist Stanley Falkow was the first person to deliver FMT through capsules, back in 1957. At the time, his hospital was besieged by a rogue strain of *Staphylococcus*, and all patients had to take pre-emptive antibiotics before an operation. Unfortunately, these drugs also annihilated their gut bacteria, leaving them with diarrhoea and indigestion. Having realised what was happening, Falkow asked visiting patients to bring a stool sample with them. He then piped the sample into capsules, which he asked the patients to swallow once they re-emerged from surgery. 'The chief hospital administrator discovered what was up,' Falkow later wrote. 'He confronted me and exclaimed: Falkow, is it true you've been feeding the patients shit! I responded: Yes I had been a participant in a clinical study that involved the patients ingesting their own feces.' He was fired, but re-hired two days later (Falkow, 2013).
39. Smith et al., 2014.
40. One team recently reported a case of weight gain after an FMT, but it was unclear if the procedure led to the extra pounds (Alang and Kelly, 2015).
41. The Power of Poop (thepowerofpoop.com) collects stories from DIY faecal-transplanters and campaigns for doctors to take the procedure seriously.

42. While I was writing this chapter, a stranger emailed me to ask if she needed FMT because she had been drinking diet soda. For the record, the answer is no.
43. Signatories include people we've met already, such as Jeff Gordon, Rob Knight and Martin Blaser: Hecht et al., 2014.
44. RePOOPulate: Petrof et al., 2013; other studies that created defined cocktails of mcirobes: Buffie et al., 2014; Lawley et al., 2012.
45. Khoruts doesn't quite agree with this. 'The full spectrum of microbes harvested from donors has been designed by nature, and has a proven safety track record in the original host,' he says. 'That's a very hard benchmark to improve upon with any kind of synthetic.' If he needed a transplant himself, he'd do it the old-fashioned way
46. Haiser et al., 2013.
47. As reviewed in Carmody and Turnbaugh, 2014; Clayton et al., 2009; Vétizou et al., 2015.
48. Dobson et al., 2015; Smith et al., 2015.
49. As reviewed in Haiser and Turnbaugh, 2012; Holmes et al., 2012; Lemon et al., 2012; Sonnenburg and Fischbach, 2011.
50. A review of Hazen's work on TMAO: Tang and Hazen, 2014; the team finds a chemical that stops bacteria from making TMAO: Wang et al., 2015.
51. I wrote about these smart probiotics for *New Scientist* in 2015 (Yong, 2015c).
52. Kotula et al., 2014.
53. Chang's work on *E. coli*: Saeidi et al., 2011. Jim Collins has co-founded a start-up called Synlogic to take these microbes to market, and he thinks they're just a couple of years away from their first clinical trials.
54. Rutherford, 2013.
55. As reviewed in Claesen and Fischbach, 2015; Sonnenburg and Fischbach, 2011.
56. Timothy Lu published the first paper on programming B-theta (Mimee et al., 2015); Sonnenburg's group is not far behind.
57. Olle, 2013.
58. As reviewed in Iturbe-Ormaetxe et al., 2011 and LePage and Bordenstein, 2013.
59. His original idea was to genetically engineer *Wolbachia*, equipping it with genes that would produce anti-dengue antibodies. If that worked, the bacterium should spread quickly through a population, as it is wont to do, and take its dengue-blocking antibodies with it. But engineering *Wolbachia* was not easy; O'Neill gave up after six years, and no one has since succeeded in his place.
60. The first mention of the popcorn strain: Min and Benzer, 1997; Conor McMeniman stably infects eggs with *Wolbachia:* McMeniman et al., 2009.

61. The simulations were done by Michael Turelli at the University of California, Davis (Bull and Turelli, 2013) and later confirmed in a field trial. When the team released mosquitoes carrying the popcorn strain on a small island in Vietnam, neither the insects nor their symbionts managed to gain a foothold.
62. Karyn Johnson and Luis Teixeira show that *Wolbachia* makes flies resistant to viruses: Hedges et al., 2008; Teixeira et al., 2008; O'Neill's team, including Luciano Moreira, shows that the same is true for mosquitoes: Moreira et al., 2009.
63. Tom Walker loaded wMel into *Aedes* eggs, while Ary Hoffmann and Scott Ritchie co-led the cage trials with O'Neill (Walker et al., 2011).
64. O'Neill knew what could happen if scientists ignored local communities. In 1969, World Health Organization scientists travelled to India to try a variety of new techniques for controlling mosquitoes, including genetic modification, irradiation, and *Wolbachia* to try and control mosquito populations (*Nature*, 1975). The project was secretive and people grew suspicious. Newspapers started accusing the scientists, some of whom were American, of using India as a test-bed for experiments that were too dangerous for US soil, and even developing biological weapons. The team responded by not responding at all. 'It was a PR nightmare,' says O'Neill. 'They were thrown out of the country, and the controversy made the genetic modification of mosquitoes taboo for twenty years.' O'Neill wanted to avoid making the same mistake.
65. Hoffmann et al., 2011.
66. The Eliminate Dengue projects: www.eliminatedengue. com; O'Neill and Kate Retzki discussed the Townsville project with me; Bekti Andari and Ana Cristina Patino Taborda talked about the Indonesia and Colombia projects with me.
67. Chrostek et al., 2013; McGraw and O'Neill, 2013.
68. Adding *Wolbachia* to *Anopheles* mosquitoes: Bian et al., 2013; using *Wolbachia* to control other insect pests: Doudoumis et al., 2013; certain gut bacteria in mosquitoes can also block *Plasmodium* parasites, and could be fed to the insects as an anti-malarial probiotic: Hughes et al., 2014.

10. TOMORROW THE WORLD

1. Our microbial cloud: Meadow et al., 2015; estimates of aerosolised bacteria: Qian et al., 2012.
2. Lax et al., 2014.
3. For the record, it's called an axilla.
4. Van Bonn et al., 2015.
5. The Hospital Microbiome Project: Westwood et al., 2014; on hospital microbes and infections: Lax and Gilbert, 2015.

6 Gibbons et al., 2015.

7. Green's work on hospital windows: Kembel et al., 2012; Florence Nightingale's writings: Nightingale, 1859.

8. On the microbiome of the indoor environment: Adams et al., 2015; Jessica Green's work on Lillis Hall: Kembel et al., 2014; Green's TED talk and review on bioinformed design: Green, 2011, 2014.

9. Gilbert et al., 2010; Jansson and Prosser, 2013; Svoboda, 2015.

10. Alivisatos et al., 2015.

BIBLIOGRAPHY

Abbott, A.C. (1894) *The Principles of Bacteriology* (Philadelphia: Lea Bros & Co.).

Acuna, R., Padilla, B.E., Florez-Ramos, C.P., Rubio, J.D., Herrera, J.C., Benavides, P., Lee, S-J., Yeats, T.H., Egan, A.N., Doyle, J.J., et al. (2012) 'Adaptive horizontal transfer of a bacterial gene to an invasive insect pest of coffee', *Proc. Natl. Acad. Sci.* 109, 4197–4202.

Adams, A.S., Aylward, F.O., Adams, S.M., Erbilgin, N., Aukema, B.H., Currie, C.R., Suen, G., and Raffa, K.F. (2013) 'Mountain pine beetles colonizing historical and naive host trees are associated with a bacterial community highly enriched in genes contributing to terpene metabolism', *Appl. Environ. Microbiol.* 79, 3468–3475.

Adams, R.I., Bateman, A.C., Bik, H.M., and Meadow, J.F. (2015) 'Microbiota of the indoor environment: a meta-analysis', *Microbiome* 3. doi: 10.1186/s40168-015-0108-3.

Alang, N. and Kelly, C.R. (2015) 'Weight gain after fecal microbiota transplantation', *Open Forum Infect. Dis.* 2, ofv004–ofv004.

Alcaide, M., Messina, E., Richter, M., Bargiela, R., Peplies, J., Huws, S.A., Newbold, C.J., Golyshin, P.N., Simón, M.A., López, G., et al. (2012) 'Gene sets for utilization of primary and secondary nutrition supplies in the distal gut of endangered Iberian lynx', *PLoS ONE* 7, e51521.

Alcock, J., Maley, C.C., and Aktipis, C.A. (2014) 'Is eating behavior manipulated by the gastrointestinal microbiota? Evolutionary pressures and potential mechanisms', *BioEssays* 36, 940–949.

Alegado, R.A. and King, N. (2014) 'Bacterial influences on animal origins', *Cold Spring Harb. Perspect. Biol.* 6, a016162–a016162.

Alegado, R.A., Brown, L.W., Cao, S., Dermenjian, R.K., Zuzow, R., Fairclough, S.R., Clardy, J., and King, N. (2012) 'A bacterial sulfonolipid triggers multicellular development in the closest living relatives of animals', *Elife* 1, e00013.

AlFaleh, K. and Anabrees, J. (2014) 'Probiotics for prevention of necrotizing enterocolitis in preterm infants', in *Cochrane Database of Systematic Reviews*, The Cochrane Collaboration (Chichester, UK: John Wiley & Sons).

Alivisatos, A.P., Blaser, M.J., Brodie, E.L., Chun, M., Dangl, J.L., Donohue, T.J., Dorrestein, P.C., Gilbert, J.A., Green, J.L., Jansson, J.K., et al. (2015) 'A unified initiative to harness Earth's microbiomes', *Science* 350, 507–508.

Allen, S.J., Martinez, E.G., Gregorio, G.V., and Dans, L.F. (2010) 'Probiotics for treating acute infectious diarrhoea', in *Cochrane Database of Systematic Reviews*, The Cochrane Collaboration (Chichester, UK: John Wiley & Sons).

Allison, M.J., Mayberry, W.R., Mcsweeney, C.S., and Stahl, D.A. (1992) '*Synergistes jonesii, gen. nov., sp.nov.*: a rumen bacterium that degrades toxic pyridinediols', *Syst. Appl. Microbiol.* 15, 522–529.

The Allium (2014) 'New Salmonella diet achieves 'amazing' weight-loss for microbiologist'.

Altman, L.K. (April 1993) 'Dr. Denis Burkitt is dead at 82; thesis changed diets of millions', *New York Times.*

Amato, K.R., Leigh, S.R., Kent, A., Mackie, R.I., Yeoman, C.J., Stumpf, R.M., Wilson, B.A., Nelson, K.E., White, B.A., and Garber, P.A. (2015) 'The gut microbiota appears to compensate for seasonal diet variation in the wild black howler monkey (*Alouatta pigra*)', *Microb. Ecol.* 69, 434–443.

American Chemical Society (1999) Alexander Fleming Discovery and Development of Penicillin. http://www.acs.org/content/acs/en/education/whatischemistry/landmarks/flemingpenicillin.html#alexander-fleming-penicillin.

Amphibian Ark (2012) Chytrid fungus – causing global amphibian mass extinction. http:\\www.amphibianark.org/the-crisis/chytrid-fungus/.

Anderson, D. (2014) 'Still going strong: Leeuwenhoek at eighty', *Antonie Van Leeuwenhoek* 106, 3–26.

Anderson, J.L., Edney, R.J., and Whelan, K. (2012) 'Systematic review: faecal microbiota transplantation in the management of inflammatory bowel disease', *Aliment. Pharmacol. Ther.* 36, 503–516.

Anukam, K.C. and Reid, G. (2007) 'Probiotics: 100 years (1907–2007) after Elie Metchnikoff's observation', in *Communicating Current Research and Educational Topics and Trends in Applied Microbiology* (FORMATEX).

Archibald. J. (2014) *One Plus One Equals One: Symbiosis and the Evolution of Complex Life* (Oxford: Oxford University Press).

Archie, E.A. and Theis, K.R. (2011) 'Animal behaviour meets microbial ecology', *Anim. Behav.* 82, 425–436.

Aroniadis, O.C. and Brandt, L.J. (2014) 'Intestinal microbiota and the efficacy of fecal microbiota transplantation in gastrointestinal disease', *Gastroenterol. Hepatol.* 10, 230–237.

Arrieta, M-C., Stiemsma, L.T., Dimitriu, P.A., Thorson, L., Russell, S., Yurist-Doutsch, S., Kuzeljevic, B., Gold, M.J., Britton, H.M., Lefebvre, D.L., et al. (2015) 'Early infancy microbial and metabolic alterations affect risk of childhood asthma', *Sci. Transl. Med.* 7, 307ra152.

Asano, Y., Hiramoto, T., Nishino, R., Aiba, Y., Kimura, T., Yoshihara, K., Koga, Y., and Sudo, N. (2012) 'Critical role of gut microbiota in the production of biologically active, free catecholamines in the gut lumen of mice', *AJP Gastrointest. Liver Physiol.* 303, G1288–G1295.

Atarashi, K., Tanoue, T., Shima, T., Imaoka, A., Kuwahara, T., Momose, Y., Cheng, G., Yamasaki, S., Saito, T., Ohba, Y., et al. (2011) 'Induction of colonic regulatory T cells by indigenous *Clostridium* species', *Science* 331, 337–341.

Atarashi, K., Tanoue, T., Oshima, K., Suda, W., Nagano, Y., Nishikawa, H., Fukuda, S., Saito, T., Narushima, S., Hase, K., et al. (2013) 'Treg induction by a rationally selected mixture of *Clostridia* strains from the human microbiota', *Nature* 500, 232–236.

Aung, A. (2007) *Feeding of Leucaena Mimosine on Small Ruminants: Investigation on the Control of its Toxicity in Small Ruminants* (Göttingen: Cuvillier Verlag).

Bäckhed, F., Ding, H., Wang, T., Hooper, L.V., Koh, G.Y., Nagy, A., Semenkovich, C.F., and Gordon, J.I. (2004) 'The gut microbiota as an environmental factor that regulates fat storage', *Proc. Natl. Acad. Sci. U. S. A.* 101, 15718–15723.

Bäckhed, F., Fraser, C.M., Ringel, Y., Sanders, M.E., Sartor, R.B., Sherman, P.M., Versalovic, J., Young, V., and Finlay, B.B. (2012) 'Defining a healthy human gut microbiome: current concepts, future directions, and clinical applications', *Cell Host Microbe* 12, 611–622.

Bäckhed, F., Roswall, J., Peng, Y., Feng, Q., Jia, H., Kovatcheva-Datchary, P., Li, Y., Xia, Y., Xie, H., Zhong, H., et al. (2015) 'Dynamics and stabilization of the human gut microbiome during the first year of life', *Cell Host Microbe* 17, 690–703.

Ballal, S.A., Veiga, P., Fenn, K., Michaud, M., Kim, J.H., Gallini, C.A., Glickman, J.N., Quéré, G., Garault, P., Béal, C., et al. (2015) 'Host lysozyme-mediated lysis of *Lactococcus lactis* facilitates delivery of colitis-attenuating superoxide dismutase to inflamed colons', *Proc. Natl. Acad. Sci.* 112, 7803–7808.

Barott, K.L., and Rohwer, F.L. (2012) 'Unseen players shape benthic competition on coral reefs', *Trends Microbiol.* 20, 621–628.

Barr, J.J., Auro, R., Furlan, M., Whiteson, K.L., Erb, M.L., Pogliano, J., Stotland, A., Wolkowicz, R., Cutting, A.S., and Doran, K.S. (2013) 'Bacteriophage adhering to mucus provide a non–host-derived immunity', *Proc. Natl. Acad. Sci.* 110, 10771–10776.

Bates, J.M., Mittge, E., Kuhlman, J., Baden, K.N., Cheesman, S.E., and Guillemin, K. (2006) 'Distinct signals from the microbiota promote different aspects of zebrafish gut differentiation', *Dev. Biol.* 297, 374–386.

Baumann, P., Lai, C., Baumann, L., Rouhbakhsh, D., Moran, N.A., and Clark, M.A. (1995) 'Mutualistic associations of aphids and prokaryotes: biology of the genus *Buchnera*', *Appl. Environ. Microbiol.* 61, 1–7.

BBC (23 January 2015) *The 25 biggest turning points in Earth's history.*

Beasley, D.E., Koltz, A.M., Lambert, J.E., Fierer, N., and Dunn, R.R. (2015) 'The evolution of stomach acidity and its relevance to the human microbiome', *PloS One* 10, e0134116.

Beaumont, W. (1838) *Experiments and Observations on the Gastric Juice, and the Physiology of Digestion* (Edinburgh: Maclachlan & Stewart).

Becerra, J.X., Venable, G.X., and Saeidi, V. (2015) '*Wolbachia*-free heteropterans do not produce defensive chemicals or alarm pheromones', *J. Chem. Ecol.* 41, 593–601.

Becker, M.H., Walke, J.B., Cikanek, S., Savage, A.E., Mattheus, N., Santiago, C.N., Minbiole, K.P.C., Harris, R.N., Belden, L.K. and Gratwicke, B. (2015) 'Composition of symbiotic bacteria predicts survival in Panamanian golden frogs infected with a lethal fungus', *Proc. R. Soc. B Biol. Sci.* 282, doi: 10.1098/rspb.2014.2881.

Belkaid, Y. and Hand, T.W. (2014) 'Role of the microbiota in immunity and inflammation; *Cell* 157, 121–141.

Bennett, G.M. and Moran, N.A. (2013) 'Small, smaller, smallest: the origins and evolution of ancient dual symbioses in a phloem-feeding insect', *Genome Biol. Evol.* 5, 1675–1688.

Bennett, G.M. and Moran, N.A. (2015) 'Heritable symbiosis: the advantages and perils of an evolutionary rabbit hole', *Proc. Natl. Acad. Sci.* 112, 10169–10176.

Benson, A.K., Kelly, S.A., Legge, R., Ma, F., Low, S.J., Kim, J., Zhang, M., Oh, P.L., Nehrenberg, D., Hua, K., et al. (2010) 'Individuality in gut microbiota composition is a complex polygenic trait shaped by multiple environmental and host genetic factors', *Proc. Natl. Acad. Sci.* 107, 18933–18938.

Bercik, P., Denou, E., Collins, J., Jackson, W., Lu, J., Jury, J., Deng, Y., Blennerhassett, P., Macri, J., McCoy, K.D., et al. (2011) 'The intestinal microbiota affect central levels of brain-derived neurotropic factor and behavior in mice', *Gastroenterology* 141, 599–609.e3.

Berer, K., Mues, M., Koutrolos, M., Rasbi, Z.A., Boziki, M., Johner, C., Wekerle, H., and Krishnamoorthy, G. (2011) 'Commensal microbiota and myelin autoantigen cooperate to trigger autoimmune demyelination', *Nature* 479, 538–541.

Bergman, E.N. (1990) 'Energy contributions of volatile fatty acids from the gastrointestinal tract in various species', *Physiol. Rev.* 70, 567–590.

Bevins, C.L. and Salzman, N.H. (2011) 'The potter's wheel: the host's role in sculpting its microbiota', *Cell. Mol. Life Sci.* 68, 3675–3685.

Bezier, A., Annaheim, M., Herbiniere, J., Wetterwald, C., Gyapay, G., Bernard-Samain, S., Wincker, P., Roditi, I., Heller, M., Belghazi, M., et al. (2009) 'Polydnaviruses of braconid wasps derive from an ancestral nudivirus', *Science* 323, 926–930.

Bian, G., Joshi, D., Dong, Y., Lu, P., Zhou, G., Pan, X., Xu, Y., Dimopoulos, G., and Xi, Z. (2013). 'Wolbachia invades Anopheles stephensi populations and induces refractoriness to Plasmodium infection', *Science* 340, 748–751.

Bindels, L.B., Delzenne, N.M., Cani, P.D., and Walter, J. (2015) 'Towards a more comprehensive concept for prebiotics', *Nat. Rev. Gastroenterol. Hepatol.* 12, 303–310.

Blakeslee, S. (15 October 1996) 'Microbial life's steadfast champion', *New York Times.*

Blaser, M. (1 February 2005) 'An endangered species in the stomach; *Sci. Am.*

Blaser, M. (2010) 'Helicobacter pylori and esophageal disease: wake-up call?', *Gastroenterology* 139, 1819–1822.

Blaser, M. (2014) *Missing Microbes: How the Overuse of Antibiotics Is Fueling Our Modern Plagues* (New York: Henry Holt & Co.).

Blaser, M. and Falkow, S. (2009) 'What are the consequences of the disappearing human microbiota?' *Nat. Rev. Microbiol.* 7, 887–894.

Blazejak, A., Erseus, C., Amann, R., and Dubilier, N. (2005) 'Coexistence of bacterial sulfide oxidizers, sulfate reducers, and spirochetes in a gutless worm (*Oligochaeta*) from the Peru Margin', *Appl. Environ. Microbiol.* 71, 1553–1561.

Bletz, M.C., Loudon, A.H., Becker, M.H., Bell, S.C., Woodhams, D.C., Minbiole, K.P.C., and Harris, R.N. (2013) 'Mitigating amphibian chytridiomycosis with bioaugmentation: characteristics of effective probiotics and strategies for their selection and use', *Ecol. Lett.* 16, 807–820.

Blumberg, R. and Powrie, F. (2012) 'Microbiota, disease, and back to health: a metastable journey', *Sci. Transl. Med.* 4, 137rv7–rv137rv7.

Bode, L. (2012) 'Human milk oligosaccharides: every baby needs a sugar mama', *Glycobiology* 22, 1147–1162.

Bode, L., Kuhn, L., Kim, H-Y., Hsiao, L., Nissan, C., Sinkala, M., Kankasa, C., Mwiya, M., Thea, D.M., and Aldrovandi, G.M. (2012) 'Human milk oligosaccharide concentration and risk of postnatal transmission of HIV through breastfeeding', *Am. J. Clin. Nutr.* 96, 831–839.

Bohnhoff, M., Miller, C.P., and Martin, W.R. (1964) 'Resistance of the mouse's intestinal tract to experimental *Salmonella* infection', *J. Exp. Med.* 120, 817–828.

Boone, C.K., Keefover-Ring, K., Mapes, A.C., Adams, A.S., Bohlmann, J., and Raffa, K.F. (2013) 'Bacteria associated with a tree-killing insect reduce concentrations of plant defense compounds', *J. Chem. Ecol.* 39, 1003–1006.

Bordenstein, S.R. and Theis, K.R. (2015) 'Host biology in light of the microbiome: ten principles of holobionts and hologenomes', *PLoS Biol.* 13, e1002226.

Bordenstein, S.R., O'Hara, F.P., and Werren, J.H. (2001) 'Wolbachia-induced incompatibility precedes other hybrid incompatibilities in *Nasonia*', *Nature* 409, 707–710.

Bosch, T.C. (2012) 'What *Hydra* has to say about the role and origin of symbiotic interactions', *Biol. Bull.* 223, 78–84.

Boto, L. (2014) 'Horizontal gene transfer in the acquisition of novel traits by metazoans', *Proc. R. Soc. B Biol. Sci.* 281, doi: 10.1098/rspb.2013.2450.

Bouskra, D., Brézillon, C., Bérard, M., Werts, C., Varona, R., Boneca, I.G., and Eberl, G. (2008) 'Lymphoid tissue genesis induced by commensals through NOD1 regulates intestinal homeostasis', *Nature* 456, 507–510.

Bouslimani, A., Porto, C., Rath, C.M., Wang, M., Guo, Y., Gonzalez, A., Berg-Lyon, D., Ackermann, G., Moeller Christensen, G.J., Nakatsuji, T. et al. (2015) 'Molecular cartography of the human skin surface in 3D', *Proc. Natl. Acad. Sci. U. S. A.* 112, E2120–E2129.

Braniste, V., Al-Asmakh, M., Kowal, C., Anuar, F., Abbaspour, A., Tóth, M., Korecka, A., Bakocevic, N., Ng, L.G., Kundu, P. et al. (2014) 'The gut microbiota influences blood-brain barrier permeability in mice', *Sci. Transl. Med.* 6, 263ra158.

Bravo, J.A., Forsythe, P., Chew, M.V., Escaravage, E., Savignac, H.M., Dinan, T.G., Bienenstock, J., and Cryan, J.F. (2011) 'Ingestion of *Lactobacillus* strain regulates emotional behavior and central GABA receptor expression in a mouse via the vagus nerve', *Proc. Natl. Acad. Sci.* 108, 16050–16055.

Broderick, N.A., Raffa, K.F., and Handelsman, J. (2006) 'Midgut bacteria required for *Bacillus thuringiensis* insecticidal activity', *Proc. Natl. Acad. Sci.* 103, 15196–15199.

Brown, C.T., Hug, L.A., Thomas, B.C., Sharon, I., Castelle, C.J., Singh, A., Wilkins, M.J., Wrighton, K.C., Williams, K.H., and Banfield, J.F. (2015) 'Unusual biology across a group comprising more than 15% of domain bacteria', *Nature* 523, 208–211.

Brown, E.M., Arrieta, M-C., and Finlay, B.B. (2013) 'A fresh look at the hygiene hypothesis: how intestinal microbial exposure drives immune effector responses in atopic disease', *Semin. Immunol.* 25, 378–387.

Bruce-Keller, A.J., Salbaum, J.M., Luo, M., Blanchard, E., Taylor, C.M., Welsh, D.A., and Berthoud, H-R. (2015) 'Obese-type gut microbiota induce neurobehavioral changes in the absence of obesity', *Biol. Psychiatry* 77, 607–615.

Brucker, R.M. and Bordenstein, S.R. (2013) 'The hologenomic basis of speciation: gut bacteria cause hybrid lethality in the genus *Nasonia*', *Science* 341, 667–669.

Brucker, R.M., and Bordenstein, S.R. (2014) Response to Comment on 'The hologenomic basis of speciation: gut bacteria cause hybrid lethality in the genus *Nasonia*', Science 345, 1011–1011.

Bshary, R. (2002) 'Biting cleaner fish use altruism to deceive image-scoring client reef fish', *Proc. Biol. Sci.* 269, 2087–2093.

Buchner, P. (1965) *Endosymbiosis of Animals with Plant Microorganisms* (New York: Interscience Publishers / John Wiley).

Buffie, C.G., Bucci, V., Stein, R.R., McKenney, P.T., Ling, L., Gobourne, A., No, D., Liu, H., Kinnebrew, M., Viale, A., et al. (2014) 'Precision microbiome reconstitution restores bile acid mediated resistance to *Clostridium difficile*', *Nature* 517, 205–208.

Bull, J.J. and Turelli, M. (2013) 'Wolbachia versus dengue: evolutionary forecasts', *Evol. Med. Public Health* 2013, 197–201.

Bulloch, W. (1938) *The History of Bacteriology* (Oxford: Oxford University Press).

Cadwell, K., Patel, K.K., Maloney, N.S., Liu, T-C., Ng, A.C.Y., Storer, C.E., Head, R.D., Xavier, R., Stappenbeck, T.S., and Virgin, H.W. (2010) 'Virus-plus-susceptibility gene interaction determines Crohn's Disease gene Atg16L1 phenotypes in intestine', *Cell* 141, 1135–1145.

Cafaro, M.J., Poulsen, M., Little, A.E.F., Price, S.L., Gerardo, N.M., Wong, B., Stuart, A.E., Larget, B., Abbot, P., and Currie, C.R. (2011) 'Specificity in the symbiotic association between fungus-growing ants and protective *Pseudonocardia* bacteria', *Proc. R. Soc. B Biol. Sci.* 278, 1814–1822.

Campbell, M.A., Leuven, J.T.V., Meister, R.C., Carey, K.M., Simon, C., and McCutcheon, J.P. (2015), 'Genome expansion via lineage splitting and genome reduction in the cicada endosymbiont Hodgkinia', *Proc. Natl. Acad. Sci.* 112, 10192–10199.

Caporaso, J.G., Lauber, C.L., Costello, E.K., Berg-Lyons, D., Gonzalez, A., Stombaugh, J., Knights, D., Gajer, P., Ravel, J., and Fierer, N. (2011) 'Moving pictures of the human microbiome', *Genome Biol.* 12, R50.

Carmody, R.N. and Turnbaugh, P.J. (2014) 'Host–microbial interactions in the metabolism of therapeutic and diet-derived xenobiotics', *J. Clin. Invest.* 124, 4173–4181.

Caspi-Fluger, A., Inbar, M., Mozes-Daube, N., Katzir, N., Portnoy, V., Belausov, E., Hunter, M.S., and Zchori-Fein, E. (2012) 'Horizontal transmission of the insect symbiont *Rickettsia* is plant-mediated', *Proc. R. Soc. B Biol. Sci.* 279, 1791–1796.

Cavanaugh, C.M., Gardiner, S.L., Jones, M.L., Jannasch, H.W., and Waterbury, J.B. (1981) 'Prokaryotic cells in the hydrothermal vent tube worm *Riftia pachyptila Jones*: possible chemoautotrophic symbionts', *Science* 213, 340–342.

Ceja-Navarro, J.A., Vega, F.E., Karaoz, U., Hao, Z., Jenkins, S., Lim, H.C., Kosina, P., Infante, F., Northen, T.R., and Brodie, E.L. (2015) 'Gut microbiota mediate caffeine detoxification in the primary insect pest of coffee', *Nat. Commun.* 6, 7618.

Chandler, J.A. and Turelli, M. (2014) Comment on 'The hologenomic basis of speciation: gut bacteria cause hybrid lethality in the genus *Nasonia*', *Science* 345, 1011–1011.

Chassaing, B., Koren, O., Goodrich, J.K., Poole, A.C., Srinivasan, S., Ley, R.E., and Gewirtz, A.T. (2015) 'Dietary emulsifiers impact the mouse gut microbiota promoting colitis and metabolic syndrome', *Nature* 519, 92–96.

Chau, R., Kalaitzis, J.A., and Neilan, B.A. (2011) 'On the origins and biosynthesis of tetrodotoxin', *Aquat. Toxicol. Amst. Neth.* 104, 61–72.

Cheesman, S.E. and Guillemin, K. (2007) 'We know you are in there: conversing with the indigenous gut microbiota', *Res. Microbiol.* 158, 2–9.

Chen, Y., Segers, S., and Blaser, M.J. (2013) 'Association between *Helicobacter pylori* and mortality in the NHANES III study', *Gut* 62, 1262–1269.

Chichlowski, M., German, J.B., Lebrilla, C.B., and Mills, D.A. (2011) 'The influence of milk oligosaccharides on microbiota of infants: opportunities for formulas', *Annu. Rev. Food Sci. Technol.* 2, 331–351.

Cho, I. and Blaser, M.J. (2012) 'The human microbiome: at the interface of health and disease', *Nat. Rev. Genet.* 13, 260–270.

Cho, I., Yamanishi, S., Cox, L., Methé, B.A., Zavadil, J., Li, K., Gao, Z., Mahana, D., Raju, K., Teitler, I., et al. (2012) 'Antibiotics in early life alter the murine colonic microbiome and adiposity', *Nature* 488, 621–626.

Chou, S., Daugherty, M.D., Peterson, S.B., Biboy, J., Yang, Y., Jutras, B.L., Fritz-Laylin, L.K., Ferrin, M.A., Harding, B.N., Jacobs-Wagner, C., et al. (2014) 'Transferred interbacterial antagonism genes augment eukaryotic innate immune function', *Nature* 518, 98–101.

Chrostek, E., Marialva, M.S.P., Esteves, S.S., Weinert, L.A., Martinez, J., Jiggins, F.M., and Teixeira, L. (2013) 'Wolbachia variants induce differential protection to viruses in *Drosophila melanogaster*: a phenotypic and phylogenomic analysis', *PLoS Genet.* 9, e1003896.

Chu, C-C., Spencer, J.L., Curzi, M.J., Zavala, J.A., and Seufferheld, M.J. (2013) 'Gut bacteria facilitate adaptation to crop rotation in the western corn rootworm', *Proc. Natl. Acad. Sci.* 110, 11917–11922.

Chung, K-T. and Bryant, M.P. (1997) 'Robert E. Hungate: pioneer of anaerobic microbial ecology', *Anaerobe* 3, 213–217.

Chung, K-T. and Ferris, D.H. (1996) 'Martinus Willem Beijerinck', *ASM News* 62, 539–543.

Chung, H., Pamp, S.J., Hill, J.A., Surana, N.K., Edelman, S.M., Troy, E.B., Reading , N.C., Villablanca, E.J., Wang, S., Mora, J.R., et al. (2012) 'Gut immune maturation depends on colonization with a host-specific microbiota', *Cell* 149, 1578–1593.

Chung, S.H., Rosa, C., Scully, E.D., Peiffer, M., Tooker, J.F., Hoover, K., Luthe, D.S., and Felton, G.W. (2013) 'Herbivore exploits orally secreted bacteria to suppress plant defenses', *Proc. Natl. Acad. Sci. U. S. A.* 110, 15728–15733.

Ciorba, M.A. (2012) 'A gastroenterologist's guide to probiotics', *Clin. Gastroenterol. Hepatol.* 10, 960–968.

Claesen, J. and Fischbach, M.A. (2015) 'Synthetic microbes as drug delivery systems', *ACS Synth. Biol.* 4, 358–364.

Clayton, A.L., Oakeson, K.F., Gutin, M., Pontes, A., Dunn, D.M., von Niederhausern, A.C., Weiss, R.B., Fisher, M., and Dale, C. (2012) 'A novel human-infection-derived bacterium provides insights into the evolutionary origins of mutualistic insect–bacterial symbioses', *PLoS Genet.* 8, e1002990.

Clayton, T.A., Baker, D., Lindon, J.C., Everett, J.R., and Nicholson, J.K. (2009) 'Pharmacometabonomic identification of a significant host–microbiome metabolic interaction affecting human drug metabolism', *Proc. Natl. Acad. Sci. U. S. A.* 106, 14728–14733.

Clemente, J.C., Pehrsson, E.C., Blaser, M.J., Sandhu, K., Gao, Z., Wang, B., Magris, M., Hidalgo, G., Contreras, M., Noya-Alarcon, O., et al. (2015) 'The microbiome of uncontacted Amerindians', *Sci. Adv.* 1, e1500183.

Cobb, M. (3 June 2013) 'Oswald T. Avery, the unsung hero of genetic science', *The Guardian.*

Cockburn, S.N., Haselkorn, T.S., Hamilton, P.T., Landzberg, E., Jaenike, J., and Perlman, S.J. (2013) 'Dynamics of the continent-wide spread of a *Drosophila* defensive symbiont', *Ecol. Lett.* 16, 609–616.

Collins, S.M., Surette, M., and Bercik, P. (2012) 'The interplay between the intestinal microbiota and the brain', *Nat. Rev. Microbiol.* 10, 735–742.

Coon, K.L., Vogel, K.J., Brown, M.R., and Strand, M.R. (2014) 'Mosquitoes rely on their gut microbiota for development', *Mol. Ecol.* 23, 2727–2739.

Costello, E.K., Lauber, C.L., Hamady, M., Fierer, N., Gordon, J.I., and Knight, R. (2009) 'Bacterial community variation in human body habitats across space and time', *Science* 326, 1694–1697.

Cox, L.M. and Blaser, M.J. (2014) 'Antibiotics in early life and obesity', *Nat. Rev. Endocrinol.* 11, 182–190.

Cox, L.M., Yamanishi, S., Sohn, J., Alekseyenko, A.V., Leung, J.M., Cho, I., Kim, S.G., Li, H., Gao, Z., Mahana, D., et al. (2014) 'Altering the intestinal microbiota during a critical developmental window has lasting metabolic consequences', *Cell* 158, 705–721.

Cryan, J.F. and Dinan, T.G. (2012) 'Mind-altering microorganisms: the impact of the gut microbiota on brain and behaviour', *Nat. Rev. Neurosci.* 13, 701–712.

CSIROpedia Leucaena toxicity solution.

Dalal, S.R., and Chang, E.B. (2014) 'The microbial basis of inflammatory bowel diseases', *J. Clin. Invest.* 124, 4190–4196.

Dale, C. and Moran, N.A. (2006) 'Molecular interactions between bacterial symbionts and their hosts', *Cell* 126, 453–465.

Danchin, E.G.J. and Rosso, M-N. (2012) 'Lateral gene transfers have polished animal genomes: lessons from nematodes', *Front. Cell. Infect. Microbiol.* 2. doi: 10.3389/fcimb.2012.00027.

Danchin, E.G.J., Rosso, M-N., Vieira, P., de Almeida-Engler, J., Coutinho, P.M., Henrissat, B., and Abad, P. (2010) 'Multiple lateral gene transfers and duplications have promoted plant parasitism ability in nematodes', *Proc. Natl. Acad. Sci.* 107, 17651–17656.

Darmasseelane, K., Hyde, M.J., Santhakumaran, S., Gale, C., and Modi, N. (2014) 'Mode of delivery and offspring body mass index, overweight and obesity in adult life: a systematic review and meta-analysis', *PloS One* 9, e87896.

David, L.A., Maurice, C.F., Carmody, R.N., Gootenberg, D.B., Button, J.E., Wolfe, B.E., Ling, A.V., Devlin, A.S., Varma, Y., Fischbach, M.A., et al. (2013) 'Diet rapidly and reproducibly alters the human gut microbiome', *Nature* 505, 559–563.

David, L.A., Materna, A.C., Friedman, J., Campos-Baptista, M.I., Blackburn, M.C., Perrotta, A., Erdman, S.E., and Alm, E.J. (2014) 'Host lifestyle affects human microbiota on daily timescales', *Genome Biol.* 15, R89.

Dawkins, Richard (1982) *The Extended Phenotype* (Oxford: Oxford University Press).

De Filippo, C., Cavalieri, D., Di Paola, M., Ramazzotti, M., Poullet, J.B., Massart, S., Collini, S., Pieraccini, G., and Lionetti, P. (2010) 'Impact of diet in shaping gut microbiota revealed by a comparative study in children from Europe and rural Africa', *Proc. Natl. Acad. Sci.* 107, 14691–14696.

Delsuc, F., Metcalf, J.L., Wegener Parfrey, L., Song, S.J., González, A., and Knight, R. (2014) 'Convergence of gut microbiomes in myrmecophagous mammals', *Mol. Ecol.* 23, 1301–1317.

Delzenne, N.M., Neyrinck, A.M., and Cani, P.D. (2013) 'Gut microbiota and metabolic disorders: how prebiotic can work?' *Br. J. Nutr.* 109, S81–S85.

Derrien, M., and van Hylckama Vlieg, J.E.T. (2015) 'Fate, activity, and impact of ingested bacteria within the human gut microbiota', *Trends Microbiol.* 23, 354–366.

Dethlefsen, L. and Relman, D.A. (2011) 'Incomplete recovery and individualized responses of the human distal gut microbiota to repeated antibiotic perturbation', *Proc. Natl. Acad. Sci.* 108, 4554–4561.

Dethlefsen, L., McFall-Ngai, M., and Relman, D.A. (2007) 'An ecological and evolutionary perspective on human–microbe mutualism and disease', *Nature* 449, 811–818.

Dethlefsen, L., Huse, S., Sogin, M.L., and Relman, D.A. (2008) 'The pervasive effects of an antibiotic on the human gut microbiota, as revealed by deep 16S rRNA sequencing', *PLoS Biol.* 6, e280.

Devkota, S., Wang, Y., Musch, M.W., Leone, V., Fehlner-Peach, H., Nadimpalli, A., Antonopoulos, D.A., Jabri, B., and Chang, E.B. (2012) 'Dietary-fat-induced taurocholic acid promotes pathobiont expansion and colitis in Il10−/− mice', *Nature* 487, 104–108.

Dill-McFarland, K.A., Weimer, P.J., Pauli, J.N., Peery, M.Z., and Suen, G. (2015) 'Diet specialization selects for an unusual and simplified gut microbiota in two- and three-toed sloths', *Environ. Microbiol.* 509, 357–360.

Dillon, R.J., Vennard, C.T., and Charnley, A.K. (2000) 'Pheromones: exploitation of gut bacteria in the locust', *Nature* 403, 851.

Ding, T. and Schloss, P.D. (2014) 'Dynamics and associations of microbial community types across the human body', *Nature* 509, 357–360.

Dinsdale, E.A., Pantos, O., Smriga, S., Edwards, R.A., Angly, F., Wegley, L., Hatay, M., Hall, D., Brown, E., Haynes, M., et al. (2008) 'Microbial ecology of four coral atolls in the Northern Line Islands', *PLoS ONE* 3, e1584.

Dobell, C. (1932) *Antony Van Leeuwenhoek and His 'Little Animals'* (New York: Dover Publications).

Dobson, A.J., Chaston, J.M., Newell, P.D., Donahue, L., Hermann, S.L., Sannino, D.R., Westmiller, S., Wong, A.C-N., Clark, A.G., Lazzaro, B.P., et al. (2015) 'Host genetic determinants of microbiota-dependent nutrition revealed by genome-wide analysis of *Drosophila melanogaster*', *Nat. Commun.* 6, 6312.

Dodd, D.M.B. (1989) 'Reproductive isolation as a consequence of adaptive divergence in *Drosophila pseudoobscura*', *Evolution* 43, 1308–1311.

Dominguez-Bello, M.G., Costello, E.K., Contreras, M., Magris, M., Hidalgo, G., Fierer, N., and Knight, R. (2010) 'Delivery mode shapes the acquisition and structure of the initial microbiota across multiple body habitats in newborns', *Proc. Natl. Acad. Sci.* 107, 11971–11975.

Dorrestein, P.C., Mazmanian, S.K., and Knight, R. (2014) 'Finding the missing links among metabolites, microbes, and the host', *Immunity* 40, 824–832.

Doudoumis, V., Alam, U., Aksoy, E., Abd-Alla, A.M.M., Tsiamis, G., Brelsfoard, C., Aksoy, S., and Bourtzis, K. (2013) 'Tsetse–*Wolbachia* symbiosis: comes of age and has great potential for pest and disease control', *J. Invertebr. Pathol.* 112, S94–S103.

Douglas, A.E. (2006) 'Phloem-sap feeding by animals: problems and solutions', *J. Exp. Bot.* 57, 747–754.

Douglas, A.E. (2008) 'Conflict, cheats and the persistence of symbioses', *New Phytol.* 177, 849–858.

Dubilier, N., Mülders, C., Ferdelman, T., de Beer, D., Pernthaler, A., Klein, M., Wagner, M., Erséus, C., Thiermann, F., Krieger, J., et al. (2001) 'Endosymbiotic sulphate-reducing and sulphide-oxidizing bacteria in an oligochaete worm', *Nature* 411, 298–302.

Dubilier, N., Bergin, C., and Lott, C. (2008) 'Symbiotic diversity in marine animals: the art of harnessing chemosynthesis', *Nat. Rev. Microbiol.* 6, 725–740.

Dubos, R.J. (1965) *Man Adapting* (New Haven and London: Yale University Press).

Dubos, R.J. (1987) *Mirage of Health: Utopias, Progress, and Biological Change* (New Brunswick, NJ: Rutgers University Press).

Dunlap, P.V. and Nakamura, M. (2011) 'Functional morphology of the luminescence system of *Siphamia versicolor* (*Perciformes: Apogonidae*), a bacterially luminous coral reef fish', *J. Morphol.* 272, 897–909.

Dunning-Hotopp, J.C. (2011) 'Horizontal gene transfer between bacteria and animals', *Trends Genet.* 27, 157–163.

Eakin, E. (1 December 2014) 'The excrement experiment', *New Yorker.*

Eckburg, P.B. (2005) 'Diversity of the human intestinal microbial flora', *Science* 308, 1635–1638.

Eisen, J. (2014) Overselling the microbiome award: *Time* Magazine & Martin Blaser for 'antibiotics are extinguishing our microbiome'. http://phylogenomics.blogspot.co.uk/2014/05/overselling-microbiome-award-time.html.

Elahi, S., Ertelt, J.M., Kinder, J.M., Jiang, T.T., Zhang, X., Xin, L., Chaturvedi, V., Strong, B.S., Qualls, J.E., Steinbrecher, K.A., et al. (2013) 'Immunosuppressive CD71+ erythroid cells compromise neonatal host defence against infection', *Nature* 504, 158–162.

Ellis, M.L., Shaw, K.J., Jackson, S.B., Daniel, S.L., and Knight, J. (2015) 'Analysis of commercial kidney stone probiotic supplements', *Urology* 85, 517–521.

Eskew, E.A. and Todd, B.D. (2013) 'Parallels in amphibian and bat declines from pathogenic fungi', *Emerg. Infect. Dis.* 19, 379–385.

Everard, A., Belzer, C., Geurts, L., Ouwerkerk, J.P., Druart, C., Bindels, L.B., Guiot, Y., Derrien, M., Muccioli, G.G., Delzenne, N.M., et al. (2013) 'Cross-talk between *Akkermansia muciniphila* and intestinal epithelium controls diet-induced obesity', *Proc. Natl. Acad. Sci*, 110, 9066–9071.

Ezenwa, V.O. and Williams, A.E. (2014) 'Microbes and animal olfactory communication: where do we go from here?', *BioEssays* 36, 847–854.

Faith, J.J., Guruge, J.L., Charbonneau, M., Subramanian, S., Seedorf, H., Goodman, A.L., Clemente, J.C., Knight, R., Heath, A.C., and Leibel, R.L. (2013) 'The long-term stability of the human gut microbiota', *Science* 341. doi: 10.1126/science.1237439.

Falkow, S. (2013) Fecal Transplants in the 'Good Old Days'. http://schaechter.asmblog.org/schaechter/2013/05/fecal-transplants-in-the-good-old-days.html.

Feldhaar, H. (2011) 'Bacterial symbionts as mediators of ecologically important traits of insect hosts', *Ecol. Entomol.* 36, 533–543.

Fierer, N., Hamady, M., Lauber, C.L., and Knight, R. (2008) 'The influence of sex, handedness, and washing on the diversity of hand surface bacteria', *Proc. Natl. Acad. Sci. U. S. A.* 105, 17994–17999.

Finucane, M.M., Sharpton, T.J., Laurent, T.J., and Pollard, K.S. (2014) 'A taxonomic signature of obesity in the microbiome? Getting to the guts of the matter', *PLoS ONE* 9, e84689.

Fischbach, M.A. and Sonnenburg, J.L. (2011) 'Eating for two: how metabolism establishes interspecies interactions in the gut', *Cell Host Microbe* 10, 336–347.

Folsome, C. (1985) *Microbes,* in *The Biosphere Catalogue* (Fort Worth, Texas: Synergistic Press).

Franzenburg, S., Walter, J., Kunzel, S., Wang, J., Baines, J.F., Bosch, T.C.G., and Fraune, S. (2013) 'Distinct antimicrobial peptide expression determines host species-specific bacterial associations', *Proc. Natl. Acad. Sci.* 110, E3730–E3738.

Fraune, S. and Bosch, T.C. (2007) 'Long-term maintenance of species-specific bacterial microbiota in the basal metazoan *Hydra*', *Proc. Natl. Acad. Sci.* 104, 13146–13151.

Fraune, S. and Bosch, T.C.G. (2010) 'Why bacteria matter in animal development and evolution', *BioEssays* 32, 571–580.

Fraune, S., Abe, Y., and Bosch, T.C.G. (2009) 'Disturbing epithelial homeostasis in the metazoan *Hydra* leads to drastic changes in associated microbiota', *Environ. Microbiol.* 11, 2361–2369.

Fraune, S., Augustin, R., Anton-Erxleben, F., Wittlieb, J., Gelhaus, C., Klimovich, V.B., Samoilovich, M.P., and Bosch, T.C.G. (2010) 'In an early branching metazoan, bacterial colonization of the embryo is controlled by maternal antimicrobial peptides', *Proc. Natl. Acad. Sci.* 107, 18067–18072.

Freeland, W.J. and Janzen, D.H. (1974) 'Strategies in herbivory by mammals: the role of plant secondary compounds', *Am. Nat.* 108, 269–289.

Frese, S.A., Benson, A.K., Tannock, G.W., Loach, D.M., Kim, J., Zhang, M., Oh, P.L., Heng, N.C.K., Patil, P.B., Juge, N., et al. (2011) 'The evolution of host specialization in the vertebrate gut symbiont *Lactobacillus reuteri*', *PLoS Genet.* 7, e1001314.

Fujimura, K.E. and Lynch, S.V. (2015) 'Microbiota in allergy and asthma and the emerging relationship with the gut microbiome', *Cell Host Microbe* 17, 592–602.

Fujimura, K.E., Demoor, T., Rauch, M., Faruqi, A.A., Jang, S., Johnson, C.C., Boushey, H.A., Zoratti, E., Ownby, D., Lukacs, N.W., et al. (2014) 'House dust exposure mediates gut microbiome *Lactobacillus* enrichment and airway immune defense against allergens and virus infection', *Proc. Natl. Acad. Sci.* 111, 805–810.

Funkhouser, L.J. and Bordenstein, S.R. (2013) 'Mom knows best: the universality of maternal microbial transmission', *PLoS Biol.* 11, e1001631.

Furusawa, Y., Obata, Y., Fukuda, S., Endo, T.A., Nakato, G., Takahashi, D., Nakanishi, Y., Uetake, C., Kato, K., Kato, T., et al. (2013) 'Commensal microbe-derived butyrate induces the differentiation of colonic regulatory T cells', *Nature* 504, 446–450.

Gajer, P., Brotman, R.M., Bai, G., Sakamoto, J., Schutte, U.M.E., Zhong, X., Koenig, S.S.K., Fu, L., Ma, Z., Zhou, X., et al. (2012) 'Temporal dynamics of the human vaginal microbiota', *Sci. Transl. Med.* 4, 132ra52–ra132ra52.

Garcia, J.R. and Gerardo, N.M. (2014) 'The symbiont side of symbiosis: do microbes really benefit?' *Front. Microbiol.* 5. doi: 10.3389/fmicb.2014.00510.

Gareau, M.G., Sherman, P.M., and Walker, W.A. (2010) 'Probiotics and the gut microbiota in intestinal health and disease', *Nat. Rev. Gastroenterol. Hepatol.* 7, 503–514.

Garrett, W.S., Lord, G.M., Punit, S., Lugo-Villarino, G., Mazmanian, S.K., Ito, S., Glickman, J.N., and Glimcher, L.H. (2007) 'Communicable ulcerative colitis induced by T-bet deficiency in the innate immune system', Cell 131, 33–45.

Garrett, W.S., Gallini, C.A., Yatsunenko, T., Michaud, M., DuBois, A., Delaney, M.L., Punit, S., Karlsson, M., Bry, L., Glickman, J.N., et al. (2010) 'Enterobacteriaceae act in concert with the gut microbiota to induce spontaneous and maternally transmitted colitis', *Cell Host Microbe* 8, 292–300.

Gehrer, L. and Vorburger, C. (2012) 'Parasitoids as vectors of facultative bacterial endosymbionts in aphids', *Biol. Lett.* 8, 613–615.

Gerrard, J.W., Geddes, C.A., Reggin, P.L., Gerrard, C.D., and Horne, S. (1976) 'Serum IgE levels in white and Metis communities in Saskatchewan', *Ann. Allergy* 37, 91–100.

Gerritsen, J., Smidt, H., Rijkers, G.T., and Vos, W.M. (2011) 'Intestinal microbiota in human health and disease: the impact of probiotics', *Genes Nutr.* 6, 209–240.

Gevers, D., Kugathasan, S., Denson, L.A., Vázquez-Baeza, Y., Van Treuren, W., Ren, B., Schwager, E., Knights, D., Song, S.J., Yassour, M., et al. (2014) 'The treatment-naive microbiome in new-onset Crohn's Disease', *Cell Host Microbe* 15, 382–392.

Gibbons, S.M., Schwartz, T., Fouquier, J., Mitchell, M., Sangwan, N., Gilbert, J.A., and Kelley, S.T. (2015) 'Ecological succession and viability of human-associated microbiota on restroom surfaces', *Appl. Environ. Microbiol.* 81, 765–773.

Gilbert, J.A. and Neufeld, J.D. (2014) 'Life in a world without microbes', *PLoS Biol.* 12, e1002020.

Gilbert, J.A., Meyer, F., Antonopoulos, D., Balaji, P., Brown, C.T., Desai, N., Eisen, J.A., Evers, D., Field, D., et al. (2010) 'Meeting Report: The Terabase Metagenomics Workshop and the Vision of an Earth Microbiome Project', *Stand. Genomic Sci.* 3, 243–248.

Gilbert, S.F., Sapp, J., and Tauber, A.I. (2012) 'A symbiotic view of life: we have never been individuals', *Q. Rev. Biol.* 87, 325–341.

Godoy-Vitorino, F., Goldfarb, K.C., Karaoz, U., Leal, S., Garcia-Amado, M.A., Hugenholtz, P., Tringe, S.G., Brodie, E.L., and Dominguez-Bello, M.G. (2012) 'Comparative analyses of foregut and hindgut bacterial communities in hoatzins and cows', *ISME J.* 6, 531–541.

Goldenberg, J.Z., Ma, S.S., Saxton, J.D., Martzen, M.R., Vandvik, P.O., Thorlund, K., Guyatt, G.H., and Johnston, B.C. (2013) 'Probiotics for the prevention of

Clostridium difficile-associated diarrhea in adults and children', in *Cochrane Database of Systematic Reviews, The Cochrane Collaboration*, ed. (Chichester, UK: John Wiley & Sons).

Gomez, A., Petrzelkova, K., Yeoman, C.J., Burns, M.B., Amato, K.R., Vlckova, K., Modry, D., Todd, A., Robbinson, C.A.J., Remis, M., et al. (2015) 'Ecological and evolutionary adaptations shape the gut microbiome of BaAka African rainforest hunter-gatherers', bioRxiv 019232.

Goodrich, J.K., Waters, J.L., Poole, A.C., Sutter, J.L., Koren, O., Blekhman, R., Beaumont, M., Van Treuren, W., Knight, R., Bell, J.T., et al. (2014) 'Human genetics shape the gut microbiome', *Cell* 159, 789–799.

Graham, D.Y. (1997) 'The only good *Helicobacter pylori* is a dead *Helicobacter pylori*', *Lancet* 350, 70–71; author reply 72.

Green, J. (2011). Are we filtering the wrong microbes? TED https://www.ted.com/talks/jessica_green_are_we_filtering_the_wrong_microbes.

Green, J.L. (2014) 'Can bioinformed design promote healthy indoor ecosystems?' *Indoor Air* 24, 113–115.

Gruber-Vodicka, H.R., Dirks, U., Leisch, N., Baranyi, C., Stoecker, K., Bulgheresi, S., Heindl, N.R., Horn, M., Lott, C., Loy, A., et al. (2011) 'Paracatenula, an ancient symbiosis between thiotrophic *Alphaproteobacteria* and catenulid flatworms', *Proc. Natl. Acad. Sci.* 108, 12078–12083.

Hadfield, M.G. (2011) 'Biofilms and marine invertebrate larvae: what bacteria produce that larvae use to choose settlement sites', *Annu. Rev. Mar. Sci.* 3, 453–470.

Haiser, H.J. and Turnbaugh, P.J. (2012) 'Is it time for a metagenomic basis of therapeutics?' *Science* 336, 1253–1255.

Haiser, H.J., Gootenberg, D.B., Chatman, K., Sirasani, G., Balskus, E.P., and Turnbaugh, P.J. (2013) 'Predicting and manipulating cardiac drug inactivation by the human gut bacterium *Eggerthella lenta*', *Science* 341, 295–298.

Hamilton, M.J., Weingarden, A.R., Unno, T., Khoruts, A., and Sadowsky, M.J. (2013) 'High-throughput DNA sequence analysis reveals stable engraftment of gut microbiota following transplantation of previously frozen fecal bacteria', *Gut Microbes* 4, 125–135.

Handelsman, J. (2007) 'Metagenomics and microbial communities', in *Encyclopedia of Life Sciences* (Chichester, UK: John Wiley & Sons).

Harley, I.T.W. and Karp, C.L. (2012) 'Obesity and the gut microbiome: striving for causality', *Mol. Metab.* 1, 21–31.

Harris, R.N., James, T.Y., Lauer, A., Simon, M.A., and Patel, A. (2006) 'Amphibian pathogen *Batrachochytrium dendrobatidis* is inhibited by the cutaneous bacteria of amphibian species', *EcoHealth* 3, 53–56.

Harris, R.N., Brucker, R.M., Walke, J.B., Becker, M.H., Schwantes, C.R., Flaherty, D.C., Lam, B.A., Woodhams, D.C., Briggs, C.J., Vredenburg, V.T., et al. (2009)

'Skin microbes on frogs prevent morbidity and mortality caused by a lethal skin fungus', *ISME J.* 3, 818–824.

Haselkorn, T.S., Cockburn, S.N., Hamilton, P.T., Perlman, S.J., and Jaenike, J. (2013) 'Infectious adaptation: potential host range of a defensive endosymbiont in *Drosophila*: host range of *Spiroplasma* in *Drosophila*', *Evolution* 67, 934–945.

Hecht, G.A., Blaser, M.J., Gordon, J., Kaplan, L.M., Knight, R., Laine, L., Peek, R., Sanders, M.E., Sartor, B., Wu, G.D., et al. (2014) 'What is the value of a food and drug administration investigational new drug application for fecal microbiota transplantation to treat *Clostridium difficile* infection?' *Clin. Gastroenterol. Hepatol. Off. Clin. Pract. J. Am. Gastroenterol. Assoc.* 12, 289–291.

Hedges, L.M., Brownlie, J.C., O'Neill, S.L., and Johnson, K.N. (2008) 'Wolbachia and virus protection in insects', *Science* 322, 702.

Hehemann, J-H., Correc, G., Barbeyron, T., Helbert, W., Czjzek, M., and Michel, G. (2010) 'Transfer of carbohydrate-active enzymes from marine bacteria to Japanese gut microbiota', *Nature* 464, 908–912.

Heijtz, R.D., Wang, S., Anuar, F., Qian, Y., Bjorkholm, B., Samuelsson, A., Hibberd, M.L., Forssberg, H., and Pettersson, S. (2011) 'Normal gut microbiota modulates brain development and behavior', *Proc. Natl. Acad. Sci.* 108, 3047–3052.

Heil, M., Barajas-Barron, A., Orona-Tamayo, D., Wielsch, N., and Svatos, A. (2014) 'Partner manipulation stabilises a horizontally transmitted mutualism', *Ecol. Lett.* 17, 185–192.

Henry, L.M., Peccoud, J., Simon, J-C., Hadfield, J.D., Maiden, M.J.C., Ferrari, J., and Godfray, H.C.J. (2013) 'Horizontally transmitted symbionts and host colonization of ecological niches', *Curr. Biol.* 23, 1713–1717.

Herbert, E.E. and Goodrich-Blair, H. (2007) 'Friend and foe: the two faces of *Xenorhabdus nematophila*', *Nat. Rev. Microbiol.* 5, 634–646.

Herniou, E.A., Huguet, E., Thézé, J., Bézier, A., Periquet, G., and Drezen, J-M. (2013) 'When parasitic wasps hijacked viruses: genomic and functional evolution of polydnaviruses', *Philos. Trans. R. Soc. Lond. B Biol. Sci.* 368, 20130051.

Hilgenboecker, K., Hammerstein, P., Schlattmann, P., Telschow, A., and Werren, J.H. (2008) 'How many species are infected with *Wolbachia*? – a statistical analysis of current data: *Wolbachia* infection rates', *FEMS Microbiol. Lett.* 281, 215–220.

Hill, C., Guarner, F., Reid, G., Gibson, G.R., Merenstein, D.J., Pot, B., Morelli, L., Canani, R.B., Flint, H.J., Salminen, S., et al. (2014) 'Expert consensus document: The International Scientific Association for Probiotics and Prebiotics consensus statement on the scope and appropriate use of the term probiotic', *Nat. Rev. Gastroenterol. Hepatol.* 11, 506–514.

Himler, A.G., Adachi-Hagimori, T., Bergen, J.E., Kozuch, A., Kelly, S.E., Tabashnik, B.E., Chiel, E., Duckworth, V.E., Dennehy, T.J., Zchori-Fein, E., et al. (2011) 'Rapid spread of a bacterial symbiont in an invasive whitefly is driven by fitness benefits and female bias', *Science* 332, 254–256.

Hird, S.M., Carstens, B.C., Cardiff, S.W., Dittmann, D.L., and Brumfield, R.T. (2014) 'Sampling locality is more detectable than taxonomy or ecology in the gut microbiota of the brood-parasitic Brown-headed Cowbird (*Molothrus ater*)', *PeerJ 2*, e321.

Hiss, P.H. and Zinsser, H. (1910) *A Text-book of Bacteriology: a Practical Treatise for Students and Practitioners of Medicine* (New York and London: D. Appleton & Co.).

Hoerauf, A., Volkmann, L., Hamelmann, C., Adjei, O., Autenrieth, I.B., Fleischer, B., and Büttner, D.W. (2000) 'Endosymbiotic bacteria in worms as targets for a novel chemotherapy in filariasis', *Lancet* 355, 1242–1243.

Hoerauf, A., Mand, S., Adjei, O., Fleischer, B., and Büttner, D.W. (2001) 'Depletion of *Wolbachia* endobacteria in *Onchocerca volvulus* by doxycycline and microfilaridermia after ivermectin treatment', *Lancet* 357, 1415–1416.

Hof, C., Araújo, M.B., Jetz, W., and Rahbek, C. (2011) 'Additive threats from pathogens, climate and land-use change for global amphibian diversity', *Nature* 480, 516–519.

Hoffmann, A.A., Montgomery, B.L., Popovici, J., Iturbe-Ormaetxe, I., Johnson, P.H., Muzzi, F., Greenfield, M., Durkan, M., Leong, Y.S., Dong, Y., et al. (2011) 'Successful establishment of *Wolbachia* in *Aedes* populations to suppress dengue transmission', *Nature* 476, 454–457.

Holmes, E., Kinross, J., Gibson, G., Burcelin, R., Jia, W., Pettersson, S., and Nicholson, J. (2012) 'Therapeutic modulation of microbiota–host metabolic interactions', *Sci. Transl. Med.* 4, 137rv6.

Honda, K., and Littman, D.R. (2012). 'The Microbiome in Infectious Disease and Inflammation', *Annu. Rev. Immunol.* 30, 759–795.

Hongoh, Y. (2011) 'Toward the functional analysis of uncultivable, symbiotic microorganisms in the termite gut', *Cell. Mol. Life Sci.* 68, 1311–1325.

Hooper, L.V. (2001) 'Molecular analysis of commensal host-microbial relationships in the intestine', *Science* 291, 881–884.

Hooper, L.V., Stappenbeck, T.S., Hong, C.V., and Gordon, J.I. (2003) 'Angiogenins: a new class of microbicidal proteins involved in innate immunity', *Nat. Immunol.* 4, 269–273.

Hooper, L.V., Littman, D.R., and Macpherson, A.J. (2012) 'Interactions between the microbiota and the immune system', *Science* 336, 1268–1273.

Hornett, E.A., Charlat, S., Wedell, N., Jiggins, C.D., and Hurst, G.D.D. (2009) 'Rapidly shifting sex ratio across a species range', *Curr. Biol.* 19, 1628–1631.

Hosokawa, T., Kikuchi, Y., Shimada, M., and Fukatsu, T. (2008) 'Symbiont acquisition alters behaviour of stinkbug nymphs', *Biol. Lett.* 4, 45–48.

Hosokawa, T., Koga, R., Kikuchi, Y., Meng, X.-Y., and Fukatsu, T. (2010). '*Wolbachia* as a bacteriocyte-associated nutritional mutualist', *Proc. Natl. Acad. Sci.* 107, 769–774.

Hosokawa, T., Hironaka, M., Mukai, H., Inadomi, K., Suzuki, N., and Fukatsu, T. (2012) 'Mothers never miss the moment: a fine-tuned mechanism for vertical symbiont transmission in a subsocial insect', *Anim. Behav.* 83, 293–300.

Hotopp, J.C.D., Clark, M.E., Oliveira, D.C.S.G., Foster, J.M., Fischer, P., Torres, M.C.M., Giebel, J.D., Kumar, N., Ishmael, N., Wang, S., et al. (2007) 'Widespread lateral gene transfer from intracellular bacteria to multicellular eukaryotes', *Science* 317, 1753–1756.

Hsiao, E.Y., McBride, S.W., Hsien, S., Sharon, G., Hyde, E.R., McCue, T., Codelli, J.A., Chow, J., Reisman, S.E., Petrosino, J.F., et al. (2013) 'Microbiota modulate behavioral and physiological abnormalities associated with neurodevelopmental disorders', *Cell* 155, 1451–1463.

Huang, L., Chen, Q., Zhao, Y., Wang, W., Fang, F., and Bao, Y. (2015) 'Is elective Cesarean section associated with a higher risk of asthma? A meta-analysis', *J. Asthma Off. J. Assoc. Care Asthma* 52, 16–25.

Hughes, G.L., Dodson, B.L., Johnson, R.M., Murdock, C.C., Tsujimoto, H., Suzuki, Y., Patt, A.A., Cui, L., Nossa, C.W., Barry, R.M., et al. (2014) 'Native microbiome impedes vertical transmission of *Wolbachia* in *Anopheles* mosquitoes', *Proc. Natl. Acad. Sci.* 111, 12498–12503.

Husnik, F., Nikoh, N., Koga, R., Ross, L., Duncan, R.P., Fujie, M., Tanaka, M., Satoh, N., Bachtrog, D., Wilson, A.C.C., et al. (2013) 'Horizontal gene transfer from diverse bacteria to an insect genome enables a tripartite nested mealybug symbiosis', *Cell* 153, 1567–1578.

Huttenhower, C., Gevers, D., Knight, R., Abubucker, S., Badger, J.H., Chinwalla, A.T., Creasy, H.H., Earl, A.M., FitzGerald, M.G., Fulton, R.S., et al. (2012) 'Structure, function and diversity of the healthy human microbiome', *Nature* 486, 207–214.

Huttenhower, C., Kostic, A.D., and Xavier, R.J. (2014) 'Inflammatory bowel disease as a model for translating the microbiome', *Immunity* 40, 843–854.

Iturbe-Ormaetxe, I., Walker, T., and O' Neill, S.L. (2011) '*Wolbachia* and the biological control of mosquito-borne disease', *EMBO Rep.* 12, 508–518.

Ivanov, I.I., Atarashi, K., Manel, N., Brodie, E.L., Shima, T., Karaoz, U., Wei, D., Goldfarb, K.C., Santee, C.A., Lynch, S.V., et al. (2009) 'Induction of intestinal Th17 cells by segmented filamentous bacteria', *Cell* 139, 485–498.

Jaenike, J., Polak, M., Fiskin, A., Helou, M., and Minhas, M. (2007) 'Interspecific transmission of endosymbiotic *Spiroplasma* by mites', *Biol. Lett.* 3, 23–25.

Jaenike, J., Unckless, R., Cockburn, S.N., Boelio, L.M., and Perlman, S.J. (2010) 'Adaptation via symbiosis: recent spread of a *Drosophila* defensive symbiont', *Science* 329, 212–215.

Jakobsson, H.E., Jernberg, C., Andersson, A.F., Sjölund-Karlsson, M., Jansson, J.K., and Engstrand, L. (2010) 'Short-term antibiotic treatment has differing long-term impacts on the human throat and gut microbiome', *PLoS ONE* 5, e9836.

Jansson, J.K. and Prosser, J.I. (2013) 'Microbiology: the life beneath our feet', *Nature* 494, 40–41.

Jefferson, R. (2010). The hologenome theory of evolution – Science as Social Enterprise. http://blogs.cambia.org/raj/2010/11/16/the-hologenome-theory-of-evolution/.

Jernberg, C., Lofmark, S., Edlund, C., and Jansson, J.K. (2010) 'Long-term impacts of antibiotic exposure on the human intestinal microbiota', *Microbiology* 156, 3216–3223.

Jiggins, F.M. and Hurst, G.D.D. (2011) 'Rapid insect evolution by symbiont transfer', *Science* 332, 185–186.

Johnston, K.L., Ford, L., and Taylor, M.J. (2014) 'Overcoming the challenges of drug discovery for neglected tropical diseases: the A·WoL experience', *J. Biomol. Screen.* 19, 335–343.

Jones, R.J. and Megarrity, R.G. (1986) 'Successful transfer of DHP-degrading bacteria from Hawaiian goats to Australian ruminants to overcome the toxicity of *Leucaena*', *Aust. Vet. J.* 63, 259–262.

Kaiser, W., Huguet, E., Casas, J., Commin, C., and Giron, D. (2010) 'Plant green-island phenotype induced by leaf-miners is mediated by bacterial symbionts', *Proc. R. Soc. B Biol. Sci.* 277, 2311–2319.

Kaiwa, N., Hosokawa, T., Nikoh, N., Tanahashi, M., Moriyama, M., Meng, X-Y., Maeda, T., Yamaguchi, K., Shigenobu, S., Ito, M., et al. (2014) 'Symbiont-supplemented maternal investment underpinning host's ecological adaptation', *Curr. Biol.* 24, 2465–2470.

Kaltenpoth, M., Göttler, W., Herzner, G., and Strohm, E. (2005) 'Symbiotic bacteria protect wasp larvae from fungal infestation', *Curr. Biol.* 15, 475–479.

Kaltenpoth, M., Roeser-Mueller, K., Koehler, S., Peterson, A., Nechitaylo, T.Y., Stubblefield, J.W., Herzner, G., Seger, J., and Strohm, E. (2014) 'Partner choice and fidelity stabilize coevolution in a Cretaceous-age defensive symbiosis', *Proc. Natl. Acad. Sci.* 111, 6359–6364.

Kane, M., Case, L.K., Kopaskie, K., Kozlova, A., MacDearmid, C., Chervonsky, A.V., and Golovkina, T.V. (2011) 'Successful transmission of a retrovirus depends on the commensal microbiota', *Science* 334, 245–249.

Karasov, W.H., Martínez del Rio, C., and Caviedes-Vidal, E. (2011) 'Ecological physiology of diet and digestive systems', *Annu. Rev. Physiol.* 73, 69–93.

Katan, M.B. (2012) 'Why the European Food Safety Authority was right to reject health claims for probiotics', *Benef. Microbes* 3, 85–89.

Kau, A.L., Planer, J.D., Liu, J., Rao, S., Yatsunenko, T., Trehan, I., Manary, M.J., Liu, T-C., Stappenbeck, T.S., Maleta, K.M., et al. (2015) 'Functional characterization of IgA-targeted bacterial taxa from undernourished Malawian children that produce diet-dependent enteropathy', *Sci. Transl. Med.* 7, 276ra24–ra276ra24.

Keeling, P.J. and Palmer, J.D. (2008) 'Horizontal gene transfer in eukaryotic evolution', *Nat. Rev. Genet.* 9, 605–618.

Kelly, L.W., Barott, K.L., Dinsdale, E., Friedlander, A.M., Nosrat, B., Obura, D., Sala, E., Sandin, S.A., Smith, J.E., and Vermeij, M.J. (2012) 'Black reefs: iron-induced phase shifts on coral reefs', *ISME J.* 6, 638–649.

Kembel, S.W., Jones, E., Kline, J., Northcutt, D., Stenson, J., Womack, A.M., Bohannan, B.J., Brown, G.Z., and Green, J.L. (2012) 'Architectural design influences the diversity and structure of the built environment microbiome', *ISME J.* 6, 1469–1479.

Kembel, S.W., Meadow, J.F., O'Connor, T.K., Mhuireach, G., Northcutt, D., Kline, J., Moriyama, M., Brown, G.Z., Bohannan, B.J.M., and Green, J.L. (2014) 'Architectural design drives the biogeography of indoor bacterial communities', *PLoS ONE* 9, e87093.

Kendall, A.I. (1909) 'Some observations on the study of the intestinal bacteria', *J. Biol. Chem.* 6, 499–507.

Kendall, A.I. (1921) *Bacteriology, General, Pathological and Intestinal* (Philadelphia and New York: Lea & Febiger).

Kendall, A.I. (1923) *Civilization and the Microbe* (Boston: Houghton Mifflin).

Kernbauer, E., Ding, Y., and Cadwell, K. (2014) 'An enteric virus can replace the beneficial function of commensal bacteria', *Nature* 516, 94–98.

Khoruts, A. (2013) 'Faecal microbiota transplantation in 2013: developing human gut microbiota as a class of therapeutics', *Nat. Rev. Gastroenterol. Hepatol.* 11, 79–80.

Kiers, E.T. and West, S.A. (2015) 'Evolving new organisms via symbiosis', *Science* 348, 392–394.

Kikuchi, Y., Hayatsu, M., Hosokawa, T., Nagayama, A., Tago, K., and Fukatsu, T. (2012) 'Symbiont-mediated insecticide resistance', *Proc. Natl. Acad. Sci.* 109, 8618–8622.

Kilpatrick, A.M., Briggs, C.J., and Daszak, P. (2010) 'The ecology and impact of chytridiomycosis: an emerging disease of amphibians', *Trends Ecol. Evol.* 25, 109–118.

Kirk, R.G. (2012) '"Life in a germ-free world": isolating life from the laboratory animal to the bubble boy', *Bull. Hist. Med.* 86, 237–275.

Koch, H. and Schmid-Hempel, P. (2011) 'Socially transmitted gut microbiota protect bumble bees against an intestinal parasite', *Proc. Natl. Acad. Sci.* 108, 19288–19292.

Kohl, K.D., Weiss, R.B., Cox, J., Dale, C., and Denise Dearing, M. (2014) 'Gut microbes of mammalian herbivores facilitate intake of plant toxins', *Ecol. Lett.* 17, 1238–1246.

Koren, O., Goodrich, J.K., Cullender, T.C., Spor, A., Laitinen, K., Kling Bäckhed, H., Gonzalez, A., Werner, J.J., Angenent, L.T., Knight, R., et al. (2012) 'Host remodeling of the gut microbiome and metabolic changes during pregnancy', *Cell* 150, 470–480.

Koropatkin, N.M., Cameron, E.A., and Martens, E.C. (2012) 'How glycan metabolism shapes the human gut microbiota', *Nat. Rev. Microbiol.* 10, 323–335.

Koropatnick, T.A., Engle, J.T., Apicella, M.A., Stabb, E.V., Goldman, W.E., and McFall-Ngai, M.J. (2004) 'Microbial factor-mediated development in a host-bacterial mutualism', *Science* 306, 1186–1188.

Kostic, A.D., Gevers, D., Siljander, H., Vatanen, T., Hyötyläinen, T., Hämäläinen, A-M., Peet, A., Tillmann, V., Pöhö, P., Mattila, I., et al. (2015) 'The dynamics of the human infant gut microbiome in development and in progression toward Type 1 Diabetes', *Cell Host Microbe* 17, 260–273.

Kotula, J.W., Kerns, S.J., Shaket, L.A., Siraj, L., Collins, J.J., Way, J.C., and Silver, P.A. (2014) 'Programmable bacteria detect and record an environmental signal in the mammalian gut', *Proc. Natl. Acad. Sci.* 111, 4838–4843.

Kozek, W.J. (1977) 'Transovarially-transmitted intracellular microorganisms in adult and larval stages of *Brugia malayi*', *J. Parasitol.* 63, 992–1000.

Kozek, W.J., and Rao, R.U. (2007) 'The Discovery of Wolbachia in arthropods and nematodes – a historical perspective', in *Wolbachia: A Bug's Life in another Bug*, A. Hoerauf and R.U. Rao, eds., pp. 1–14 (Basel: Karger).

Kremer, N., Philipp, E.E.R., Carpentier, M-C., Brennan, C.A., Kraemer, L., Altura, M.A., Augustin, R., Häsler, R., Heath-Heckman, E.A.C., Peyer, S.M., et al. (2013) 'Initial symbiont contact orchestrates host-organ-wide transcriptional changes that prime tissue colonization', *Cell Host Microbe* 14, 183–194.

Kroes, I., Lepp, P.W., and Relman, D.A. (1999) 'Bacterial diversity within the human subgingival crevice', *Proc. Natl. Acad. Sci.* 96, 14547–14552.

Kruif, P.D. (2002) *Microbe Hunters* (Boston: Houghton Mifflin Harcourt).

Kueneman, J.G., Parfrey, L.W., Woodhams, D.C., Archer, H.M., Knight, R., and McKenzie, V.J. (2014) 'The amphibian skin-associated microbiome across species, space and life history stages', *Mol. Ecol.* 23, 1238–1250.

Kunz, C. (2012) 'Historical aspects of human milk oligosaccharides', *Adv. Nutr. Int. Rev. J.* 3, 430S – 439S.

Kunzig, R. (2000) *Mapping the Deep: The Extraordinary Story of Ocean Science* (New York: W. W. Norton & Co.).

Kuss, S.K., Best, G.T., Etheredge, C.A., Pruijssers, A.J., Frierson, J.M., Hooper, L.V., Dermody, T.S., and Pfeiffer, J.K. (2011) 'Intestinal microbiota promote enteric virus replication and systemic pathogenesis', *Science* 334, 249–252.

Kwong, W.K. and Moran, N.A. (2015) 'Evolution of host specialization in gut microbes: the bee gut as a model', *Gut Microbes* 6, 214–220.

Lander, E.S., Linton, L.M., Birren, B., Nusbaum, C., Zody, M.C., Baldwin, J., Devon, K., Dewar, K., Doyle, M., FitzHugh, W., et al. (2001) 'Initial sequencing and analysis of the human genome', *Nature* 409, 860–921.

Lane, N. (2015a) *The Vital Question: Why Is Life the Way It Is?* (London: Profile Books).

Lane, N. (2015b) 'The unseen world: reflections on Leeuwenhoek (1677) "Concerning little animals"' *Philos. Trans. R. Soc. B Biol. Sci.* 370, doi: 10.1098/rstb. 2014. 0344.

Lang, J.M., Eisen, J.A., and Zivkovic, A.M. (2014) 'The microbes we eat: abundance and taxonomy of microbes consumed in a day's worth of meals for three diet types', *PeerJ 2*, e659.

Lawley, T.D., Clare, S., Walker, A.W., Stares, M.D., Connor, T.R., Raisen, C., Goulding, D., Rad, R., Schreiber, F., Brandt, C., et al. (2012) 'Targeted restoration of the intestinal microbiota with a simple, defined bacteriotherapy resolves relapsing *Clostridium difficile* disease in mice', *PLoS Pathog.* 8, e1002995.

Lax, S. and Gilbert, J.A. (2015) 'Hospital-associated microbiota and implications for nosocomial infections', *Trends Mol. Med.* 21, 427–432.

Lax, S., Smith, D.P., Hampton-Marcell, J., Owens, S.M., Handley, K.M., Scott, N.M., Gibbons, S.M., Larsen, P., Shogan, B.D., Weiss, S., et al. (2014) 'Longitudinal analysis of microbial interaction between humans and the indoor environment', *Science* 345, 1048–1052.

Le Chatelier, E., Nielsen, T., Qin, J., Prifti, E., Hildebrand, F., Falony, G., Almeida, M., Arumugam, M., Batto, J-M., Kennedy, S., et al. (2013) 'Richness of human gut microbiome correlates with metabolic markers', *Nature* 500, 541–546.

Le Clec'h, W., Chevalier, F.D., Genty, L., Bertaux, J., Bouchon, D., and Sicard, M. (2013) 'Cannibalism and predation as paths for horizontal passage of *Wolbachia* between terrestrial isopods', *PLoS ONE* 8, e60232.

Lee, Y.K. and Mazmanian, S.K. (2010) 'Has the microbiota played a critical role in the evolution of the adaptive immune system?', *Science* 330, 1768–1773.

Lee, B.K., Magnusson, C., Gardner, R.M., Blomström, Å., Newschaffer, C.J., Burstyn, I., Karlsson, H., and Dalman, C. (2015) 'Maternal hospitalization with infection during pregnancy and risk of autism spectrum disorders', *Brain. Behav. Immun.* 44, 100–105.

Leewenhoeck, A. van (1677) 'Observation, communicated to the publisher by Mr. Antony van Leewenhoeck, in a Dutch letter of the 9 Octob. 1676 here English'd: concerning little animals by him observed in rain-well-sea and snow water; as also in water wherein pepper had lain infused', *Phil. Trans.* 12, 821–831.

Leewenhook, A. van (1674), More Observations from Mr. Leewenhook, in a Letter of Sept. 7, 1674. sent to the Publisher', *Phil Trans* 12, 178–182.

Lemon, K.P., Armitage, G.C., Relman, D.A., and Fischbach, M.A. (2012) 'Microbiota-targeted therapies: an ecological perspective', *Sci. Transl. Med.* 4, 137rv5–rv137rv5.

LePage, D., and Bordenstein, S.R. (2013) '*Wolbachia*: can we save lives with a great pandemic?', *Trends Parasitol.* 29, 385–393.

Leroi, A.M. (2014) *The Lagoon: How Aristotle Invented Science* (New York: Viking Books).

Leroy, P.D., Sabri, A., Heuskin, S., Thonart, P., Lognay, G., Verheggen, F.J., Francis, F., Brostaux, Y., Felton, G.W., and Haubruge, E. (2011) 'Microorganisms from aphid honeydew attract and enhance the efficacy of natural enemies', *Nat. Commun.* 2, 348.

Ley, R.E., Bäckhed, F., Turnbaugh, P., Lozupone, C.A., Knight, R.D., and Gordon, J.I. (2005) 'Obesity alters gut microbial ecology', *Proc. Natl. Acad. Sci. U. S. A.* 102, 11070–11075.

Ley, R.E., Peterson, D.A., and Gordon, J.I. (2006) 'Ecological and evolutionary forces shaping microbial diversity in the human intestine', *Cell* 124, 837–848.

Ley, R.E., Hamady, M., Lozupone, C., Turnbaugh, P.J., Ramey, R.R., Bircher, J.S., Schlegel, M.L., Tucker, T.A., Schrenzel, M.D., Knight, R., et al. (2008a) 'Evolution of mammals and their gut microbes', *Science* 320, 1647–1651.

Ley, R.E., Lozupone, C.A., Hamady, M., Knight, R., and Gordon, J.I. (2008b) 'Worlds within worlds: evolution of the vertebrate gut microbiota', *Nat. Rev. Microbiol.* 6, 776–788.

Li, J., Jia, H., Cai, X., Zhong, H., Feng, Q., Sunagawa, S., Arumugam, M., Kultima, J.R., Prifti, E., Nielsen, T., et al. (2014) 'An integrated catalog of reference genes in the human gut microbiome', *Nat. Biotechnol.* 32, 834–841.

Linz, B., Balloux, F., Moodley, Y., Manica, A., Liu, H., Roumagnac, P., Falush, D., Stamer, C., Prugnolle, F., van der Merwe, S.W., et al. (2007) 'An African origin for the intimate association between humans and *Helicobacter pylori*', *Nature* 445, 915–918.

Liou, A.P., Paziuk, M., Luevano, J.-M., Machineni, S., Turnbaugh, P.J., and Kaplan, L.M. (2013) 'Conserved shifts in the gut microbiota due to gastric bypass reduce host weight and adiposity', *Sci. Transl. Med.* 5, 178ra41.

Login, F.H. and Heddi, A. (2013) 'Insect immune system maintains long-term resident bacteria through a local response', *J. Insect Physiol.* 59, 232–239.

Lombardo, M.P. (2008) 'Access to mutualistic endosymbiotic microbes: an underappreciated benefit of group living', *Behav. Ecol. Sociobiol.* 62, 479–497.

Lyte, M., Varcoe, J.J., and Bailey, M.T. (1998) 'Anxiogenic effect of subclinical bacterial infection in mice in the absence of overt immune activation', *Physiol. Behav.* 65, 63–68.

Ma, B., Forney, L.J., and Ravel, J. (2012) 'Vaginal microbiome: rethinking health and disease,' *Annu. Rev. Microbiol.* 66, 371–389.

Malkova, N.V., Yu, C.Z., Hsiao, E.Y., Moore, M.J., and Patterson, P.H. (2012) 'Maternal immune activation yields offspring displaying mouse versions of the three core symptoms of autism', *Brain. Behav. Immun.* 26, 607–616.

Manichanh, C., Borruel, N., Casellas, F., and Guarner, F. (2012) 'The gut microbiota in IBD', *Nat. Rev. Gastroenterol. Hepatol.* 9, 599–608.

Marcobal, A., Barboza, M., Sonnenburg, E.D., Pudlo, N., Martens, E.C., Desai, P., Lebrilla, C.B., Weimer, B.C., Mills, D.A., German, J.B., et al. (2011) 'Bacteroides in the infant gut consume milk oligosaccharides via mucus-utilization pathways', *Cell Host Microbe* 10, 507–514.

Margulis, L., and Fester, R. (1991) *Symbiosis as a Source of Evolutionary Innovation: Speciation and Morphogenesis* (Cambridge, Mass: The MIT Press).

Margulis, L. and Sagan, D. (2002) *Acquiring Genomes: A Theory of the Origin of Species* (New York: Perseus Books Group).

Martel, A., Sluijs, A.S. der, Blooi, M., Bert, W., Ducatelle, R., Fisher, M.C., Woeltjes, A., Bosman, W., Chiers, K., Bossuyt, F., et al. (2013) '*Batrachochytrium salamandrivorans sp. nov.* causes lethal chytridiomycosis in amphibians', *Proc. Natl. Acad. Sci.* 110, 15325–15329.

Martens, E.C., Kelly, A.G., Tauzin, A.S., and Brumer, H. (2014) 'The devil lies in the details: how variations in polysaccharide fine-structure impact the physiology and evolution of gut microbes', *J. Mol. Biol.* 426, 3851–3865.

Martínez, I., Stegen, J.C., Maldonado-Gómez, M.X., Eren, A.M., Siba, P.M., Greenhill, A.R., and Walter, J. (2015) 'The gut microbiota of rural Papua New Guineans: composition, diversity patterns, and ecological processes', *Cell Rep.* 11, 527–538.

Mayer, E.A., Tillisch, K., and Gupta, A. (2015) 'Gut/brain axis and the microbiota', *J. Clin. Invest.* 125, 926–938.

Maynard, C.L., Elson, C.O., Hatton, R.D., and Weaver, C.T. (2012) 'Reciprocal interactions of the intestinal microbiota and immune system', *Nature* 489, 231–241.

Mazmanian, S.K., Liu, C.H., Tzianabos, A.O., and Kasper, D.L. (2005) 'An immunomodulatory molecule of symbiotic bacteria directs maturation of the host immune system', *Cell* 122, 107–118.

Mazmanian, S.K., Round, J.L., and Kasper, D.L. (2008) 'A microbial symbiosis factor prevents intestinal inflammatory disease', *Nature* 453, 620–625.

McCutcheon, J.P. (2013) 'Genome evolution: a bacterium with a Napoleon Complex', *Curr. Biol.* 23, R657–R659.

McCutcheon, J.P. and Moran, N.A. (2011) 'Extreme genome reduction in symbiotic bacteria', *Nat. Rev. Microbiol.* 10, 13–26.

McDole, T., Nulton, J., Barott, K.L., Felts, B., Hand, C., Hatay, M., Lee, H., Nadon, M.O., Nosrat, B., Salamon, P., et al. (2012) 'Assessing coral reefs on a Pacific-wide scale using the microbialization score', *PLoS ONE* 7, e43233.

McFall-Ngai, M.J. (1998) 'The development of cooperative associations between animals and bacteria: establishing detente among domains', *Integr. Comp. Biol.* 38, 593–608.

McFall-Ngai, M. (2007) 'Adaptive immunity: care for the community', *Nature* 445, 153.

McFall-Ngai, M. (2014) 'Divining the essence of symbiosis: insights from the Squid-Vibrio Model', *PLoS Biol.* 12, e1001783.

McFall-Ngai, M.J. and Ruby, E.G. (1991) 'Symbiont recognition and subsequent morphogenesis as early events in an animal–bacterial mutualism', *Science* 254, 1491–1494.

McFall-Ngai, M., Hadfield, M.G., Bosch, T.C., Carey, H.V., Domazet-Lošo, T., Douglas, A.E., Dubilier, N., Eberl, G., Fukami, T., and Gilbert, S.F. (2013) 'Animals in a bacterial world, a new imperative for the life sciences', *Proc. Natl. Acad. Sci.* 110, 3229–3236.

McFarland, L.V. (2014) 'Use of probiotics to correct dysbiosis of normal microbiota following disease or disruptive events: a systematic review', *BMJ Open* 4, e005047.

McGraw, E.A. and O'Neill, S.L. (2013) 'Beyond insecticides: new thinking on an ancient problem', *Nat. Rev. Microbiol.* 11, 181–193.

McKenna, M. (2010) *Superbug: The Fatal Menace of MRSA* (New York: Free Press).

McKenna, M. (2013) Imagining the Post-Antibiotics Future. https://medium.com/@fernnews/imagining-the-post-antibiotics-future-892b57499e77.

Mclaren, D.J., Worms, M.J., Laurence, B.R., and Simpson, M.G. (1975) 'Micro-organisms in filarial larvae (*Nematoda*)', *Trans. R. Soc. Trop. Med. Hyg.* 69, 509–514.

McMaster, J. (2004). How Did Life Begin? http:www.pbs.org/wgbn/nova/evolution/how-did-life-begin.html.

McMeniman, C.J., Lane, R.V., Cass, B.N., Fong, A.W.C., Sidhu, M., Wang, Y-F., and O'Neill, S.L. (2009) 'Stable introduction of a life-shortening *Wolbachia* infection into the mosquito *Aedes aegypti*', *Science* 323, 141–144.

McNulty, N.P., Yatsunenko, T., Hsiao, A., Faith, J.J., Muegge, B.D., Goodman, A.L., Henrissat, B., Oozeer, R., Cools-Portier, S., Gobert, G., et al. (2011) 'The impact of a consortium of fermented milk strains on the gut microbiome of gnotobiotic mice and monozygotic twins', *Sci. Transl. Med.* 3, 106ra106.

Meadow, J.F., Bateman, A.C., Herkert, K.M., O'Connor, T.K., and Green, J.L. (2013) 'Significant changes in the skin microbiome mediated by the sport of roller derby', *PeerJ* 1, e53.

Meadow, J.F., Altrichter, A.E., Bateman, A.C., Stenson, J., Brown, G.Z., Green, J.L., and Bohannan, B.J.M. (2015) 'Humans differ in their personal microbial cloud', *PeerJ* 3, e1258.

Metcalf, J.A., Funkhouser-Jones, L.J., Brileya, K., Reysenbach, A-L., and Bordenstein, S.R. (2014) 'Antibacterial gene transfer across the tree of life', eLife 3.

Miller, A.W., Kohl, K.D., and Dearing, M.D. (2014) 'The gastrointestinal tract of the white-throated woodrat (*Neotoma albigula*) harbors distinct consortia of oxalate-degrading bacteria', *Appl. Environ. Microbiol.* 80, 1595–1601.

Mimee, M., Tucker, A.C., Voigt, C.A., and Lu, T.K. (2015) 'Programming a human commensal bacterium, *Bacteroides thetaiotaomicron*, to sense and respond to stimuli in the murine gut microbiota', *Cell Syst.* 1, 62–71.

Min, K.-T., and Benzer, S. (1997) 'Wolbachia, normally a symbiont of Drosophila, can be virulent, causing degeneration and early death', *Proc. Natl. Acad. Sci. U. S. A.* 94, 10792–10796.

Moberg, S. (2005) *René Dubos, Friend of the Good Earth: Microbiologist, Medical Scientist, Environmentalist* (Washington, DC: ASM Press).

Moeller, A.H., Li, Y., Mpoudi Ngole, E., Ahuka-Mundeke, S., Lonsdorf, E.V., Pusey, A.E., Peeters, M., Hahn, B.H., and Ochman, H. (2014) 'Rapid changes in the gut microbiome during human evolution', *Proc. Natl. Acad. Sci. U. S. A.* 111, 16431–16435.

Montgomery, M.K. and McFall-Ngai, M. (1994) 'Bacterial symbionts induce host organ morphogenesis during early postembryonic development of the squid *Euprymna scolopes*', *Dev. Camb. Engl.* 120, 1719–1729.

Moran, N.A. and Dunbar, H.E. (2006) 'Sexual acquisition of beneficial symbionts in aphids', *Proc. Natl. Acad. Sci.* 103, 12803–12806.

Moran, N.A. and Sloan, D.B. (2015) 'The Hologenome Concept: helpful or hollow?' *PLoS Biol.* 13, e1002311.

Moran, N.A., Degnan, P.H., Santos, S.R., Dunbar, H.E., and Ochman, H. (2005) 'The players in a mutualistic symbiosis: insects, bacteria, viruses, and virulence genes', *Proc. Natl. Acad. Sci. U. S. A.* 102, 16919–16926.

Moreira, L.A., Iturbe-Ormaetxe, I., Jeffery, J.A., Lu, G., Pyke, A.T., Hedges, L.M., Rocha, B.C., Hall-Mendelin, S., Day, A., Riegler, M., et al. (2009) 'A *Wolbachia* symbiont in *Aedes aegypti* limits infection with dengue, chikungunya, and plasmodium', *Cell* 139, 1268–1278.

Morell, V. (1997) 'Microbial biology: microbiology's scarred revolutionary', *Science* 276, 699–702.

Morgan, X.C., Tickle, T.L., Sokol, H., Gevers, D., Devaney, K.L., Ward, D.V., Reyes, J.A., Shah, S.A., LeLeiko, N., Snapper, S.B., et al. (2012) 'Dysfunction of the intestinal microbiome in inflammatory bowel disease and treatment', *Genome Biol.* 13, R79.

Mukherjee, S. (2011) *The Emperor of All Maladies* (London:Fourth Estate).

Mullard, A. (2008) 'Microbiology: the inside story', *Nature* 453, 578–580.

National Research Council (US) Committee on Metagenomics (2007) *The New Science of Metagenomics: Revealing the Secrets of Our Microbial Planet* (Washington, DC: National Academies Press (US)).

Nature (1975) 'Oh, New Delhi; oh, Geneva', *Nature* 256, 355–357.

Nature (2013) 'Culture shock', *Nature* 493, 133–134.

Nelson, B. (2014). Medicine's dirty secret. http://mosaicscience.com/story/medicine%E2%80%99s-dirty-secret.

Neufeld, K.M., Kang, N., Bienenstock, J., and Foster, J.A. (2011) 'Reduced anxiety-like behavior and central neurochemical change in germ-free mice: behavior in germ-free mice', *Neurogastroenterol. Motil.* 23, 255–e119.

Newburg, D.S., Ruiz-Palacios, G.M., and Morrow, A.L. (2005) 'Human milk glycans protect infants against enteric pathogens', *Annu. Rev. Nutr.* 25, 37–58.

New York Times (12 February 1985) 'Science watch: miracle plant tested as cattle fodder'.

Nicholson, J.K., Holmes, E., Kinross, J., Burcelin, R., Gibson, G., Jia, W., and Pettersson, S. (2012) 'Host–Gut Microbiota Metabolic Interactions', *Science* 336, 1262–1267.

Nightingale, F. (1859) *Notes on Nursing: What It Is, and What It Is Not* (New York: D. Appleton & Co.).

Nougué, O., Gallet, R., Chevin, L-M., and Lenormand, T. (2015) 'Niche limits of symbiotic gut microbiota constrain the salinity tolerance of brine shrimp', *Am. Nat.* 186, 390–403.

Nováková, E., Hypša, V., Klein, J., Foottit, R.G., von Dohlen, C.D., and Moran, N.A. (2013) 'Reconstructing the phylogeny of aphids (*Hemiptera: Aphididae*) using DNA of the obligate symbiont *Buchnera aphidicola', Mol. Phylogenet. Evol.* 68, 42–54.

Obregon-Tito, A.J., Tito, R.Y., Metcalf, J., Sankaranarayanan, K., Clemente, J.C., Ursell, L.K., Zech Xu, Z., Van Treuren, W., Knight, R., Gaffney, P.M., et al. (2015) 'Subsistence strategies in traditional societies distinguish gut microbiomes', *Nat. Commun.* 6, 6505.

Ochman, H., Lawrence, J.G., and Groisman, E.A. (2000) 'Lateral gene transfer and the nature of bacterial innovation', *Nature* 405, 299–304.

Ohbayashi, T., Takeshita, K., Kitagawa, W., Nikoh, N., Koga, R., Meng, X-Y., Tago, K., Hori, T., Hayatsu, M., Asano, K., et al. (2015) 'Insect's intestinal organ for symbiont sorting', *Proc. Natl. Acad. Sci.* 112, E5179–E5188.

Oliver, K.M., Moran, N.A., and Hunter, M.S. (2005) 'Variation in resistance to parasitism in aphids is due to symbionts not host genotype', *Proc. Natl. Acad. Sci. U. S. A.* 102, 12795–12800.

Oliver, K.M., Campos, J., Moran, N.A., and Hunter, M.S. (2008) 'Population dynamics of defensive symbionts in aphids', *Proc. R. Soc. B Biol. Sci.* 275, 293–299.

Olle, B. (2013) 'Medicines from microbiota', *Nat. Biotechnol.* 31, 309–315.

Olszak, T., An, D., Zeissig, S., Vera, M.P., Richter, J., Franke, A., Glickman, J.N., Siebert, R., Baron, R.M., Kasper, D.L., et al. (2012) 'Microbial exposure during early life has persistent effects on natural killer T cell function', *Science* 336, 489–493.

O'Malley, M.A. (2009) 'What did Darwin say about microbes, and how did microbiology respond?', *Trends Microbiol.* 17, 341–347.

Osawa, R., Blanshard, W., and Ocallaghan, P. (1993) 'Microbiological studies of the intestinal microflora of the Koala, *Phascolarctos-Cinereus* .2. Pap, a special maternal feces consumed by juvenile koalas', *Aust. J. Zool.* 41, 611–620.

Ott, S.J., Musfeldt, M., Wenderoth, D.F., Hampe, J., Brant, O., Fölsch, U.R., Timmis, K.N., and Schreiber, S. (2004) 'Reduction in diversity of the colonic mucosa associated bacterial microflora in patients with active inflammatory bowel disease', *Gut* 53, 685–693.

Ott, B.M., Rickards, A., Gehrke, L., and Rio, R.V.M. (2015) 'Characterization of shed medicinal leech mucus reveals a diverse microbiota', *Front. Microbiol.* 5. doi: 10.3389/fmicb.2014.00757.

Pace, N.R., Stahl, D.A., Lane, D.J., and Olsen, G.J. (1986) 'The analysis of natural microbial populations by ribosomal RNA Sequences', in *Advances in Microbial Ecology*, K.C. Marshall, ed. (New York: Springer US), pp. 1–55.

Paine, R.T., Tegner, M.J., and Johnson, E.A. (1998) 'Compounded perturbations yield ecological surprises', *Ecosystems* 1, 535–545.

Pais, R., Lohs, C., Wu, Y., Wang, J., and Aksoy, S. (2008) 'The obligate mutualist *Wigglesworthia glossinidia* influences reproduction, digestion, and immunity processes of its host, the tsetse fly', *Appl. Environ. Microbiol.* 74, 5965–5974.

Pannebakker, B.A., Loppin, B., Elemans, C.P., Humblot, L., and Vavre, F. (2007) 'Parasitic inhibition of cell death facilitates symbiosis', *Proc. Natl. Acad. Sci.* 104, 213–215.

Payne, A.S. (1970) *The Cleere Observer. A Biography of Antoni Van Leeuwenhoek* (London: Macmillan).

Petrof, E.O. and Khoruts, A. (2014) 'From stool transplants to next-generation microbiota therapeutics', *Gastroenterology* 146, 1573–1582.

Petrof, E., Gloor, G., Vanner, S., Weese, S., Carter, D., Daigneault, M., Brown, E., Schroeter, K., and Allen-Vercoe, E. (2013) 'Stool substitute transplant therapy for the eradication of *Clostridium difficile* infection: 'RePOOPulating' the gut', *Microbiome 2013*, 3.

Petschow, B., Doré, J., Hibberd, P., Dinan, T., Reid, G., Blaser, M., Cani, P.D., Degnan, F.H., Foster, J., Gibson, G., et al. (2013) 'Probiotics, prebiotics, and the host microbiome: the science of translation', *Ann. N. Y. Acad. Sci.* 1306, 1–17.

Pickard, J.M., Maurice, C.F., Kinnebrew, M.A., Abt, M.C., Schenten, D., Golovkina, T.V., Bogatyrev, S.R., Ismagilov, R.F., Pamer, E.G., Turnbaugh, P.J., et al. (2014) 'Rapid fucosylation of intestinal epithelium sustains host–commensal symbiosis in sickness', *Nature* 514, 638–641.

Poulsen, M., Hu, H., Li, C., Chen, Z., Xu, L., Otani, S., Nygaard, S., Nobre, T., Klaubauf, S., Schindler, P.M., et al. (2014) 'Complementary symbiont contributions to plant decomposition in a fungus-farming termite', *Proc. Natl. Acad. Sci.* 111, 14500–14505.

Qian, J., Hospodsky, D., Yamamoto, N., Nazaroff, W.W., and Peccia, J. (2012). 'Size-resolved emission rates of airborne bacteria and fungi in an occupied classroom: size-resolved bioaerosol emission rates', *Indoor Air* 22, 339–351.

Quammen, D. (1997) *The Song of the Dodo: Island Biogeography in an Age of Extinction* (New York: Scribner).

Rawls, J.F., Samuel, B.S., and Gordon, J.I. (2004) 'Gnotobiotic zebrafish reveal evolutionarily conserved responses to the gut microbiota', *Proc. Natl. Acad. Sci. U. S. A.* 101, 4596–4601.

Rawls, J.F., Mahowald, M.A., Ley, R.E., and Gordon, J.I. (2006) 'Reciprocal gut microbiota transplants from zebrafish and mice to germ-free recipients reveal host habitat selection', *Cell* 127, 423–433.

Redford, K.H., Segre, J.A., Salafsky, N., del Rio, C.M., and McAloose, D. (2012) 'Conservation and the Microbiome: Editorial. *Conserv. Biol.* 26, 195–197.

Reid, G. (2011) 'Opinion paper: Quo vadis – EFSA?', *Benef. Microbes* 2, 177–181.

Relman, D.A. (2008), '"Til death do us part": coming to terms with symbiotic relationships', Foreword. *Nat. Rev. Microbiol.* 6, 721–724.

Relman, D.A. (2012) 'The human microbiome: ecosystem resilience and health', *Nutr. Rev.* 70, S2–S9.

Ridaura, V.K., Faith, J.J., Rey, F.E., Cheng, J., Duncan, A.E., Kau, A.L., Griffin, N.W., Lombard, V., Henrissat, B., Bain, J.R., et al. (2013). 'Gut microbiota from twins discordant for obesity modulate metabolism in mice', *Science* 341, 1241214.

Rigaud, T., and Juchault, P. (1992). Heredity – Abstract of article: 'Genetic control of the vertical transmission of a cytoplasmic sex factor in *Armadillidium vulgare Latr.* (Crustacea, Oniscidea)', *Heredity* 68, 47–52.

Riley, D.R., Sieber, K.B., Robinson, K.M., White, J.R., Ganesan, A., Nourbakhsh, S., and Dunning Hotopp, J.C. (2013) 'Bacteria–human somatic cell lateral gene transfer is enriched in cancer samples', *PLoS Comput. Biol.* 9, e1003107.

Roberts, C.S. (1990) 'William Beaumont, the man and the opportunity', in *Clinical Methods: The History, Physical, and Laboratory Examinations*, H.K. Walker, W.D. Hall, and J.W. Hurst, eds (Boston: Butterworths).

Roberts, S.C., Gosling, L.M., Spector, T.D., Miller, P., Penn, D.J., and Petrie, M. (2005) 'Body Odor Similarity in Noncohabiting Twins', *Chem. Senses* 30, 651–656.

Rogier, E.W., Frantz, A.L., Bruno, M.E., Wedlund, L., Cohen, D.A., Stromberg, A.J., and Kaetzel, C.S. (2014) 'Secretory antibodies in breast milk promote long-term intestinal homeostasis by regulating the gut microbiota and host gene expression', *Proc. Natl. Acad. Sci.* 111, 3074–3079.

Rohwer, F. and Youle, M. (2010) *Coral Reefs in the Microbial Seas* (United States: Plaid Press).

Rook, G.A.W., Lowry, C.A., and Raison, C.L. (2013) 'Microbial 'Old Friends', immunoregulation and stress resilience', *Evol. Med. Public Health* 2013, 46–64.

Rosebury, T. (1962) *Microorganisms Indigenous to Man* (New York: McGraw-Hill).

Rosebury, T. (1969) *Life on Man* (New York: Viking Press).

Rosenberg, E., Sharon, G., and Zilber-Rosenberg, I. (2009) 'The hologenome theory of evolution contains Lamarckian aspects within a Darwinian framework', *Environ. Microbiol.* 11, 2959–2962.

Rosner, J. (2014) 'Ten times more microbial cells than body cells in humans?', *Microbe* 9, 47.

Round, J.L., and Mazmanian, S.K. (2009) 'The gut microbiota shapes intestinal immune responses during health and disease', *Nat. Rev. Immunol.* 9, 313–323.

Round, J.L. and Mazmanian, S.K. (2010) 'Inducible Foxp3+ regulatory T-cell development by a commensal bacterium of the intestinal microbiota', *Proc. Natl. Acad. Sci. U. S. A.* 107, 12204–12209.

Russell, C.W., Bouvaine, S., Newell, P.D., and Douglas, A.E. (2013a) 'Shared metabolic pathways in a coevolved insect–bacterial symbiosis', *Appl. Environ. Microbiol.* 79, 6117–6123.

Russell, J.A., Funaro, C.F., Giraldo, Y.M., Goldman-Huertas, B., Suh, D., Kronauer, D.J.C., Moreau, C.S., and Pierce, N.E. (2012) 'A veritable menagerie of heritable bacteria from ants, butterflies, and beyond: broad molecular surveys and a systematic review', *PLoS ONE* 7, e51027.

Russell, J.A., Weldon, S., Smith, A.H., Kim, K.L., Hu, Y., Łukasik, P., Doll, S., Anastopoulos, I., Novin, M., and Oliver, K.M. (2013b) 'Uncovering symbiont-driven genetic diversity across North American pea aphids', *Mol. Ecol.* 22, 2045–2059.

Rutherford, A. (2013). *Creation: The Origin of Life / The Future of Life* (London: Penguin).

Sachs, J.L., Skophammer, R.G., and Regus, J.U. (2011) 'Evolutionary transitions in bacterial symbiosis', *Proc. Natl. Acad. Sci.* 108, 10800–10807.

Sacks, O. (23 April 2015) 'A General Feeling of Disorder.' *N. Y. Rev. Books.*

Saeidi, N., Wong, C.K., Lo, T-M., Nguyen, H.X., Ling, H., Leong, S.S.J., Poh, C.L., and Chang, M.W. (2011) 'Engineering microbes to sense and eradicate *Pseudomonas aeruginosa*, a human pathogen', *Mol. Syst. Biol.* 7, 521.

Sagan, L. (1967) 'On the origin of mitosing cells', *J. Theor. Biol.* 14, 255–274.

Salter, S.J., Cox, M.J., Turek, E.M., Calus, S.T., Cookson, W.O., Moffatt, M.F., Turner, P., Parkhill, J., Loman, N.J., and Walker, A.W. (2014) 'Reagent and laboratory contamination can critically impact sequence-based microbiome analyses', *BMC Biol.* 12, 87.

Salzberg, S.L. (2001) 'Microbial genes in the human genome: lateral transfer or gene loss?', *Science* 292, 1903–1906.

Salzberg, S.L., Hotopp, J.C., Delcher, A.L., Pop, M., Smith, D.R., Eisen, M.B., and Nelson, W.C. (2005) 'Serendipitous discovery of *Wolbachia* genomes in multiple *Drosophila* species', *Genome Biol.* 6, R23.

Sanders, J.G., Beichman, A.C., Roman, J., Scott, J.J., Emerson, D., McCarthy, J.J., and Girguis, P.R. (2015) 'Baleen whales host a unique gut microbiome with similarities to both carnivores and herbivores', *Nat. Commun.* 6, 8285.

Sangodeyi, F.I. (2014) 'The Making of the Microbial Body, 1900s–2012.' Harvard University.

Sapp, J. (1994) *Evolution by Association: A History of Symbiosis* (New York: Oxford University Press).

Sapp, J. (2002) 'Paul Buchner (1886–1978) and hereditary symbiosis in insects', *Int. Microbiol.* 5, 145–150.

Sapp, J. (2009) *The New Foundations of Evolution: On the Tree of Life* (Oxford and New York: Oxford University Press).

Savage, D.C. (2001) 'Microbial biota of the human intestine: a tribute to some pioneering scientists', *Curr. Issues Intest. Microbiol.* 2, 1–15.

Schilthuizen, M.O. and Stouthamer, R. (1997) Horizontal transmission of parthenogenesis-inducing microbes in *Trichogramma* wasps', *Proc. R. Soc. Lond. B Biol. Sci.* 264, 361–366.

Schluter, J. and Foster, K.R. (2012) 'The evolution of mutualism in gut microbiota via host epithelial selection', *PLoS Biol.* 10, e1001424.

Schmidt, C. (2013) 'The startup bugs', *Nat. Biotechnol.* 31, 279–281.

Schmidt, T.M., DeLong, E.F., and Pace, N.R. (1991) 'Analysis of a marine picoplankton community by 16S rRNA gene cloning and sequencing', *J. Bacteriol.* 173, 4371–4378.

Schnorr, S.L., Candela, M., Rampelli, S., Centanni, M., Consolandi, C., Basaglia, G., Turroni, S., Biagi, E., Peano, C., Severgnini, M., et al. (2014) 'Gut microbiome of the Hadza hunter-gatherers', *Nat. Commun.* 5, 3654.

Schubert, A.M., Sinani, H., and Schloss, P.D. (2015) 'Antibiotic-induced alterations of the murine gut microbiota and subsequent effects on colonization resistance against *Clostridium difficile*', mBio 6, e00974–15.

Sela, D.A. and Mills, D.A. (2014) 'The marriage of nutrigenomics with the microbiome: the case of infant-associated bifidobacteria and milk', *Am. J. Clin. Nutr.* 99, 697S–703S.

Sela, D.A., Chapman, J., Adeuya, A., Kim, J.H., Chen, F., Whitehead, T.R., Lapidus, A., Rokhsar, D.S., Lebrilla, C.B., and German, J.B. (2008) 'The genome sequence of *Bifidobacterium longum subsp. infantis* reveals adaptations for milk utilization within the infant microbiome', *Proc. Natl. Acad. Sci.* 105, 18964–18969.

Selosse, M-A., Bessis, A., and Pozo, M.J. (2014) 'Microbial priming of plant and animal immunity: symbionts as developmental signals', *Trends Microbiol.* 22, 607–613.

Shanahan, F. (2010) 'Probiotics in perspective', *Gastroenterology* 139, 1808–1812.

Shanahan, F. (2012) 'The microbiota in inflammatory bowel disease: friend, bystander, and sometime-villain', *Nutr. Rev.* 70, S31–S37.

Shanahan, F. and Quigley, E.M.M. (2014) 'Manipulation of the microbiota for treatment of IBS and IBD – challenges and controversies', *Gastroenterology* 146, 1554–1563.

Sharon, G., Segal, D., Ringo, J.M., Hefetz, A., Zilber-Rosenberg, I., and Rosenberg, E. (2010) 'Commensal bacteria play a role in mating preference of *Drosophila melanogaster*', *Proc. Natl. Acad. Sci.* 107, 20051–20056.

Sharon, G., Garg, N., Debelius, J., Knight, R., Dorrestein, P.C., and Mazmanian, S.K. (2014) 'Specialized metabolites from the microbiome in health and disease. *Cell Metab.* 20, 719–730.

Shikuma, N.J., Pilhofer, M., Weiss, G.L., Hadfield, M.G., Jensen, G.J., and Newman, D.K. (2014) 'Marine tubeworm metamorphosis induced by arrays of bacterial phage tail-Like structures', *Science* 343, 529–533.

Six, D.L. (2013) 'The Bark Beetle holobiont: why microbes matter', *J. Chem. Ecol.* 39, 989–1002.

Sjögren, K., Engdahl, C., Henning, P., Lerner, U.H., Tremaroli, V., Lagerquist, M.K., Bäckhed, F., and Ohlsson, C. (2012) 'The gut microbiota regulates bone mass in mice', *J. Bone Miner. Res. Off. J. Am. Soc. Bone Miner. Res.* 27, 1357–1367.

Slashinski, M.J., McCurdy, S.A., Achenbaum, L.S., Whitney, S.N., and McGuire, A.L. (2012) "Snake-oil,' 'quack medicine,' and 'industrially cultured organisms:' biovalue and the commercialization of human microbiome research', *BMC Med. Ethics* 13, 28.

Slatko, B.E., Taylor, M.J., and Foster, J.M. (2010) 'The *Wolbachia* endosymbiont as an anti-filarial nematode target', *Symbiosis* 51, 55–65.

Smillie, C.S., Smith, M.B., Friedman, J., Cordero, O.X., David, L.A., and Alm, E.J. (2011) 'Ecology drives a global network of gene exchange connecting the human microbiome', *Nature* 480, 241–244.

Smith, C.C., Snowberg, L.K., Gregory Caporaso, J., Knight, R., and Bolnick, D.I. (2015) 'Dietary input of microbes and host genetic variation shape among-population differences in stickleback gut microbiota', *ISME J.* 9, 2515–2526.

Smith, J.E., Shaw, M., Edwards, R.A., Obura, D., Pantos, O., Sala, E., Sandin, S.A., Smriga, S., Hatay, M., and Rohwer, F.L. (2006) 'Indirect effects of algae on coral: algae-mediated, microbe-induced coral mortality', *Ecol. Lett.* 9, 835–845.

Smith, M., Kelly, C., and Alm, E. (2014) 'How to regulate faecal transplants', *Nature* 506, 290–291.

Smith, M.I., Yatsunenko, T., Manary, M.J., Trehan, I., Mkakosya, R., Cheng, J., Kau, A.L., Rich, S.S., Concannon, P., Mychaleckyj, J.C., et al. (2013a) 'Gut microbiomes of Malawian twin pairs discordant for kwashiorkor', *Science* 339, 548–554.

Smith, P.M., Howitt, M.R., Panikov, N., Michaud, M., Gallini, C.A., Bohlooly-Y, M., Glickman, J.N., and Garrett, W.S. (2013b) 'The microbial metabolites, short-chain fatty acids, regulate colonic Treg cell homeostasis', *Science* 341, 569–573.

Smithsonian National Museum of Natural History (2010) Giant Tube Worm: *Riftia pachyptila.* http://www.mnh.si.edu/onehundredyears/featured-objects/Riftia.html.

Sneed, J.M., Sharp, K.H., Ritchie, K.B., and Paul, V.J. (2014) 'The chemical cue tetrabromopyrrole from a biofilm bacterium induces settlement of multiple Caribbean corals', *Proc. R. Soc. B Biol. Sci.* 281, 20133086.

Sokol, H., Pigneur, B., Watterlot, L., Lakhdari, O., Bermúdez-Humarán, L.G., Gratadoux, J-J., Blugeon, S., Bridonneau, C., Furet, J-P., Corthier, G., et al. (2008) '*Faecalibacterium prausnitzii* is an anti-inflammatory commensal bacterium identified by gut microbiota analysis of Crohn disease patients', *Proc. Natl. Acad. Sci.*

Soler, J.J., Martín-Vivaldi, M., Ruiz-Rodríguez, M., Valdivia, E., Martín-Platero, A.M., Martínez-Bueno, M., Peralta-Sánchez, J.M., and Méndez, M. (2008) 'Symbiotic association between hoopoes and antibiotic-producing bacteria that live in their uropygial gland', *Funct. Ecol.* 22, 864–871.

Sommer, F. and Bäckhed, F. (2013) 'The gut microbiota – masters of host development and physiology', *Nat. Rev. Microbiol.* 11, 227–238.

Sonnenburg, E.D. and Sonnenburg, J.L. (2014) 'Starving our microbial self: the deleterious consequences of a diet deficient in microbiota-accessible carbohydrates', *Cell Metab.* 20, 779–786.

Sonnenburg, E.D., Smits, S.A., Tikhonov, M., Higginbottom, S.K., Wingreen, N.S., and Sonnenburg, J.L. (2016) 'Diet-induced extinctions in the gut microbiota compound over generations', *Nature* 529, 212–215.

Sonnenburg, J.L., and Fischbach, M.A. (2011) 'Community health care: therapeutic opportunities in the human microbiome', *Sci. Transl. Med.* 3, 78ps12.

Sonnenburg, J. and Sonnenburg, E. (2015) *The Good Gut: Taking Control of Your Weight, Your Mood, and Your Long-Term Health* (New York: The Penguin Press).

Spor, A., Koren, O., and Ley, R. (2011) 'Unravelling the effects of the environment and host genotype on the gut microbiome', *Nat. Rev. Microbiol.* 9, 279–290.

Stahl, D.A., Lane, D.J., Olsen, G.J., and Pace, N.R. (1985) 'Characterization of a Yellowstone hot spring microbial community by 5S rRNA sequences', *Appl. Environ. Microbiol.* 49, 1379–1384.

Stappenbeck, T.S., Hooper, L.V., and Gordon, J.I. (2002) 'Developmental regulation of intestinal angiogenesis by indigenous microbes via Paneth cells', *Proc. Natl. Acad. Sci. U. S. A.* 99, 15451–15455.

Stefka, A.T., Feehley, T., Tripathi, P., Qiu, J., McCoy, K., Mazmanian, S.K., Tjota, M.Y., Seo, G-Y., Cao, S., Theriault, B.R., et al. (2014) 'Commensal bacteria protect against food allergen sensitization', *Proc. Natl. Acad. Sci.* 111, 13145–13150.

Stevens, C.E. and Hume, I.D. (1998) 'Contributions of microbes in vertebrate gastrointestinal tract to production and conservation of nutrients', *Physiol. Rev.* 78, 393–427.

Stewart, F.J. and Cavanaugh, C.M. (2006) 'Symbiosis of thioautotrophic bacteria with *Riftia pachyptila*', *Prog. Mol. Subcell. Biol.* 41, 197–225.

Stilling, R.M., Dinan, T.G., and Cryan, J.F. (2015) 'The brain's Geppetto – microbes as puppeteers of neural function and behaviour?', *J. Neurovirol.* doi: 10.3389/fcimb.2014.00147.

Stoll, S., Feldhaar, H., Fraunholz, M.J., and Gross, R. (2010) 'Bacteriocyte dynamics during development of a holometabolous insect, the carpenter ant *Camponotus floridanus*', *BMC Microbiol.* 10, 308.

Strachan, D.P. (1989) 'Hay fever, hygiene, and household size', *BMJ* 299, 1259–1260.

Strachan, D.P. (2015). Re: 'The 'hygiene hypothesis' for allergic disease is a misnomer.' *BMJ* 349, g5267.

Strand, M.R. and Burke, G.R. (2012) 'Polydnaviruses as symbionts and gene delivery systems', *PLoS Pathog.* 8, e1002757.

Subramanian, S., Huq, S., Yatsunenko, T., Haque, R., Mahfuz, M., Alam, M.A., Benezra, A., DeStefano, J., Meier, M.F., Muegge, B.D., et al. (2014) 'Persistent gut microbiota immaturity in malnourished Bangladeshi children', *Nature* 510, 417–421.

Sudo, N., Chida, Y., Aiba, Y., Sonoda, J., Oyama, N., Yu, X-N., Kubo, C., and Koga, Y. (2004) 'Postnatal microbial colonization programs the hypothalamic-pituitary–adrenal system for stress response in mice', *J. Physiol.* 558, 263–275.

Sundset, M.A., Barboza, P.S., Green, T.K., Folkow, L.P., Blix, A.S., and Mathiesen, S.D. (2010) 'Microbial degradation of usnic acid in the reindeer rumen', *Naturwissenschaften* 97, 273–278.

Svoboda, E. (2015) How Soil Microbes Affect the Environment. http://www.quantamagazine.org/20150616-soil-microbes-bacteria-climate-change/.

Tang, W.H.W. and Hazen, S.L. (2014) 'The contributory role of gut microbiota in cardiovascular disease', *J. Clin. Invest.* 124, 4204–4211.

Taylor, M.J. and Hoerauf, A. (1999) '*Wolbachia* bacteria of filarial nematodes', *Parasitol. Today* 15, 437–442.

Taylor, M.J., Makunde, W.H., McGarry, H.F., Turner, J.D., Mand, S., and Hoerauf, A. (2005) 'Macrofilaricidal activity after doxycycline treatment of *Wuchereria bancrofti*: a double-blind, randomised placebo-controlled trial', *Lancet* 365, 2116–2121.

Taylor, M.J., Hoerauf, A., and Bockarie, M. (2010) 'Lymphatic filariasis and onchocerciasis', *Lancet* 376, 1175–1185.

Taylor, M.J., Voronin, D., Johnston, K.L., and Ford, L. (2013) '*Wolbachia* filarial interactions: *Wolbachia* filarial cellular and molecular interactions', *Cell. Microbiol.* 15, 520–526.

Taylor, M.J., Hoerauf, A., Townson, S., Slatko, B.E., and Ward, S.A. (2014) 'Anti-*Wolbachia* drug discovery and development: safe macrofilaricides for onchocerciasis and lymphatic filariasis', *Parasitology* 141, 119–127.

Teixeira, L., Ferreira, Á., and Ashburner, M. (2008) 'The bacterial symbiont *Wolbachia* induces resistance to RNA viral infections in *Drosophila melanogaster*', *PLoS Biol.* 6, e1000002.

Thacker, R.W. and Freeman, C.J. (2012) 'Sponge–microbe symbioses', in *Advances in Marine Biology* (Philadelphia: Elsevier), pp. 57–111.

Thaiss, C.A., Zeevi, D., Levy, M., Zilberman-Schapira, G., Suez, J., Tengeler, A.C., Abramson, L., Katz, M.N., Korem, T., Zmora, N., et al. (2014) 'Transkingdom control of microbiota diurnal oscillations promotes metabolic homeostasis', *Cell* 159, 514–529.

Theis, K.R., Venkataraman, A., Dycus, J.A., Koonter, K.D., Schmitt-Matzen, E.N., Wagner, A.P., Holekamp, K.E., and Schmidt, T.M. (2013) 'Symbiotic bacteria appear to mediate hyena social odors', *Proc. Natl. Acad. Sci.* 110, 19832–19837.

Thurber, R.L.V., Barott, K.L., Hall, D., Liu, H., Rodriguez-Mueller, B., Desnues, C., Edwards, R.A., Haynes, M., Angly, F.E., Wegley, L., et al. (2008) 'Metagenomic analysis indicates that stressors induce production of herpes-like viruses in the coral *Porites compressa*', *Proc. Natl. Acad. Sci.* 105, 18413–18418.

Thurber, R.V., Willner-Hall, D., Rodriguez-Mueller, B., Desnues, C., Edwards, R.A., Angly, F., Dinsdale, E., Kelly, L., and Rohwer, F. (2009) 'Metagenomic analysis of stressed coral holobionts', *Environ. Microbiol.* 11, 2148–2163.

Tillisch, K., Labus, J., Kilpatrick, L., Jiang, Z., Stains, J., Ebrat, B., Guyonnet, D., Legrain-Raspaud, S., Trotin, B., Naliboff, B., et al. (2013) 'Consumption of fermented milk product with probiotic modulates brain activity', *Gastroenterology* 144, 1394–1401.e4.

Tito, R.Y., Knights, D., Metcalf, J., Obregon-Tito, A.J., Cleeland, L., Najar, F., Roe, B., Reinhard, K., Sobolik, K., Belknap, S., et al. (2012) 'Insights from "Characterizing Extinct Human Gut Microbiomes"', *PLoS ONE* 7, e51146.

Trasande, L., Blustein, J., Liu, M., Corwin, E., Cox, L.M., and Blaser, M.J. (2013) 'Infant antibiotic exposures and early-life body mass', *Int. J. Obes.* 2005 37, 16–23.

Tung, J., Barreiro, L.B., Burns, M.B., Grenier, J-C., Lynch, J., Grieneisen, L.E., Altmann, J., Alberts, S.C., Blekhman, R., and Archie, E.A. (2015) 'Social networks predict gut microbiome composition in wild baboons', *eLife* 4.

Turnbaugh, P.J., Ley, R.E., Mahowald, M.A., Magrini, V., Mardis, E.R., and Gordon, J.I. (2006) 'An obesity-associated gut microbiome with increased capacity for energy harvest', *Nature* 444, 1027–1131.

Underwood, M.A., Salzman, N.H., Bennett, S.H., Barman, M., Mills, D.A., Marcobal, A., Tancredi, D.J., Bevins, C.L., and Sherman, M.P. (2009) 'A randomized placebo-controlled comparison of 2 prebiotic/probiotic combinations in preterm infants: impact on weight gain, intestinal microbiota, and fecal short-chain fatty acids', *J. Pediatr. Gastroenterol. Nutr.* 48, 216–225.

University of Utah (2012). How Insects Domesticate Bacteria. http://archive.unews.utah.edu/news-releases/how-insects-domesticate-bacteria/.

Vaishnava, S., Behrendt, C.L., Ismail, A.S., Eckmann, L., and Hooper, L.V. (2008) 'Paneth cells directly sense gut commensals and maintain homeostasis at the intestinal host-microbial interface', *Proc. Natl. Acad. Sci.* 105, 20858–20863.

Van Bonn, W., LaPointe, A., Gibbons, S.M., Frazier, A., Hampton-Marcell, J., and Gilbert, J. (2015) 'Aquarium microbiome response to ninety-percent system water change: clues to microbiome management', *Zoo Biol.* 34, 360–367.

Van Leuven, J.T., Meister, R.C., Simon, C., and McCutcheon, J.P. (2014) 'Sympatric speciation in a bacterial endosymbiont results in two genomes with the functionality of one', *Cell* 158, 1270–1280.

Van Nood, E., Vrieze, A., Nieuwdorp, M., Fuentes, S., Zoetendal, E.G., de Vos, W.M., Visser, C.E., Kuijper, E.J., Bartelsman, J.F.W.M., Tijssen, J.G.P., et al. (2013) 'Duodenal infusion of donor feces for recurrent *Clostridium difficile*', *N. Engl. J. Med.* 368, 407–415.

Verhulst, N.O., Qiu, Y.T., Beijleveld, H., Maliepaard, C., Knights, D., Schulz, S., Berg-Lyons, D., Lauber, C.L., Verduijn, W., Haasnoot, G.W., et al. (2011) 'Composition of human skin microbiota affects attractiveness to malaria mosquitoes', *PLoS ONE* 6, e28991.

Vétizou, M., Pitt, J.M., Daillère, R., Lepage, P., Waldschmitt, N., Flament, C., Rusakiewicz, S., Routy, B., Roberti, M.P., Duong, C.P.M., et al. (2015) 'Anticancer immunotherapy by CTLA-4 blockade relies on the gut microbiota', *Science* 350, 1079–1084.

Vigneron, A., Masson, F., Vallier, A., Balmand, S., Rey, M., Vincent-Monégat, C., Aksoy, E., Aubailly-Giraud, E., Zaidman-Rémy, A., and Heddi, A. (2014) 'Insects recycle endosymbionts when the benefit is over', *Curr. Biol.* 24, 2267–2273.

Voronin, D., Cook, D.A.N., Steven, A., and Taylor, M.J. (2012) 'Autophagy regulates Wolbachia populations across diverse symbiotic associations', *Proc. Natl. Acad. Sci.* 109, E1638–E1646.

Vrieze, A., Van Nood, E., Holleman, F., Salojärvi, J., Kootte, R.S., Bartelsman, J.F.W.M., Dallinga-Thie, G.M., Ackermans, M.T., Serlie, M.J., Oozeer, R., et al. (2012) 'Transfer of intestinal microbiota from lean donors increases insulin sensitivity in individuals with metabolic syndrome', *Gastroenterology* 143, 913–916.e7.

Wada-Katsumata, A., Zurek, L., Nalyanya, G., Roelofs, W.L., Zhang, A., and Schal, C. (2015) 'Gut bacteria mediate aggregation in the German cockroach', *Proc. Natl. Acad. Sci* doi: 10.1073/pnas.1504031112.

Wahl, M., Goecke, F., Labes, A., Dobretsov, S., and Weinberger, F. (2012) 'The second skin: ecological role of epibiotic biofilms on marine organisms', *Front. Microbiol.* 3 doi: 10.3389/fmicb.2012.00292.

Walke, J.B., Becker, M.H., Loftus, S.C., House, L.L., Cormier, G., Jensen, R.V., and Belden, L.K. (2014) 'Amphibian skin may select for rare environmental microbes', *ISME J.* 8, 2207–2217.

Walker, T., Johnson, P.H., Moreira, L.A., Iturbe-Ormaetxe, I., Frentiu, F.D., McMeniman, C.J., Leong, Y.S., Dong, Y., Axford, J., Kriesner, P., et al. (2011) 'The wMel Wolbachia strain blocks dengue and invades caged Aedes aegypti populations', *Nature* 476, 450–453.

Wallace, A.R. (1855) 'On the law which has regulated the introduction of new species', *Ann. Mag. Nat. Hist.* 16, 184–196.

Wallin, I.E. (1927) *Symbionticism and the Origin of Species* (Baltimore: Williams & Wilkins Co.).

Walter, J. and Ley, R. (2011) 'The human gut microbiome: ecology and recent evolutionary changes', *Annu. Rev. Microbiol.* 65, 411–429.

Walters, W.A., Xu, Z., and Knight, R. (2014) 'Meta-analyses of human gut microbes associated with obesity and IBD', *FEBS Lett.* 588, 4223–4233.

Wang, Z., Roberts, A.B., Buffa, J.A., Levison, B.S., Zhu, W., Org, E., Gu, X., Huang, Y., Zamanian-Daryoush, M., Culley, M.K., et al. (2015) 'Non-lethal inhibition of gut microbial trimethylamine production for the treatment of atherosclerosis. *Cell* 163, 1585–1595.

Ward, R.E., Ninonuevo, M., Mills, D.A., Lebrilla, C.B., and German, J.B. (2006) 'In vitro fermentation of breast milk oligosaccharides by *Bifidobacterium infantis* and *Lactobacillus gasseri*', *Appl. Environ. Microbiol.* 72, 4497–4499.

Weeks, P. (2000) 'Red-billed oxpeckers: vampires or tickbirds?', *Behav. Ecol.* 11, 154–160.

Wells, H.G., Huxley, J., and Wells, G.P. (1930) *The Science of Life* (London: Cassell).

Wernegreen, J.J. (2004) 'Endosymbiosis: lessons in conflict resolution', *PLoS Biol.* 2, e68.

Wernegreen, J.J. (2012) 'Mutualism meltdown in insects: bacteria constrain thermal adaptation', *Curr. Opin. Microbiol.* 15, 255–262.

Wernegreen, J.J., Kauppinen, S.N., Brady, S.G., and Ward, P.S. (2009) 'One nutritional symbiosis begat another: phylogenetic evidence that the ant tribe *Camponotini* acquired *Blochmannia* by tending sap-feeding insects', *BMC Evol. Biol.* 9, 292.

Werren, J.H., Baldo, L., and Clark, M.E. (2008) '*Wolbachia*: master manipulators of invertebrate biology', *Nat. Rev. Microbiol.* 6, 741–751.

West, S.A., Fisher, R.M., Gardner, A., and Kiers, E.T. (2015) 'Major evolutionary transitions in individuality', *Proc. Natl. Acad. Sci. U. S. A.* 112, 10112–10119.

Westwood, J., Burnett, M., Spratt, D., Ball, M., Wilson, D.J., Wellsteed, S., Cleary, D., Green, A., Hutley, E., Cichowska, A., et al. (2014). The Hospital Microbiome Project: meeting report for the UK science and innovation network UK–USA workshop 'Beating the superbugs: hospital microbiome studies for tackling antimicrobial resistance', 14 October 2013. *Stand. Genomic Sci.* 9, 12.

The Wilde Lecture (1901) 'The Wilde Medal and Lecture of the Manchester Literary and Philosophical Society.' *Br. Med. J.* 1, 1027–1028.

Willingham, E. (2012). Autism, immunity, inflammation, and the *New York Times*. http://www.emilywillinghamphd.com/2012/08/autism-immunity-inflammation-and-new.html.

Wilson, A.C.C., Ashton, P.D., Calevro, F., Charles, H., Colella, S., Febvay, G., Jander, G., Kushlan, P.F., Macdonald, S.J., Schwartz, J.F., et al. (2010) 'Genomic insight into the amino acid relations of the pea aphid, *Acyrthosiphon pisum*, with its symbiotic bacterium *Buchnera aphidicola*', *Insect Mol. Biol.* 19 Suppl. 2, 249–258.

Wlodarska, M., Kostic, A.D., and Xavier, R.J. (2015) 'An integrative view of microbiome-host interactions in inflammatory bowel diseases', *Cell Host Microbe* 17, 577–591.

Woese, C.R. and Fox, G.E. (1977) 'Phylogenetic structure of the prokaryotic domain: the primary kingdoms', *Proc. Natl. Acad. Sci. U. S. A.* 74, 5088–5090.

Woodhams, D.C., Vredenburg, V.T., Simon, M-A., Billheimer, D., Shakhtour, B., Shyr, Y., Briggs, C.J., Rollins-Smith, L.A., and Harris, R.N. (2007) 'Symbiotic bacteria contribute to innate immune defenses of the threatened mountain yellow-legged frog, *Rana muscosa*', *Biol. Conserv.* 138, 390–398.

Woodhams, D.C., Brandt, H., Baumgartner, S., Kielgast, J., Küpfer, E., Tobler, U., Davis, L.R., Schmidt, B.R., Bel, C., Hodel, S., et al. (2014) 'Interacting symbionts and immunity in the amphibian skin mucosome predict disease risk and probiotic effectiveness', *PLoS ONE* 9, e96375.

Wu, H., Tremaroli, V., and Bäckhed, F. (2015) 'Linking microbiota to human diseases: a systems biology perspective', *Trends Endocrinol. Metab.* 26, 758–770.

Wybouw, N., Dermauw, W., Tirry, L., Stevens, C., Grbić, M., Feyereisen, R., and Van Leeuwen, T. (2014) 'A gene horizontally transferred from bacteria protects arthropods from host plant cyanide poisoning', *eLife* 3.

Yatsunenko, T., Rey, F.E., Manary, M.J., Trehan, I., Dominguez-Bello, M.G., Contreras, M., Magris, M., Hidalgo, G., Baldassano, R.N., Anokhin, A.P., et al. (2012) 'Human gut microbiome viewed across age and geography', *Nature* 486 (7402), 222–227.

Yong, E. (2014a) The Unique Merger That Made You (and Ewe, and Yew). http://nautil.us/issue/10/mergers-acquisitions/the-unique-merger-that-made-you-and-ewe-and-yew.

Yong, E. (2014b) Zombie roaches and other parasite tales. https://www.ted.com/talks/ed_yong_suicidal_wasps_zombie_roaches_and_other_tales_of_parasites?language=en.

Yong, E. (2014c) 'There is no 'healthy' microbiome', *N. Y. Times*.

Yong, E. (2015a) 'A visit to Amsterdam's Microbe Museum', *New Yorker*.

Yong, E. (2015b) 'Microbiology: here's looking at you, squid', *Nature* 517, 262–264.

Yong, E. (2015c) 'Bugs on patrol', *New Sci.* 226, 40–43.

Yoshida, N., Oeda, K., Watanabe, E., Mikami, T., Fukita, Y., Nishimura, K., Komai, K., and Matsuda, K. (2001) 'Protein function: chaperonin turned insect toxin', *Nature* 411, 44–44.

Youngster, I., Russell, G.H., Pindar, C., Ziv-Baran, T., Sauk, J., and Hohmann, E.L. (2014) 'Oral, capsulized, frozen fecal microbiota transplantation for relapsing *Clostridium difficile* infection', *JAMA* 312, 1772.

Zhang, F., Luo, W., Shi, Y., Fan, Z., and Ji, G. (2012) 'Should we standardize the 1,700-year-old fecal microbiota transplantation?', *Am. J. Gastroenterol.* 107, 1755–1755.

Zhang, Q., Raoof, M., Chen, Y., Sumi, Y., Sursal, T., Junger, W., Brohi, K., Itagaki, K., and Hauser, C.J. (2010) 'Circulating mitochondrial DAMPs cause inflammatory responses to injury', *Nature* 464, 104–107.

Zhao, L. (2013) 'The gut microbiota and obesity: from correlation to causality', *Nat. Rev. Microbiol.* 11, 639–647.

Zilber-Rosenberg, I. and Rosenberg, E. (2008) 'Role of microorganisms in the evolution of animals and plants: the hologenome theory of evolution', *FEMS Microbiol. Rev.* 32, 723–735.

Zimmer, C. (2008) *Microcosm: E-coli and The New Science of Life* (London: William Heinemann).

Zug, R. and Hammerstein, P. (2012) 'Still a host of hosts for *Wolbachia*: analysis of recent data suggests that 40% of terrestrial arthropod species are infected', *PLoS ONE* 7, e38544.

INDEX